EVOLUTIONARY ECOLOGY

EVOLUTIONARY ECOLOGY

THE 23RD SYMPOSIUM
OF THE BRITISH ECOLOGICAL SOCIETY
LEEDS 1982

EDITED BY

B. SHORROCKS
University of Leeds

BLACKWELL SCIENTIFIC PUBLICATIONS
OXFORD LONDON EDINBURGH
BOSTON MELBOURNE

3/1985
Biol.

Editorial offices:
Osney Mead, Oxford, OX2 OEL
8 John Street, London, WC1N 2ES
9 Forrest Road, Edinburgh, EH1 2QH
52 Beacon Street, Boston
Massachusetts 02108, USA
99 Barry Street, Carlton
Victoria 3053, Australia

First published 1984

Typeset by Thomson Press (India) Limited, New Delhi and printed and bound in Great Britain by The Alden Press, Oxford.

DISTRIBUTORS

USA
Blackwell Mosby Book Distributors
11830 Westline Industrial Drive
St Louis, Missouri 63141

Canada
Blackwell Mosby Book Distributors
120 Melford Drive, Scarborough
Ontario, M1B 2X4

Australia
Blackwell Scientific Book Distributors
31 Advantage Road, Highett
Victoria 3190

British Library
Cataloguing in Publication Data

British Ecological Society. *Symposium (23rd: 1982: Leeds)*
Evolutionary ecology.
1. Ecology—Congresses
2. Evolution—Congresses
I. Shorrocks, B.
574.5 QH540

ISBN 0 632 01189 0

CONTENTS

	Preface	vii
1	The importance of evolutionary ideas in ecology—and vice versa A.D. BRADSHAW, *University of Liverpool*	1
2	The genetical analysis of ecological traits M.J. LAWRENCE, *University of Birmingham*	27
3	The definition and measurement of fitness F.B. CHRISTIANSEN, *University of Aarhus, Denmark*	65
4	Economics of ontogeny—adaptational aspects P. CALOW, *University of Glasgow*	81
5	Fitness variation among branches within trees D.E. GILL and T.G. HALVERSON, *University of Maryland, U.S.A.*	105
6	The evolutionary genetics of life-histories B. CHARLESWORTH, *University of Sussex*	117
7	What is a population? T.J. CRAWFORD, *University of York*	135
8	Density and individual fitness: assymetric competition M. BEGON, *University of Liverpool*	175
9	The population as a unit of selection J. MAYNARD SMITH, *University of Sussex*	195
10	Strong present-day competition between the *Anolis* lizard populations of St Maarten (Neth. Antilles) J. ROUGHGARDEN, *Stanford University, U.S.A.*	203
11	The evolution of mutualism J. VANDERMEER, *University of Michigan, U.S.A.*	221
12	The evolutionary ecology of predation J.J.D. GREENWOOD, *University of Dundee*	233
13	The genetics of host–parasite interaction J.A. BARRETT, *University of Liverpool*	275

14 Genetic diversity and stability in parasite-host systems 295
D. PIMENTEL, *Cornell University, U.S.A.*

15 Darwin's coffin and Dr Pangloss—do adaptionist models explain mimicry? 313
J.R.G. TURNER, *University of Leeds*

16 Genetic diversity and ecological stability 363
G.S. MANI, *University of Manchester*

Index 397

PREFACE

Most ecologists are now aware that many of the variables they are seeking to measure are genetically determined. None the less, they have frequently tended either to regard intra-population variation as negligible or, alternatively, to assume that natural selection will have maximized fitness. In addition some ecologists believe that such variation is irrelevant to ecology and make the distinction between evolutionary time scales and ecological time scales. The consequence of all such thinking is that intra-population variation is ignored and a 'population type' replaces the 'species type' of a century ago.

For the pre-Darwinians the notion of the 'ideal' or 'type' to which actual objects were imperfect approximations was central. The fact that individual cases failed to match these ideals was simply a measure of the imperfection of nature. However, Darwin called attention to this variation between individuals as the most essential and central property of natural populations. He saw evolution as the conversion of this variation between individuals to variation between populations and finally species. Rather than regarding the variation as a 'blurring' of the ideal, an annoying distraction, he made it the central and essential part of his theory of evolution.

The demonstration by population geneticists that a significant part of almost all genomes is heterozygous (5 to 10% of vertebrates and 10 to 20% of invertebrate loci) with perhaps a third of all genes segregating, coupled with the fact that selection intensities of 10% may be common in natural situations, means that populations can respond quickly to both spatial and temporal environmental pressures. The population, like the species, is polytypic and ecological time scales are no different than the ones traditionally studied by population geneticists. The best way to study populations is to integrate genetical and ecological concepts. Without one or the other, a vital dimension is missing and dangerous misconceptions can occur. Of course, ecologists and geneticists have always been interested in related population parameters and processes. However, they have often approached evolutionary problems from different starting points. For example, while population geneticists have often been interested in the phenotypic effects of major genes or chromosomal units, ecologists see the process of adaptation as involving continuous characters such as egg number, body weight or tolerance of some environmental variable such as temperature or salinity. These continuous or quantitative variations are more difficult to analyse genetically than discontinuous phenotypes and it is important that ecologists are familiar with and understand the techniques and rationale of quantitative genetics.

The symposium volume is set out as follows. The first three chapters form an introduction to the topic by suggesting why genotypes are important in ecology, how quantitative variation can be analysed genetically and how fitness should be measured. The next six chapters look at the evolutionary ecology of single species populations and the next six look at species interactions of one kind or another. The final chapter looks at the genetic diversity that may influence ecological processes and its mechanism of maintenance. I am indebted to the symposium committee (L.M. Cook Department of Zoology, University of Manchester; M. Lawrence, Department of Genetics, University of Birmingham and R. Law, Department of Botany, University of Sheffield) who helped to plan the programme and select the contributors. I would also like to thank the several referees who provided valuable comments upon the final manuscripts.

Leeds 1983 Bryan Shorrocks

1. THE IMPORTANCE OF EVOLUTIONARY IDEAS IN ECOLOGY—AND VICE VERSA

A.D. BRADSHAW
Department of Botany, University of Liverpool, Liverpool L69 3BX

INTRODUCTION

If anyone asked whether ecologists can do good work without any knowledge of evolution, the answer must surely be yes. Ecologists work mainly on situations existing at one particular moment of time or in one particular area. As a result it would appear that they can take the properties of a species as fixed and assume that evolution, which is all about changes in species and populations, has nothing to do with ecology. A perusal of any ecological journal would support this; not more than one article in twenty mentions evolution or overtly considers evolutionary processes—yet there is nothing scientifically inadequate about the rest.

Yet if we look more critically, it is patently wrong to descry evolution as a proper subject for ecologists, for three major reasons:

(i) We all tacitly assume that what we examine ecologically is the product of evolution, and that as a result of natural selection it is adapted, more or less, to its environment, in the sense that it is fitted (*aptare*) to (*ad*) it.

(ii) Species are clearly not fixed, but consist of a complex of different populations, often with extremely different ecological properties, which can change as a result of evolutionary processes in only a few generations.

(iii) Ecologists themselves study life and death, and mechanisms of fitness, which are the stuff of evolution as we understand it.

So not only does evolutionary thinking form a conceptual background to ecology, but there is a two-way relationship between ecology and evolution, in that the findings of each is very relevant to the other. I shall endeavour to explore both sides of this relationship, using mainly plant examples, not because animal examples are not equally relevant but because of my own particular experience.

BIRTH, DEATH AND SURVIVAL

One of the first major steps towards an understanding of evolutionary processes and a retreat from special creation was unwittingly taken, of course, by Malthus, a demographer. It was his obstinate arguments for the discrepancy between rates of increase of populations and the resources available to them that started the slow revolution in thinking about man's place in nature. His book, reprinted several times, forced people to think about the struggle for existence rather than to run away from the idea.

Now, demographic parameters are crucial objects of ecological study. Not only are we being given very detailed information about the apparent extravagances of nature in the processes of reproduction of species, but we are beginning see that this extravagance in plants particularly, can take place at many different places in the life cycle (Harper 1977; Begon & Mortimer 1981). Correlated with this, in accordance with Malthus, death takes its toll. It can be of whole individuals, and, in plants, not just in overcrowded young or at one particular period of 'old age', but progressively over several years. In *Plantago* (Sagar 1959) or *Anthoxanthum* (Antonovics 1972) it is better to talk of a half-life of 2 years because the depletion rate is so constant. Many plants spend a considerable portion of their life cycle as dormant seeds. Death works inexorably here too, as in many common annuals (Roberts 1964).

But death does not affect only whole individuals, or genets; it also affects single parts of an individual, or ramets. The ramets of *Ranunculus repens* have very similar patterns of survivorship to that of whole plants (Sarukhan & Harper 1973). So the quantity of a single individual which survives from one period to another can change considerably.

In the same manner, during growth the numbers of single parts of an individual can be altered. Growth is increase; but in different conditions the amount by which a plant may increase, and therefore contribute to the next generation, can vary dramatically, as is demonstrated by every experiment on plant density. Perhaps more importantly, the same can occur in conditions influenced by competition from other species; death does not necessarily occur, but large changes in growth and seed output are commonplace. Such differences of growth will be cumulative in vegetatively propagating plants, and we can observe differences in relative abundance becoming more and more pronounced with time even between genotypes (Hickey & McNeilly 1975). Differences between species, measured as Relative Replacement Rates (de Wit 1960; Van dèn Bergh 1968) can easily reach 5 or 10.

In perennial species, establishment of new seedlings can occur at the same time as older plants are dying. Since these two processes may compensate each other, the total population can appear static when, in fact, considerable change is occurring (Fig. 1.1). Rates of such turnover in perennial populations can often be as high as 100% per annum.

All this implies the possibility of natural selection, in which individuals

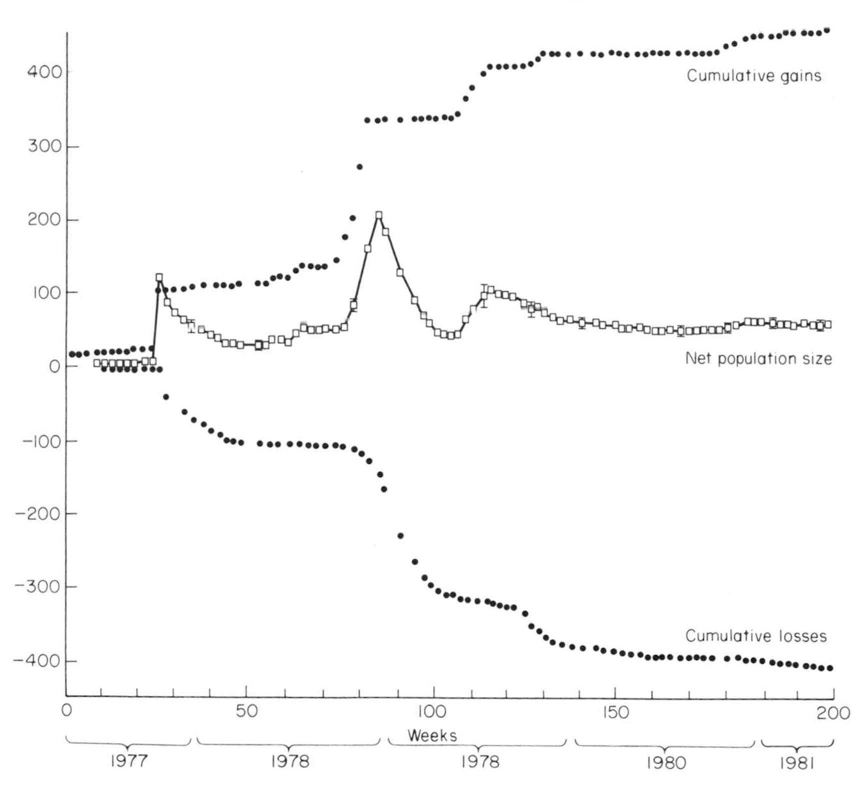

FIG. 1.1. The number of individuals, and the cumulative gains and losses, in a population of *Agrostis canina* on a copper mine at Glasdir, Dolgelley, Wales (data of S. Farrow, T. McNeilly and P.D. Putwain).

of one sort replace those of another, or at least survive better than another. Demography is therefore the key to natural selection (Solbrig 1980). What can occur has been demonstrated by many artificial mixtures of selected genotypes (reviewed by Bradshaw 1972). Perhaps the most outstanding example is in mixtures of either *Lolium* or *Dactylis* cultivars sown as swards and given various treatments (Charles 1961). In all cases the mixtures changed their composition substantially, in different ways with different treatments, within a single generation, due to differential mortality, and expansion of the survivors. Any plant community is the result of such processes; pastures can reveal that it has happened by an examination of their genotypic structure (Fig. 1.2); far fewer genotypes remain than must have been present initially.

But selection is only effective as an evolutionary force if what survives is

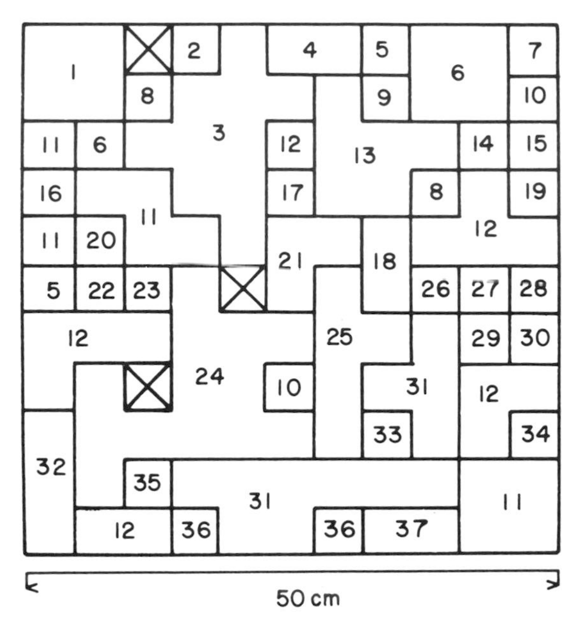

Fig. 1.2. The pattern of genotype distribution in a 9-year-old population of *Lolium perenne* in a grazed pasture (240 seeds originally sown in 50 × 50 cm area—genotypes identified by isozyme variation at PG1, GOT1 + 2, loci) (data of M. Roose and T. McNeilly).

genetically different from the original population mean. The degree of genetic change for a generation (ΔG_g) will depend on the selection differential (the difference between the phenotypes selected and the mean of the population ($P - \bar{P}$) and the degree to which that difference is passed onto the next generation [the heritability (h^2)].

$$\Delta G_g = h^2(P - \bar{P})$$

This simple concept, well described in many places (e.g. Falconer 1981), is the key to understanding evolutionary change. Clearly if the individuals selected do not differ much from the mean or if their special characteristics are only poorly inherited, then evolutionary change will be limited. This was well understood, although perhaps oversimplified, by Fisher in his fundamental theorem of natural selection—the rate of increase in fitness of a population equals its genetic variance in fitness at the time.

HERITABILITY

We must therefore ask whether characters, especially those of ecological importance are heritable. Nearly all ecologically important characters are

TABLE 1.1. The various origins of variation in a character of a population, expressed as components of variance

Components of variance (of a character in individuals of a population)

$$V_P = V_G + V_E$$
$$= V_A + V_D + V_I + V_E$$

where V_P = total variance between phenotypes
V_G = variance due to genetic effects
V_A = variance due to additive genetic effects
V_D = variance due to dominance genetic effects
V_I = variance due to interaction genetic effects
V_E = variance due to environmental effects

continuously varying and so-called metric (see subsequent discussion by Lawrence in Chapter 2, p. 29). For these characters the variation must have several origins or components (Table 1.1). The degree of genetic determination of the character (or broad sense heritability) is given by a comparison of the total genetic variance with the total variance from all sources,

$$\text{degree of genetic determination (broad sense heritability)} = \frac{V_G}{V_P}.$$

The degree to which a character is transmitted to offspring is given by the additive genetic variance,

$$\text{degree of heritability (narrow sense heritability)} = \frac{V_A}{V_P}.$$

The broad sense heritability determines the outcome of selection on individuals. It will influence changes in genotype frequency in a population, since it measures the degree to which individuals are consistently different from one another. As such it is important, and can easily be measured if individual plants can be clonally replicated. Broad sense heritability is usually very high, e.g. > 0.85 for time of ear emergence in *Lolium* (Cooper 1959). But such differences, although important to the survival of the individual plants, may be due to various genetic and other effects which are not inherited, again very clearly shown in *Lolium* (Hayward 1970).

For this reason the true (or narrow sense) heritability is needed. This can be measured in a number of ways (Falconer 1981; Lawrence in Chapter 2, p. 29), of which the regression of offspring on parents is the most obvious, but not necessarily any simpler than other methods, such as the variation between sibs or half-sibs produced by pair crosses or a polycross. Common sense must indicate to ecologists that characters are inherited, since offspring

TABLE 1.2. Narrow sense heritabilities of some ecologically important characters in plants

Plants	Characters	Heritability	References
Lolium perenne	Date of ear emergence	0.87 (parent/offspring)	(Cooper 1959)
	Rate of leaf appearance	0.73 (pair crosses)	(Cooper & Edwards 1961)
	Rates of photosynthesis		
	(high light)	0.30 (diallel)	(Wilson & Cooper 1969)
	(low light)	0.14 (diallel)	(Wilson & Cooper 1969)
	Mesophyll cell thickness	0.33 (diallel)	(Wilson & Cooper 1969)
	Leaf length	0.42 (diallel)	(Wilson & Cooper 1969)
Dactylis glomerata	Yield		
	(spring)	0.67 (polycross)	(Stratton *et al.* 1979)
	(autumn)	0.04 (polycross)	(Stratton *et al.* 1979)
	Digestibility		
	(spring)	0.62 (polycross)	(Stratton *et al.* 1979)
	(autumn)	0.49 (polycross)	(Stratton *et al.* 1979)
Oryza sativa	Competitive ability (panicle number)	0.12 (F_3 families)	(Sakai 1961)
Agrostis tenuis	Tolerance to copper	0.70 (diallel)	(Gartside & McNeilly 1974)
Geranium carolinianum	Resistance to SO_2	0.50 (parent/offspring)	(Taylor 1978)
Linum usitatissimum	Plasticity to density		
	(height)	0.74 (F_3 families)	(Khan *et al.* 1976)
	(capsule number)	0.66 (F_3 families)	(Khan *et al.* 1976)

always resemble parents. But the essence of a heritability test is to find out how far different parents give rise to different progenies, each similar to their parents, or how far the progeny of one set of parents resemble each other and are different from progeny of other parents.

In fact, heritabilities of ecologically important characters can be quite high (Table 1.2). The same range of heritabilities can be found for characters in animals. However in all cases heritabilities tend to be lower for fitness characters. When heritabilities are high for a given selection differential, progress under selection can be considerable, and, in the appropriate material, progress can be maintained for many generations, as shown by the famous Illinois corn experiment (Woodworth, Leng & Jugenheimer 1952), in which the selected lines transgressed completely the limits of the original population.

However, when heritabilities are low, little progress under selection may be possible. This is shown by a comparison of selection for egg number ($h^2 = 0.25$) and egg weight ($h^2 = 0.75$) in chickens (Lerner 1958).

For characters determined by a single gene, the concept of heritability does not apply and high selection pressures inevitably have radical effects on gene frequency (e.g. Shorrocks 1978).

NATURAL SELECTION AND EVOLUTIONARY CHANGE

High birth and death rates combined with high heritability of ecologically important characters imply that evolution should be happening everywhere, and at appreciable rates, just as it can in experimental material. Yet in practice this hardly seems true. In natural situations, change in important characters such as in horses' teeth (Simpson 1953) or many other examples, appears to occur only over geologic time.

Nevertheless, over the past few decades we have uncovered a few examples where rapid change has occurred. Metal tolerance for instance seems able to evolve in only one or two generations (Wu, Bradshaw & Thurman 1975). Sulphur dioxide tolerance in plants and industrial melanism in insects have obviously evolved since industrialization (Taylor & Murdy 1975; Bishop & Cook 1980). Similarly the population differentiation in snails and grass species in response to man-made variations of habitats (Cain & Sheppard 1954; Bradshaw 1959) must have come about in the last few hundred years. Yet for most characters in most species there is little change in either time or space.

Why should so many characters in so many species be so stable? It cannot any longer be because we have not looked for evolutionary change. Nor can it be because there are no opportunities. On the waste heaps of metal mines, for instance, there is still plenty of open space available for colonization by, more, tolerant species than those which exist at the present. Similarly, until the appearance of the new amphidiploid grass species, *Spartina anglica*, estuarine saltings have been available for angiosperm colonists ever since angiosperms first appeared in the Mesozoic. And in habitats already colonized there are still opportunities for new, more competitive species, as is demonstrated by every successful alien.

GENOSTASIS AND EVOLUTIONARY PLATEAUS

If birth and death rates can be so high and therefore generate high potential selection pressures, it follows from simple theory that natural selection may exhaust the variation available to it. In long-term selection experiments, limits to selection and selection plateaus are well known for many different characters in different organisms (Lerner 1958; Robertson 1955). These can be due to complex causes, such as epistasis in which genes do not act additively, co-adapted gene complexes in which genes have been selected to act together, pleiotropy in which there are subsidiary effects of genes which are disadvantageous, or linkage in which the genes concerned are linked to other genes which are disadvantageous (for discussion cf. Antonovics 1976 and Falconer 1981). All of these effects can cause the otherwise advantageous effects of genes, and therefore the influence of selection, to be nullified.

It is possible that species or populations have become adapted to a particular environment in such a way that any deviations in their characteristics, no matter what direction, lead to loss of adaptation. Under such conditions, stabilizing selection operates, eliminating the deviants and maintaining the population as it is in perpetuity. However it must be remembered that in such situations the possibility of evolution to become *more* adapted to the existing situation must still exist. There can be no reason why, for instance, even a turtle cannot become a better turtle by evolving mechanisms for more efficient food conversion, or for faster swimming at no other cost. The occurrence of stabilizing selection is direct evidence of some other restrictions to evolution.

We therefore come to an alternative possibility commonly overlooked. Lack of genetic change in situations of potentially high selective pressure can be due to lack of appropriate variation. Selection can only act on herit-

TABLE 1.3. Examples of the different sources of ecological adaptation in plant breeding programmes

Sources	Adaptations
From original gene pools	
Potato	Blight resistance within *S. tuberosum*
Alfalfa	Spotted aphid resistance
Sugar beet	Sugar content
Rye	Reduced height
From other gene pools—other cultivars	
Barley	Yellow dwarf resistance from Abyssinian cultivars
Wheat	Dwarfing genes from Japanese cultivars
Grapes	Root aphis resistance from American material
Cotton	Blackarm resistance from African cultivars
From other gene pools—other species	
Oats	Mildew resistance from *A. ludoviciana*
Bread wheat	Stem rust resistance from *T. dicoccum*
Bread what	Eye spot resistance from *Aegilops ventricosa*
Rice	Grassy stunt resistance from *O. nivara*
Delphinium	Red flower colour from *D. cardinale*
Potato	Blight resistance from *S. demissum*

able variation; if the appropriate variation is not present then evolutionary change is not possible. It is of no significance if other, inappropriate, variation occurs. This situation is familiar to every plant breeder, who may be lucky in finding the character he wishes within his starting material, but may also be unlucky and have to look for it in other, often distantly related, material (Table 1.3). Yet we usually forget that this must be common in more normal evolution.

This situation appears to be a dominant factor in the evolution of metal tolerance. Evolution of tolerance in individual species appears to be related to the occurrence of variability in metal tolerance in the normal populations of those species (Gartside & McNeilly 1974). This relationship has been examined by a new sensitive culture technique and appears to hold widely (Table 1.4). It would appear usually to be a property of the whole species. But there is no reason why, in cases where there is no gene exchange, it should not be a property of individual populations. In small populations, whether by drift or founder effects, there is the possibility of gene fixation and exclusion of particular genes, as in the absence of genes for copper tolerance in

TABLE 1.4. The percentage of copper tolerant individuals found in normal populations of various grass species, in relation to the presence of the species on copper polluted waste and whether the plants collected were tolerant of copper (data of C. Ingram)

Species	Percentage of occurrence of tolerant individuals	Presence of species on mines		Tolerance of collected adult plants
		On waste	Margins	
Holcus lanatus	0.16	+	+	+
Agrostis tenuis	0.13	+	+	+
Festuca ovina	0.07	–	+	–
Dactylis glomerata	0.05	+	+	+
Deschampsia flexuosa	0.03	+	+	+
Anthoxanthum odoratum	0.02	–	+	–
Festuca rubra	0.01	+	+	+
Lolium perenne	0.005	–	+	–
Poa pratensis	0.0	–	+	–
Poa trivialis	0.0	–	+	–
Phleum pratense	0.0	–	+	–
Cynosurus cristatus	0.0	–	+	–
Alopecurus pratensis	0.0	–	+	–
Bromus mollis	0.0	–	+	–
Arrhenatherum elatius	0.0	–	+	–

a modern cultivar of *Agrostis tenuis* (Humphreys & Bradshaw 1976).

Lack of variability for resistance has recently become apparent in disease outbreaks, and has lead to catastrophes. The effect of the introduction of chestnut blight into North America from China has been to cause the total elimination of the American chestnut (*Castanea dentata*), which apparently possesses no variability whatever for resistance. The same must be true for foliose lichens and SO_2 : they disappear completely in areas of high SO_2. But interestingly this then gives rise to evolution of melanism in moths, which do possess appropriate variability in pigmentation.

When there is lack of variability for a character, either because there was little initially or because natural selection has fixed the additive variation previously present, heritabilities will be low, because V_A is small compared with V_P. This is characteristic of many fitness characters, presumably because of the prior effects of natural selection; but it also appears in artificial selection experiments as additive variation is exhausted and only epistatic non-additive variation is left (Lerner 1958).

Situations in which evolution is limited by lack of appropriate genetic variation, are obviously widespread and indeed almost universal. The condition, which can be called *genostasis*, requires much wider recognition than it has previously been accorded. It may, or may not, be related to the *stasis* suggested in current discussions on long-term evolution involving punctuated equilibria (Eldredge & Gould 1972, Grant 1982). From the point of present-day evolution and ecology it is certainly crucial to appreciate when genostasis may, or may not, be occurring.

VARIATION AND EVOLUTIONARY CHANGE

When variation is available there is no doubt that evolution can occur, often very rapidly. Despite Haldane's (1957) dilemma about the cost of natural selection, if we assume the usual situation in which populations are controlled in some density-dependent manner, then the increase in numbers of individuals possessing an advantageous gene can be purely at the expense of those not possessing it (Sved 1968).

Evolution is likely to be most rapid in those situations in which selection has not already acted, because in relation to the particular selection pressure, there can be a store of hidden unselected variability. The extent of this hidden variability can be seen when a population is subject to a new environment; although in the old environment it can be very uniform, in the new it can be very variable, as in *Lolium rigidum* exposed to the unusual conditions (for it) of continuous light and warmth (Cooper 1954). The best examples of rapid

TABLE 1.5. Examples of evolution of resistance in weeds which have been exposed to triazine herbicides

Have evolved resistance*	Have not evolved resistance
Amaranthus retroflexus	*Agropyron repens*
A. powellii	*Anagallis arvenis*
A. hybridus	*Capsella bursapastoris*
Brassica campestris	*Sonchus arvensis*
Chenopodium album	*S. oleracea*
C. strictum	*Stellaria media*
Senecio vulgaris	*Taraxacum officinale*
Solanum nigrum	*Thlaspi arvense*

*From Bandeen, Stephenson & Cowett 1982; Gressel *et al.* 1982.

evolution, including those already mentioned, do in fact come, for both plants and animals, from new man-made situations such as those arising from the use of pesticides and herbicides and from industrial pollution (Bishop & Cook 1981 Bradshaw & McNeilly 1981, LeBaron & Gressel 1982).

Other material will be in a genostatic condition, which will only be relieved by mutation giving real novelties, or by hybridization or gene flow giving new gene pools, as in *Drosophila* (Lewontin & Birch 1966). The most spectacular present-day examples of the relief of a genostatic condition with startling evolutionary and ecological consequences must surely be the appearance of *Spartina anglica* from the introduction of the alien *S. alterniflora* (Marchant 1968). If we understood the genetic situation more clearly it is possible that the present havoc in Europe from Dutch elm disease (*Ceratocystis ulmi*) after its re-introduction from America has the same cause—the acquisition of virulence genes in America.

Certainly it is possible to mimic such startling evolutionary changes. In this country resistance to the widely used triazine herbicides has been absent from common weed populations until recently, and still does not occur in many species (Table 1.5). Yet if genes for resistance to triazines are introduced into a susceptible population, the change in frequency of resistant genotypes is remarkably fast (Fig. 1.3).

Yet for reasons already given, such rapid change is the exception rather than the rule. Genostasis is the common condition. Does this suggest that evolutionary and ecological time scales are really very different (Begon & Mortimer 1981) and that the disciplines can have little to do with each other?

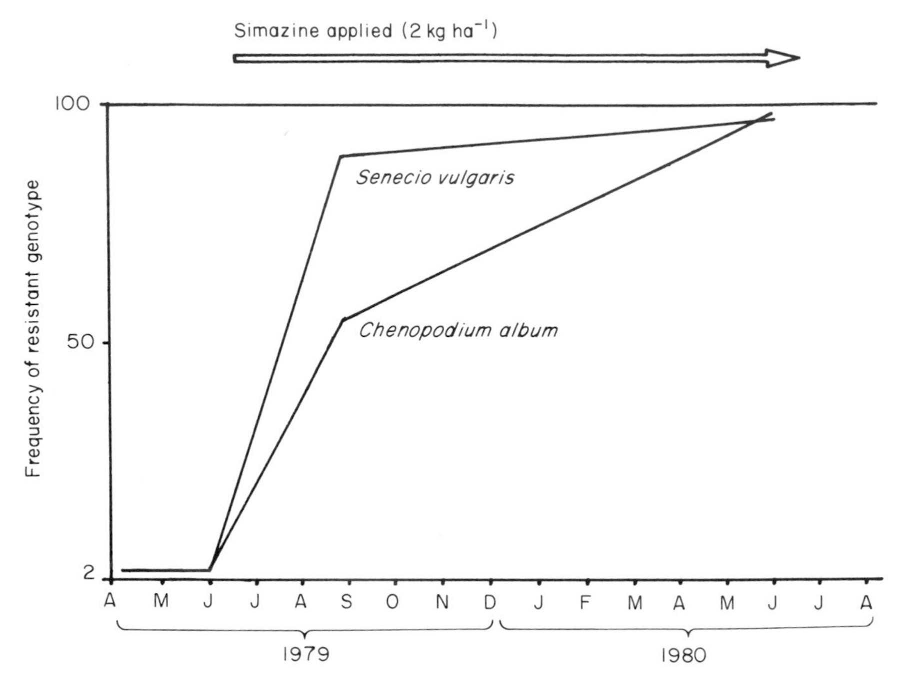

FIG. 1.3. The change of frequency of resistant genotypes in populations of *Senecio vulgaris* and *Chenopodium album* exposed to simazine, into which 2% of a resistant genotype was introduced (data of K.R. Scott and P.D. Putwain)

POPULATION DIFFERENTIATION AND ECOLOGICAL AMPLITUDE

Despite the normal differences in time scale it must be remembered that we will always be dealing with material which could, in the past, have suffered evolutionary change because of past, if not present, variability. In fact, the sorting, and favouring process caused by natural selection seems always to have proceeded some distance in all species. Population differentiation within species is universal. Sometimes, as in *Achillea* (Clausen, Keck & Heisey 1948), the amount of evolutionary adjustment is so considerable that the extreme populations of the same species cannot survive in each other's habitat. The differences, as in calcium response in *Festuca ovina* (Snaydon & Bradshaw 1961), can be so great that populations behave as differently as ecologically distinct species.

Any character can be affected. There are excellent examples of population differences in response to climate, for example in *Potentilla* (Clausen,

Keck & Heisey 1940) and in different prairie grasses (McMillan, 1959) and to soil, for example in *Trifolium repens* (Snaydon & Bradshaw 1962) and in *Dactylis glomerata* (Crossley & Bradshaw 1968). But we now realize that life history parameters can be affected, for example in *Poa annua* (Law, Bradshaw & Putwain 1977) and in *Taraxacum* (Gadgil & Solbrig 1972). Even phenotypic plasticity can be affected by natural selection, for example in *Linum usitassimum* (Khan, Antonovics & Bradshaw 1976) (where heritabilities have been determined), in *Capsella bursa-pastoris* (Sϕrensen 1954) and in *Ranunculus flammula* (Cook & Johnson 1968), so that sensitivity to environmental factors is itself under evolutionary control. The physiological processes which underlie population differences are themselves becoming understood, for instance climatic response in *Solidago virgaurea* (Bjorkman & Holmgren, 1963) and zinc tolerance in grasses (Brookes, Collins & Thurman 1981). A good review of the evidence is provided by Heslop-Harrison (1964).

At the same time it has become realized that these differences can occur very locally, well within the confines of what ecologists might treat as a single habitat, as in the cliff populations of *Agrostis stolonifera* (Aston & Bradshaw 1966), or vernal pool populations of *Veronica peregrina* (Linhart 1974). The balance between the opposing forces of gene flow and selection appears weighted in favour of selection (Jain & Bradshaw 1966), as any ecologist might have suspected.

Populations also differentiate in relation to changes occurring in time, in relation to successional changes. This is very clear from recent work on *Spartina patens* (Silander & Antonovics (1979) and on *Poa annua* and other species (Law 1979). Some species, at least, can climb their own seres.

So a species cannot be understood by the ecological behaviour of one of its populations, but by the behaviour of the sum (not the mean) of all its populations. This becomes very apparent from transplant experiments where extreme populations of a single species cannot survive in each other's habitat, whether over large geographical distances, e.g. *Achillea* (Clausen, Keck & Heisey 1948) or small, e.g. *Agrostis tenuis* (Bradshaw 1960). The ability to colonize a habitat may often be related more to evolutionary capability than anything else. There is no doubt, for instance, that only a few species can colonize metal-contaminated habitats (Bradshaw 1975) (Table 1.6), despite the fact that a wide range of species have had the opportunity to do so. This can be related to the evidence already given in Table 1.4. Ecological amplitude on a local or a wide scale has therefore a strong evolutionary component. It must be related to genetical rather than physiological flexibility.

Table 1.6. Species to be found in mown grassland in copper-contaminated and uncontaminated areas at Prescot, Lancs. (Bradshaw 1975, updated)

Copper in soil (ppm)	Species found	
< 2000	*Agrostis stolonifera*	*Festuca rubra*
Adjacent to refinery	*A. tenuis*	*Agropyron repens*
		Holcus lanatus
< 500	*Ranunculus repens*	*Achillea millefolium*
Away from refinery	*R. bulbosus*	*Hypochaeris radicata*
	Cerastium vulgatum	*Leontodon autumnale*
	Trifolium repens	*Luzula campestris*
	T. pratense	*Lolium perenne*
	Taraxacum officinale	*Poa annua*
	Rumex obtusifolius	*P. pratensis*
	Prunella vulgaris	*P. trivialis*
	Plantago lanceolata	*Dactylis glomerata*
	Bellis perennis	*Cynosurus cristatus*
		Hordeum murinum

THE NATURE OF SELECTION

While the findings of evolution can be applied to ecology, the reverse is also important. As evolutionary theory has developed there has been a notable lack of critical ecological thinking on the nature of selection and its effects. The effects of selection on continuous variation have been separated by Mather (1953) into three types: directional, stabilizing and disruptive. These at first sight seem fairly straightforward, yet examined ecologically, they perhaps have complexities.

Directional selection

Situations in which individuals at one end of a continuous range of variation are disfavoured, are basically simple and lead to normal types of evolutionary progression. Yet precise analysis of many situations has shown that the coefficients of selection generated can be immensely variable, for instance, in barley mixtures, for the variety Vaughn, 0.22–1.59 (Allard, Harding & Wehrhan 1966). So in practice evolutionary changes can be reversed from one generation to another, very apparent when the changes in gene frequency in composite cross populations of barley are followed over a number of years (Suneson & Stevens 1953). Any simple ecological experience should lead us to expect this; annual variations in climatic conditions alone can have startling effects on populations. But an examination of the precise ecological causes of such *fluctuating selection* would provide us with a much better understanding of the nature of this important aspect of selection (Felsenstein 1976).

When directional selection for one character is occurring certain genes are increasing in frequency. There is no reason why these genes should not (a) have direct effects on other characters by pleiotropy, (b) be closely linked to other genes whose frequency and therefore effects will also change, and (c) be acting on a physiological or growth system with limits, so that changes in one aspect of the system are accompanied by reciprocal or compensatory effects in another aspect. All this gives rise to *correlated response*, which is well known to plant breeders. It can cause trivial effects, e.g. changes in panicle characters in *Lolium perenne* when flowering time is being selected, due to linkage (Cooper 1960), or major effects, e.g. changes in frequency of the genes for cyanogenesis in legume species related to temperature rather than to the presence of predators, because of the pleiotropic effects of the genes on cold tolerance (Jones 1973).

It is difficult to envisage that any single gene can act in isolation from other genes; since most characters are complex many genes are usually involved together. This can lead to *co-adaptation* within gene complexes. The result is that where there is a simple geographical gradient in selection, one balanced conservative gene complex will persist over a large section of that gradient until it is suddenly replaced by another. As a result sudden changes in a character can occur without obvious relationship to similar changes in the environment. These *area effects*, reported particularly in snails *Partula* (Clarke 1968) and *Cepaea* (Cain & Currey 1963), inevitably complicate the simple effects of directional selection.

Stabilizing selection

There has been wide acceptance of the idea that individuals which deviate most, for any character, from the mean of population are likely to be less fit. It is a concept that seems ecologically very reasonable, and evidence for it continues to appear, whether for chaeta number in *Drosophila* (Kearsey & Barnes 1970), flowering time in *Phleum* (Charles 1964) or response to soil nutrients in *Dactylis* (Crossley & Bradshaw 1968).

Yet surprisingly little work has been done by ecologists to confirm that extreme individuals are usually less fit and to show the processes which lead to this. Indeed, what has for so long been considered the most effective evidence, the recovery of individuals of *Passer domesticus* knocked down by a storm, does not on re-examination now seem to be clear cut (Johnson,

Niles & Rohwer 1972). There is certainly more work needed on the fitness of the extreme.

Disruptive selection

Perhaps it is here that the ecologist has the most scope for providing understanding. The problem is that the subject is complex at the outset because several different processes are involved (Clarke 1979). However, in all the cases to be considered it is genotypes which are different that are favoured. From this is follows that genotypes which are rare must be favoured. This means that frequency-dependent fitness and disruptive selection co-occur and are two aspects of the same selective situation.

Evidence for selection by predators against the commonest, *apostatic* selection, shows that it is of widespread occurence (Clarke 1969). Wherever it occurs it will inevitably lead to the maintenance of variation and, in an extreme form, balanced polymorphism in prey populations. The detailed ecological and behavioural evidence which is available, such as for fish/corixids (Popham 1942) and passerine birds/artificial baits (Allen 1975), illustrates exactly the important contribution that ecologists can make. It is important to realize that the principle has important applications to the evolution of plant diversity, as suggested by Gillett (1962), Janzen (1970) and others. It may perhaps apply in grazing situations, e.g. sheep/*Trifolium repens* (Cahn & Harper 1976). It may also cause diversity in predators (Paulson 1973).

Parasitism can have similar effects. There is no space to discuss the recent developments in our thinking, since these have been well reviewed (Taylor & Muller 1976) in a series of contributions all of which show the importance of an ecological approach.

When two genotypes are competing for similar resources we are now well aware that they will more escape each other's competitive effects the more different they are in their requirements. The results will be a stable mixture. This ecological concept of *annidation* or *niche diversity*, elegantly developed for interaction between plant species by de Wit (1960) and others (Harper 1977), has scarcely been applied to evolutionary situations except, notably, by Antonovics (1978). Yet it is completely applicable to interactions between genotypes within species. Very clear annidation has been shown between genotypes of *Linum* which would lead to stable polymorphisms (Khan, Putwain & Bradshaw 1975). Since in every population there is liable to be competition between the genotypes that compose it, it should follow that evolution of annidation and stable diversity as a consequence, without loss of population fitness, should be commonplace. It will be difficult to detect in outbreeding populations because of the individuality of single genotypes, which explains perhaps why it has not been demonstrated so far. But the very elegant demonstration of the evolution of a stable polymorphism due to annidation in a hybrid bulk population of *Hordeum* (Allard & Adams 1969) (which is inbreeding) should have aroused much more interest than

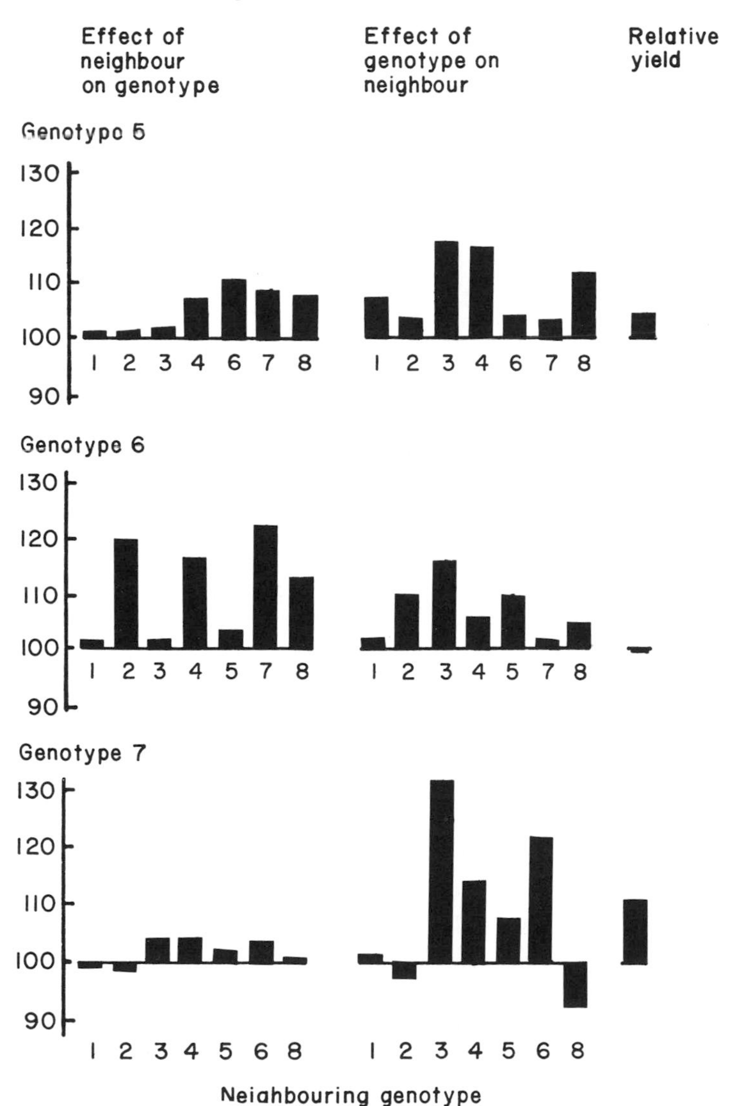

FIG. 1.4. Examples of the positive interactions, in pair mixtures, found between eight genotypes abstracted from a stable polymorphic bulk–hybrid population of *Hordeum sativum* (Allard & Adams 1969).

it has done so far, especially since the authors demonstrate that the selective forces involved can be powerful in their effects (Fig. 1.4).

Facilitation, when genotypes influence each other in a positive fashion (in contrast to annidation where there is merely the reduction of the negative effects of competition (Khan *et al.* 1975) will have similar evolutionary effects. Although the term is widely used, unequivocal examples of facilitation are rare. For this reason perhaps, and also for the more cogent reason that true facilitation is a much more complex process, there appear to be no examples so far of it contributing to evolutionary processes in populations. But this does not mean that we should not look for them.

The final possible cause of disruptive selection is *variation of environment in space* on a very localized scale—a phenomenon well known to all ecologists. Such variation grades into directional selection in different habitats when the spatial scale is large and into annidation when the spatial scale is so small that the different environments cannot be disentangled from each other. Different closely adjacent habitats are well known to be able to cause the evolutionary differentiation of distinct populations (Jain & Bradshaw 1966) despite the fact that the populations are parapatric (adjacent and able to exchange genes). The scale on which such differentiation can occur is measurable in terms of metres, e.g. in *Liatris* (Schaal 1975) or even centimetres, e.g. in *Anthoxanthum* (Snaydon & Davies 1976). In this latter experiment, where the environmental differences are due to the soil effects of the Park Grass Experiment, transplant experiments have shown that the differences in half-life between alien and native material are at least 50% (Davies & Snaydon 1976).

Localized environmental variation, over the scale of metres or fraction of a metre, can be caused by many different factors, microclimate, edaphic or biotic. They are commonplace in all habitats. As a result we have the possibility of relatively permanent, spatially fixed, very local differences in the environment experienced by a single interbreeding population of individuals, a truly multiple-niche situation, which can exert powerful disruptive, or frequent dependent, selection in a form which satisfies the conditions necessary for stable polymorphism. Yet despite this, cases in which disruptive selection due to localized environmental variation occurs have not yet been clearly established. The existence of genotypes within *Trifolium repens* showing differences in adaptation to different grass species (Turkington & Harper 1979) seem to be an example. There are, clearly, differences in performance of clones of *Agrostis stolonifera* transplanted into different areas of an old meadow (Table 1.7), but just how far local variation in environment (of what ever cause) has clearly selected different genotypes is not yet clear. It is only detailed transplant work of this sort in perennial species which is

TABLE 1.7. Specific genotype adaptation in a perennial grass in different permanent pastures in Cheshire shown by a transplant experiment with *Agrostis stolonifera*. Clones from three fields were reciprocally transplanted into three randomly chosen quadrats in three areas of each field. The table shows an analysis of variance of tiller numbers (natural logarithm) present 10 weeks after planting of single tillers. Variance ratios showing superscripts′ are quasi-*F* ratios (data of D.A. Weir and A.M. Mortimer)

Sources of variation	DF	MS	VR	
Fields (F)	2	325.41	23.42′	***
Populations (P)	2	49.49	3.23′	NS
Clones in P (C in P)	9	14.11	2.10	*
Areas in F (A in F)	6	25.03	4.21′	**
Quadrats in (A in F)	18	2.99	< 1	NS
F × P	4	10.57	1.28	NS
F × (C in P)	18	11.28	1.24	NS
P × (A in P)	12	11.94	2.31′	**
P × (Q in A in P)	36	3.97	< 1	NS
(C in P) × (A in F)	54	5.09	< 1	NS
(C in P) × (Q in A in F)	162	9.06	1.35	*
Error	308	6.71		
Total	631			

likely to elucidate a selective process which *a priori* would seem probably widespread.

THE EVOLUTIONARY ENDPOINT—ADAPTATION?

Ecologists look at characters; geneticists look at genes. Nevertheless, no matter which approach, we all tend to expect that the outcome of evolution is adaptation. But such a conclusion is simplistic and even naive. What we should only expect is that the outcome is material which is selected and more adapted than the material which preceded it. Evolution is merely a replacement system using what is better, in terms of fitness, than the rest.

In the final state, the material must have some fitness or it would not survive. So in a sense the material is adapted, better than what was there before, but certainly not in a state of grace. Evolution has to make do with what is already present and whatever new material is to hand. It is essentially a tinkering process (Jacob 1977). As a result, as we have seen, fitness can be limited by (a) genostasis—lack of variability; (b) pleiotropy and linkage—deleterious side effects; (c) phenotypic plasticity—variability hidden from selection; and (d) ancestry—limits set by the past.

Not only must we take a pluralistic approach to the agents of evolutionary change, as Gould & Lewontin (1979) have suggested, but also realize that the limitations set by the supply of appropriate variability are overriding.

These limitations can occur on a small or on a vast scale. Land plants would have had a very different scale of success if they had been able to produce a membrane which restricted the diffusion of water but not of CO_2. In grass breeding, subtle internal resource limitations, for which there seems no available genetic variability by which they can be overcome, control the improvement of leaf–area production in *Lolium* (Edwards & Cooper 1963). Ancestry can have effects which are obvious, such as no trees in the Gramineae, or subtle, such as the way defense to particular groups of insects is limited to certain families of Angiosperms (Erhlich & Raven 1964). Species can get trapped geographically, such as the Monterey pine (Cain 1944), or have lost particular genes in an evolutionary migration, such as is suggested for certain alpine Eurasiatic plants (Turesson 1931). Finally we must remember the Red Queen hypothesis (Van Valen 1973)—everything else is evolving so that any one species must itself evolve to stand still.

So we will always be dealing with imperfect organisms, certainly not showing perfect adaptation—fitting to (the present). It may indeed be better to think of them as showing "abaptation" (Harper 1982)—fitting from (the past).

CONCLUSION

In their review of genetics and ecology, Sammeta & Levins (1970) said that 'the integration of population genetics and ecology with other disciplines into a coherent evolutionary biology of populations has barely begun'. Everything the ecologist looks at is the result of evolution. We are coming to realize that species are at times both less immutable *and* at other times more immutable and capricious than we previously suspected. Ecologists cannot afford any longer to take a simplistic view of the species they work with.

At the same time nearly everything the ecologist looks at can lead to evolution. Ecology is crucial to understanding evolutionary mechanisms. But whether or not evolution will occur depends on the heritability of characters and the availability of variation.

So future progress in understanding evolution depends every much on a pluralism of approaches and a community of effort between ecologists and geneticists.

ACKNOWLEDGEMENTS

I am very grateful to Tom McNeilly, Martin Mortimer and Philip Putwain for their patient reading of earlier drafts of this paper, and to Robert Scott,

Mike Roose and Steve Farrow for allowing me to use some of their unpublished data.

REFERENCES

Allard R.W. & Adams J. (1969) Population studies in predominantly self-pollinating species XII Intergenotypic competition and population structure in barley and wheat. *American Naturalist,* **103**, 621–645.

Allard R.W., Harding J. & Wehrhan C. (1966) The estimation and use of selective values in predicting population change. *Heredity,* **21**, 547–563.

Allen J.A. (1975) Further evidence for apostatic selection by wild passerine birds: 9:1 experiments. *Heredity, London,* **36**, 173–180.

Antonovics J. (1972) Population dynamics of the grass *Anthoxanthum odoratum* on a zinc mine. *Journal of Ecology,* **60**, 351–366.

Antonovics J. (1976) The nature of limits to natural selection. *Annals of the Missouri Botanic Garden,* **63**, 224–247.

Antonovics J. (1978) Population genetics of mixtures. *Plant Relations in Pastures* (Ed. by J.R. Wilson), pp. 233–252. C.S.I.R.O., Melbourne.

Aston J. & Bradshaw A.D. (1966) Evolution in closely adjacent plant populations II *Agrostis stolonifera* in maritime habitats. *Heredity, London,* **21**, 649–664.

Bandeen J.D., Stephenson G.R. & Cowett E.R. (1982) Discovery and distribution of herbicide-resistant weeds in North America. In *Herbicide Resistance in Plants* (Ed. by H. LeBaron & J. Gressel), pp. 9–30. Wiley, New York.

Begon M.E. & Mortimer A.M. (1981) *Population Ecology.* Blackwells, Oxford.

Bishop J.A. & Cook L.M. (1980) Industrial melanism and the urban environment. *Advances in Ecological Research,* **11**, 373–404.

Bishop J.A. & Cook L.M. (1981) *Genetic Consequences of Man Made Change.* Academic Press, London.

Bjorkman D. & Holmgren P. (1963) Adaptability of the photosynthetic apparatus to light intensity in ecotypes from exposed and shaded habitats. *Physiologia Plantarum,* **16**, 889–914.

Bradshaw A.D. (1959) Population differentiation in *Agrostis tenuis* Sibth I Morphological variation. *New Phytologist,* **58**, 208–227.

Bradshaw A.D. (1960) Population differentiation in *Agrostis tenuis* Sibth III Populations in varied environments. *New Phytologist,* **59**, 92–103.

Bradshaw A.D. (1972) Some of the evolutionary consequences of being a plant. *Evolutionary Biology,* **5**, 25–47.

Bradshaw A.D. (1975) The evolution of metal tolerance and its significance for vegetation establishment on metal-contaminated sites. *Proceeding International Conference on Heavy Metals in the Environment, Toronto, 1975* (Ed. by T.C. Hutchinson), pp. 599–622. University of Toronto, Toronto.

Bradshaw A.D. & McNeilly T. (1981) *Evolution and Pollution.* Arnold, London.

Brookes A., Collins J.C. & Thurman D.A. (1981) The mechanism of zinc tolerance in grasses. *Journal of Plant Nutrition,* **3**, 695–705.

Cahn M.A. & Harper J.L. (1976) The biology of the leaf mark polymorphism in *Trifolium repens* L. II Evidence for the selection of leaf marks by rumen fistulated sheep. *Heredity, London,* **137**, 327–333.

Cain A.J. & Currey J.D. (1963) Area effects in *Cepaea. Philosophical Transactions of the Royal Society, London, B.* **246**, 1–81.

Cain A.J. & Sheppard P.M. (1954) Natural selection in *Cepaea. Genetics,* **39**, 89–116.

Cain S.A. (1944) *Foundations of Plant Geography*. Harper & Brothers, New York.

Charles, A.H. (1961) Differential survival of cultivars of *Lolium, Dactylis* and *Phleum*. *Journal of the British Grassland Society*, **16**, 69–75.

Charles A.H. (1964) Differential survival of plant types in swards. *Journal of the British Grassland Society*, **19**, 198–204.

Clarke B.C. (1968) Balanced polymorphism and regional differentiation in land snails. *Evolution and Environment* (Ed. by E.T. Drake), pp. 351–368. Yale University Press, Newhaven.

Clarke B.C. (1969) The evidence for apostatic selection. *Heredity, London*, **124**, 347–352.

Clarke B.C. (1979) The evolution of genetic diversity. *Proceedings of the Royal Society, London*, B. **205**, 453–474.

Clausen J., Keck D.D. & Heisey W.M. (1940) Experimental studies on the nature of species I The effect of varied environments on western North American plants. *Carnegie Institute of Washington Publication*, **520**.

Clausen J., Keck D.D. & Heisey W.M. (1948) Experimental studies on the nature of species III Environmental responses of climatic races of *Achillea*. *Carnegie Institute of Washington Publication*, **581**.

Cook S.A. & Johnson M.P. (1968) Adaptation to heterogeneous environments. I. Variation in heterophylly in *Ranunculus flammula* L. *Evolution*, **22**, 496–516.

Cooper J.P. (1954) Studies on growth and development in *Lolium* IV Genetic control of heading responses in local populations. *Journal of Ecology*, **42**, 521–556.

Cooper J.P. (1959) Selection and population structure in *Lolium* II Genetic control of data of ear emergence. *Heredity, London*, **13**, 445–459.

Cooper J.P. (1960) Selection and population structure in *Lolium* IV Correlated response to selection. *Heredity, London*, **14**, 229–246.

Cooper J.P. & Edwards K.J.R. (1961) The genetic control of leaf development in *Lolium*. 1 Assessment of genetic variation. *Heredity, London*, **16**, 63–82.

Crossley G.K. & Bradshaw A.D. (1968) Differences in response to mineral nutrients of populations of ryegrass, *Lolium perenne* L. and orchard grass, *Dactylis glomerata* L. *Crop Science*, **8**, 383–387.

Davies M.S. & Snaydon R.W. (1976) Rapid population differentiation in a mosaic environment III Coefficients of selection. *Heredity, London*, **36**, 56–66.

de Wit C.T. (1960) On Competition. *Verslagen van Landbouwkundige Onderzoekingen Wageningen*, **66.8**.

Edwards K.J.R. & Cooper J.P. (1963) The genetic control of leaf development in *Lolium* 2 Response to selection. *Heredity, London*, **18**, 307–317.

Eldredge N. & Gould S.I. (1972) Punctuated equilibria: an alternative to phyletic gradualism. *Models in Paleobiology* (Ed. by T.J.M. Schopf), pp. 82–115. Freeman Cooper, San Franciso.

Erhlich P.R. & Raven P.H. (1964) Butterflies and plants: a study in coevolution. *Evolution*, **18**, 586–603.

Falconer D.S. (1981) *Introduction to Quantitative Genetics* (2nd ed.) Longman, London.

Felsenstein J. (1976) The theoretical population genetics of variable selection and migration. *Annual Review of Genetics*, **10**, 253–80.

Gadgil M. & Solbrig O.T. (1972) The concept of *r*- and *K*-selection: evidence from wild flowers and some theoretical considerations. *American Naturalist*, **106**, 14–31.

Gartside D.W. & McNeilly T. (1974) The potential for evolution of heavy metal tolereance in plants II. Copper tolerance in normal populations of different plant species. *Heredity, London*, **32**, 335–48.

Gillett J.B. (1962) Pest pressure, an underestimated factor in evolution. *Taxonomy and Geography* (Ed. by D. Nichols), pp. 37–46. Systematics Association, Oxford.

Gould S.J. & Lewontin R.C. (1979) The spandrels of San Marco and the Panglossian paradigm:

a critique of the adaptationist programme. *Proceedings of the Royal Society of London, B*, **205**, 581–598.

Grant V. (1982) Punctuated equilibria: a critique. *Biologische Zentralblat*, **101**, 175–184.

Gressel J., Ammon H.U., Fogelfors H., Gasquez J., Kay Q.O.N. & Kees, H. (1982) Discovery and distribution of herbicide-resistant weeds outside North America. *Herbicide Resistance in Plants* (Ed. by H. leBaron & J. Gressel), pp. 31–55. Wiley, New York.

Haldane J.B.S. (1957) The cost of natural selection. *Journal of Genetics*, **55**, 511–524.

Harper J.L. (1977) *Population Biology of Plants*. Academic Press, London.

Harper J.L. (1982) After description. *The Plant Community as a Working Mechanism* (Ed. by E.I. Newman), pp. 11–25. Blackwells, Oxford.

Hayward M.D. (1970) Selection and survival in *Lolium perenne*. *Heredity, London*, **25**, 441–447.

Heslop-Harrison J. (1964) Forty years of genecology. *Advances in Ecological Research*, **2**, 159–247.

Hickey D.A. & McNeilly T. (1975) Competition between metal tolerant and normal plant populations; a field experiment on normal soil. *Evolution*, **29**, 458–464.

Humphreys M.D. & Bradshaw A.D. (1976) Genetic potential for solving problems of soil mineral stress: heavy metal toxicities. *Plant Adaptation to Mineral Stress in Problem Soils* (Ed. by M.J. Wright), pp. 95–105. Cornell University Press, Cornell.

Jacob F. (1977) Evolution and tinkering. *Science*, **196**, 1161–1166.

Jain S.K. & Bradshaw A.D. (1966) Evolutionary divergence among adjacent plant populations I The evidence and its theoretical analysis. *Heredity, London*, **21**, 407–421.

Janzen D.H. (1970) Herbivores and the number of tree species in tropical forests. *American Naturalist*, **104**, 501–528.

Johnson R.F., Niles D.M. & Rohwer S.A. (1972). Hermon Bumpus and natural selection in the house sparrow *Passer domesticus*. *Evolution*, **26**, 20–31.

Jones D.A. (1973) Co-evolution and cyanogenesis. *Taxonomy and Ecology* (Ed. by V.H. Heywood), pp. 213–242. Academic Press, London.

Kearsey, M.J. & Barnes, B.W. (1970) Variation for metrical characters in *Drosophila* populations II Natural selection. *Heredity, London*, **25**, 11–21.

Khan M.A., Antonovics J. & Bradshaw A.D. (1976) Adaptation to heterogeneous environments. III The inheritance of response to spacing in flax and linseed (*Linum usitatissimum*). *Australian Journal of Agricultural Research*, **27**, 649–659.

Khan M.A., Putwain P.D. & Bradshaw A.D. (1975) Population interrelationships 2. Frequency dependent fitness in *Linum*. *Heredity, London*, **34**, 145–163.

Law R. (1979) Ecological determinants in the evolution of life histories. *Population Dynamics* (Ed. by R.M. Anderson, B.D. Turner & L.R. Taylor), pp. 81–103. Blackwells, Oxford.

Law R., Bradshaw A.D. & Putwain P.D. (1977) Life history variation in *Poa annua*. *Evolution*, **31**, 233–246.

LeBaron H.M. & Gressel J. (1982) *Herbicide Resistance in Plants*. Wiley, New York.

Lerner I.M. (1958) *The Genetic Basis of Selection*. Wiley, New York.

Lewontin R.C. & Birch L.C. (1966) Hybridization as a source of variation for adaptation to new environments. *Evolution*, **20**, 315–336.

Linhart Y.B. (1974) Intra-population differentiation in annual plants I. *Veronica peregrina* L. raised under non-competitive conditions. *Evolution*, **28**, 232–243.

McMillan C. (1959) The role of ecotypic variation in the distribution of the central grassland of North America. *Ecological Monographs*, **29**, 285–308.

Marchant C.J. (1968) Evolution in *Spartina* (Gramineae) II Chromosomes, basic relationships and the problem of the *S. x townsendii* agg. *Journal of the Linnean Society* (*Botany*), **60**, 381–409.

Mather K. (1953) The genetical structure of populations. *Symposium of Society of Experimental Biology*, **7**, 66–95.

Paulson D.R. (1973) Predator polymorphism and apostatic selection. *Evolution*, **27**, 269–277.

Popham E.J. (1942) Further experimental studies on the selective action of predators. *Proceedings of the Zoological Society of London, A*. **112**, 105–117.

Roberts H.A. (1964) Emergence and longevity in cultivated soil of seeds of some annual weeds. *Weed Research*, **4**, 296–307.

Robertson A. (1955) Selection in animals: synthesis. *Cold Spring Harbor Symposium on Quantitative Biology*, **20**, 225–229.

Sagar G.R. (1959) *The Biology of Some Sympatric Species of Grassland*. D. Phil thesis, University of Oxford.

Sakai K. (1961) Competitive ability in plants: its inheritance and some related problems. *Mechanism of Biological Competition* (Ed. by F.L. Milthorpe), pp. 245–263. Society of Experimental Biology. Symp. 15. Cambridge University Press, Cambridge.

Sammeta K.P.V. & Levins R. (1970) Genetics and ecology. *Annual Review of Genetics*, **4**, 469–488.

Sarukhan J. & Harper J.L. (1973) Studies on plant demography: *Ranunculus repens* L., *R. bulbosus*, L. and *R. acris* L.I. Population flux and survivorship. *Journal of Ecology*, **61**, 675–716.

Schaal B.A. (1975) Population structure and local differentiation in *Liatris cylindracea*. *American Naturalist*, **109**, 511–528.

Shorrocks B. (1978) *The Genesis of Diversity*. Hodder and Stoughton, London.

Silander J.A. & Antonovics J. (1979) The genetic basis of the ecological amplitude of *Spartina patens*.1. Morphometric and physiological traits. *Evolution*, **33**, 1114–1127.

Simpson G.G. (1953) *The Major Features of Evolution*. Columbia University, New York.

Snaydon R.W. & Bradshaw A.D. (1961) Differential responses to calcium within the species *Festuca ovina* L. *New Phytologist*, **60**, 219–234.

Snaydon R.W. & Bradshaw A.D. (1962) Differences between natural populations of *Trifolium repens* L. in response to mineral nutrients. I. Phosphate. *Journal of Experimental Botany*, **13**, 422–434.

Snaydon R.W. & Davies M.S. (1976) Rapid population differentiation in a mosaic environment IV Populations of *Anthoxanthum odoratum* at sharp boundaries. *Heredity, London*, **37**, 9–25.

Solbrig O.T. (1980) Demography and natural selection. *Demography and Evolution in Plant Populations* (Ed. by O.T. Solbrig), pp. 1–20. Blackwell, Oxford.

Sørensen T. (1954) Adaptation of small plants to deficient nutrition and a short growing season, illustrated by cultivation experiments with *Capsella bursapastoris* (L.) *Botaniska Tidsskrift*, **51**, 339–361.

Stratton S.D. Sleper D.A. & Matches A.G. (1979) Genetic variation and inter-relationships of in-vitro dry matter disappearance and fibre content in orchard grass herbage. *Crop Science*, **19**, 329–332.

Suneson C.A. & Stevens H. (1953) Studies with bulked hybrid populations of barley. *U.S. Department of Agriculture Technical Bulletin*, **1067**.

Sved J.A. (1968) Possible rates of gene substitution in evolution. *American Naturalist*, **102**, 283–293.

Taylor A.E.R. & Muller R. (1976) *Genetic Aspects of Host-parasitic Relationship*. Blackwell, Oxford.

Taylor G.E. (1978) Genetic analysis of ecotypic differentiation of an annual plant species, *Geranium carolinianum* L. in response to sulfur dioxide. *Botanical Gazette*, **139**, 362–8.

Taylor G.E. & Murdy W.H. (1975) Population differentiation of an annual plant species, *Geranium carolinianum* L. in response to sulfur dioxide. *Botanical Gazette*, **136**, 212–215.

Turesson G. (1931) The geographical distribution of the alpine ecotype of some Eurasiatic Plants. *Hereditas*, **15**, 329–346.

Turkington R. & Harper J.L. (1979) The growth, distribution and neighbour relationship of *Trifolium repens* in a permanent pasture IV Fine scale biotic differentiation. *Journal of Ecology*, **67**, 245–254.

Van den Bergh J.P. (1968) An analysis of yields of grasses in mixed and pure stands. *Verslagen Van Landbouwkundige Onderzoekingen, Wageningen*, **714**.

Van Valen L. (1973) A new evolutionary law. *Evolutionary Theory*, **1**, 1–30.

Wilson D. & Cooper J.P. (1969) Diallel analysis of photosynthetic rate and related leaf characters among contrasting genotypes of *Lolium perenne*. *Heredity*, **24**, 633–649.

Woodworth C.M., Leng, E.R. & Jugenheimer R.W. (1952) Fifty generations of selection for protein and oil in corn. *Agronomy Journal*, **44**, 60–66.

Wu L., Bradshaw A.D. & Thurman D.A. (1975) The potential for evolution of heavy metal tolerance in plants III The rapid evolution of copper tolerance in *Agrostis stolonifera*. *Heredity, London*, **34**, 165–187.

2. THE GENETICAL ANALYSIS OF ECOLOGICAL TRAITS

M.J. LAWRENCE
Department of Genetics, University of Birmingham, Birmingham, B15 2TT

INTRODUCTION

Differences between individuals of natural populations are nearly always of a quantitative rather than a discontinuous kind and the same is true of differences between individuals from different populations, races, sub-species and, in general, species. In short, phenotype variation in natural assemblages is predominantly of the continuous rather than the discrete kind, irrespective of whether we are interested in behavioural, morphological, physiological, biochemical or fitness traits, any of which could be treated as an ecological trait.

In some circumstances we may wish only to describe the variation in one or more populations in order to find out, for example, whether there are consistent differences between the latter with respect to the trait or traits in question. However, in order to account for the presence of such variation, that is, to attempt to understand its evolutionary significance, it is necessary to satisfy ourselves that the trait is heritable. But because this variation is quantitative rather than discrete, we are forced to use the experimental procedures and analytical methods of biometrical genetics, rather than those of classical Mendelian genetics in order ot answer this question. One of the three chief purposes of this paper is to show that quite simple experiments will suffice to show that a quantitative trait is heritable.

The second problem that faces any attempt to answer the question of the evolutionary significance of variation in natural populations is that the type of selection (if any) that is maintaining this variation is usually far from obvious. It is no accident that melanism in moths, mimicry in butterflies, shell colour and pattern in snails and heavy-metal tolerance in plants are among the most throughly investigated and hence well-known case histories in ecological genetics, for in each the type of selection involved is obvious. It is not as widely known as perhaps it ought to be that a knowledge of the magnitude and direction of dominance and other kinds of non-additive genetical effects can indicate the type of selection to which the character in question is subject. Though the experimental designs required to detect and to estimate the direction and magnitude of non-additive genetical effects make greater demands on one's material than those suitable for estimates of heritability only, the additional information the former provide about the type of selection involved can make the extra effort worthwhile. Indeed, since it can almost be taken for granted that any quantitative character is at least partially heritable, though this must be demonstrated in every case, it can be argued that the chief purpose of genetical experiments with material of natural origin is the detection and estimation of non-additive, rather than additive genetical variation with which estimates of heritability are concerned. The second chief purpose of this paper is to show how non-additive genetical effects can be estimated and how information on the direction of dominance and epistasis can be used to infer the type of selection that is acting on the character in question.

THE DETECTION OF GENETICAL VARIATION

It is necessary to satisfy two requirements in all genetical analyses of quantitative characters; first, we have to be able to recognize groups of relatives and second we need to separate, in a statistical sense, the effects of the environment on phenotypic variation from those of the genotype.

The first requirement has to be satisfied in all genetical experiments, irrespective of the kind of character that is being studied. Relatives may be parents and their offspring, or preferably contemporaries such as families of full or half-sibs. However, relatives usually cannot be recognized in natural populations (with the possible exception of birds or other animals that can be ringed or marked in some other way). For this reason, genetical experiments have to be performed in a laboratory environment in which either natural progenies (from gravid females or seed from single plants) or those produced by controlled crosses are raised in the laboratory, experimental field or glasshouse.

The second requirement also can be satisfied only in a laboratory environment. In the natural habitat of the species in question, differences between individuals due to genetical causes are completely confounded with differences that are due to environmental causes; the genetical and environmental components of phenotypic variation cannot be separated therefore. These two sources of variation can be separated only in circumstances where it is possible to impose statistical control on this variation, which is achieved by replication and randomization of the material over the space occupied by the experiment.

The need to investigate the genetics of quantative characters in a laboratory environment raises two very well-known and much-discussed problems which concern respectively, the best way to sample a natural population and the extent to which inferences about variation in the natural habitat made on the basis of the performance of material raised in the laboratory are valid. This is not the place to go into the question of the best way to sample a population in any detail except to say that this will depend largely on the purpose of the investigation and on the investigator's knowledge of what is practicable with the species in question. We shall return later, however, to consider the problem of the relationship between variation in the natural and laboratory environment in more detail.

With these important preliminary matters in mind, we can turn now to consider the kinds of experiment we can use in order to accomplish the first of our objectives, namely to show that the character of interest is heritable. Details of eight experimental designs are given in the appendix (p. 44) to this paper. These vary in several ways including the number of parents involved (whether few or many), the kinds of progeny involved (whether natural or experimental) and in the type of progeny concerned when these are produced by controlled matings (whether, in the case of plants, these are produced by cross pollination only or by self-pollination also) (Table 2.1).

TABLE 2.1. A comparison of the characteristics and of the information that can be obtained from eight experimental designs

Design	a	b	c	d	e	f	g	h	i	j	k	l	m
1. Nat progenies	+	+	n	n	64	+	+						
2. BIPs	+	+	$2n$	n	128	+	+						
3. NCM1	+	+	$n_1(1+n_2)$	$n_1 n_2$	72	+	+	+		+			
4. NCM2	+	+	n_1+n_2	$n_1 n_2$	16	+	+	+	+	+	(+)		
5. Selfs	+		n	n	64	+	+		(+)	(+)	(+)		
6. ABIPs	+		$2n$	$4n$	32	+	+		+	+	+		
7. Selfing series	+	(+)	2	—	—	+	+	+	+	+	+	+	+
8. TTC	+	(+)	(2)	—	—	+	+	+	+	+	(+)	+	(+)

Key:

a suitable for self-compatible species of flowering plants.
b suitable for self-incompatible and dioecious species of flowering plants and animal species.
c number of parents used.
d number of families of offspring.
e number of parents required for sixty-four families of offspring.
f design can detect additive genetical effects.
g design can be used to estimate heritability.
h design can be used to estimate the additive genetical component of variance.
i design can detect dominance.
j design can be used to estimate the dominance component of variance.
k design can determine the direction of dominance (potence).
l design can detect non-allelic interaction (epistasis).
m design can be used to estimate the components of epistatic variation.

However, all of these designs can be used to detect genetical variation and all provide estimates of the heritability of the trait in question. In practice, one's choice of design will depend largely on the biology, particularly the reproductive biology of the species in question, such as whether, in the case of plants, the species is self-compatible or self-incompatible; and, in general, on the ease or otherwise with which crosses can be made, on the number of offspring produced per cross and so on.

An investigation of variation with respect to seven quantitative characters within, and between fourteen different natural populations of the self-incompatible poppy, *Papaver rhoeas*, will serve to illustrate the information that can be obtained from the simplest experimental design, that involving natural progenies (data from Ooi 1970). Seed was taken from between fifteen and seventy-four plants from each of these populations, these plants being chosen at random except that all parts of the population were visited (Table 2.2). The seed taken from each plant was packeted separately. In the following year, four plants were raised from the seed of each packet in each of two completely and independently randomized blocks. Each plant in the experi-

TABLE 2.2. *Papaver rhoeas*. Population means for the seven characters scored in the experiment. N = number of plants in the population from which seed was sampled

Population	N	$HT8$	$D8$	FT	HFT	PL	FD	SRN
Wool	26	5.31	20.6	28.9	43.1	25.1	8.61	11.2
Wellesbourne	62	7.33	21.9	26.4	54.1	25.8	9.84	11.6
Donnington	25	6.35	21.9	25.9	47.9	24.3	9.39	10.8
Broad Oak(1)	32	6.17	22.1	27.8	52.0	28.2	9.62	11.4
Broad Oak(2)	38	6.46	22.1	26.4	49.7	28.2	9.53	11.5
Condicote	74	6.76	21.6	25.2	49.7	26.4	9.69	10.9
Hackmans Gate(1)	52	8.05	24.2	19.5	47.6	25.5	9.50	10.5
Hackmans Gate(2)	16	7.73	22.7	19.6	42.7	24.5	9.06	10.0
Harley(1)	34	6.20	22.1	27.1	49.7	27.2	9.67	11.1
Harley(2)	50	6.55	22.5	26.1	50.9	27.4	9.84	11.3
Norwich	34	6.59	23.5	26.0	44.5	24.3	9.42	10.5
Cosford	15	7.14	24.5	23.1	44.8	25.0	9.59	10.5
Alridge	47	6.77	21.8	24.0	49.8	27.1	9.79	11.1
Alkerton	47	7.08	21.9	24.2	48.7	26.7	9.67	11.1

Key:
$HT8$ = Height of plant at 8 weeks of age (cm).
$D8$ = Diameter of plant at 8 weeks of age (cm).
FT = Flowering time (days after 31 May).
HFT = Height of plant at flowering time (cm).
PL = Pedicel length of first flower (cm).
FD = Diameter of first flower (cm).
SRN = Average stigmatic ray number of three tallest capsules on plant.

ment was scored for six quantitative characters and those of the first block for a seventh. Preliminary analysis showed that the blocks were homogeneous with respect to the six characters that were scored in both, so that these data may be treated as if they came from a single block containing a designed number of eight plants per family per population. However, for the usual reasons, not all families were in fact represented by eight plants, so that the analyses were adjusted accordingly. The purpose of this experiment was to find out whether each of these traits was heritable and also whether there were any differences between populations over and above those within them for these characters.

Suppose, first, that we are concerned with the data from one population only. Then the structure of the analysis of variance of the data for each character would be as shown on p. 50 with $r = 8$ (assuming that no plants were lost) and n equal to the number of families raised from the population in question. Since we are concerned with fourteen populations in all, there are, in principle, fourteen such analyses of variance for each character. However, it is convenient to combine the data from each population into a single three-level hierarchical analysis of variance. The analysis of variance of the data for one character, plant height at 8 weeks after sowing, is shown in Table 2.3. Since the Families' MS is significantly greater than the Individuals' MS, there is little doubt that this character is, in fact, heritable. Similarly, since the Population's MS is greater than the Families' MS, there is also little doubt that there are systematic differences between populations in respect of height at 8 weeks over and above those within them for this character. The same outcome was obtained from the analysis of each of the

TABLE 2.3. *Papaver rhoeas*. Analysis of variance of plant height at 8 weeks. The coefficients of the variance components, *n*, *r* and *s* are calculated by following a procedure given in Snedecor & Cochran (1967, pp. 291–294)

Item	df	*MS*	*EMS*
Populations	13	127.6754 ***	$\sigma_w^2 + r\sigma_f^2 + s\sigma_p^2$
Families (within populations)	532	9.4061 ***	$\sigma_w^2 + n\sigma_f^2$
Individuals (within families)	3106	3.8326	σ_w^2

*** = $P < 0.001$.

The pair of intraclass correlation coefficients that can be calculated from estimates of these variance components are:

$$t_p = \frac{\sigma_p^2}{\sigma_p^2 + \sigma_f^2 + \sigma_w^2} \text{ for populations; and}$$

$$t_f = \frac{\sigma_f^2}{\sigma_f^2 + \sigma_w^2} \text{ for families.}$$

other characters. Thus we can conclude that each of the characters scored in this experiment is heritable and that the mean expression of each trait differs from one population to another. It is worth mentioning that an analysis of covariance of the data showed that none of these characters was so highly correlated with any other as to raise serious doubts about the wisdom of continuing to treat them as seven characters rather than, say six or a smaller number.

Having shown that each of these characters is heritable, we now wish to know whether some are more heritable than others and also whether populations differ more for some characters than others. Answer to these questions can be obtained by calculating two intraclass correlation coefficients for each character, one of which measures the difference between populations relative to differences within them (t_p) and the other of which measures differences between families relative to variation within them (t_f). Since *Papaver rhoeas* is a self-incompatible species it is not unreasonable to suppose that individuals mate at random. For reasons given in the appendix (p. 51), in these circumstances the heritability of a trait is expected to be between twice and four times t_f, according to whether the members of a family are related as full-sibs (single paternity) or as half-sibs (multiple paternity). In practice, the heritability of a trait is likely to be intermediate to these extremes, with a possible bias towards the lower bound (i.e. ht $= 2t_f$), because of short-range pollen and seed dispersal and hence the likelihood of neighbourhoods of relatives over the area occupied by the population. Strictly, an estimate of heritability concerns a particular population of individuals growing in a particular environment. While the latter poses no problem, for all of the plants measured were raised in the same randomized experiment, the former does, for our estimates of t_f are, in effect, obtained by averaging over populations. On the other hand, in so far that a joint estimate of t_f is based on a greater number of degrees of freedom than one estimated from a single population, it must be regarded as more reliable. However, though there

TABLE 2.4. *Papaver rhoeas*. Intra-class correlation coefficients and heritabilities (both as percentages) of seven characters

Character	t_p	t_f	Heritability
HT8	8.06	15.91	32–64
D8	3.81	8.39	17–34
FT	15.80	23.89	48–96
HFT	9.23	21.03	42–84
PL	7.20	13.60	27–54
FD	5.63	15.61	31–62
SRN	12.34	22.88	46–92

may be some legitimate doubts about some details of this procedure, it does at least give a rough idea of the heritability of a trait relative to that of others, particularly when, as here, these comparisons are made within a single experiment.

Turning then to Table 2.4, we notice that the characters do in fact appear to have rather different heritabilities: flowering time, height at flowering time and stigmatic ray number having the highest heritabilities, and diameter at 8 weeks and pedicel length, the lowest. Though estimates of heritability have notoriously high standard errors, these differences between groups of characters are likely to be genuine both because they are, for the most part, consistent over populations and because similar estimates have been obtained from other experiments on *P. rhoeas* (Ooi 1970).

The rank-order of the intraclass correlations for populations (t_p) is, as it turns out, similar to that of t_f, with flowering time and stigmatic ray number again heading the list and diameter at 8 weeks bringing up the rear. There is no obvious genetical reason for this similarity. However, in the case of the character diameter at 8 weeks it is possible that errors of measurement have inflated the within family component (σ_w^2) and hence deflated both the between family (σ_b^2) and the between population (σ_p^2) components, since this character is not the easiest to measure. For other characters, the relationship between the magnitudes of t_p and t_f is likely to be genuine.

There are three further points worth making about the results of this experiment. First, though there is clear evidence of differentiation between both populations and families, most of the variation for a character falls to differences within families, rather than between them. Indeed, even for flowering time, $1 - t_f$, which is a measure of the proportion of variation within families, is more than three times greater than t_f and for other characters the ratio of $1 - t_f$ to t_f is even greater. This is, of course, as expected in a self-incompatible species such as *P. rhoeas*. In the closely related self-compatible species *P. dubium* however, a much higher proportion of the total phenotypic variance within a population falls to differences between families which again is as expected, because individuals of this species set three-quarters of their seed by self-pollination (Humphreys & Gale 1974). The chief point here is that the value of t_f depends not only on the heritability of the trait, but also on the genetical structure of the population and hence the breeding system of the species.

Second, though the characters that were scored in this experiment were chosen largely on the grounds of technical convenience, a case can be made that some of them are of ecological significance. For example, flowering time and stigmatic ray number are aspects of the reproductive biology of the species and height and diameter at 8 weeks are measures of the rate of establishment and early growth. However, insofar that these characters may be regarded as a random sample of metrical characters, there is no reason to suppose that their genetics is likely to be radically different from that of characters of known or suspected ecological significance.

Third, there is no evidence in these data that the differentiation between populations is of the clinal type. For example, the distance between Wool (Dorset) in the south and Alridge (Staffordshire) in the north of the area from which these samples were obtained is over 160 miles and that between Broad Oak (Herefordshire) in the west and Norwich (Norfolk) in the east is over 200 miles. Nor is there any evidence in these data that differences between populations are due to gross ecological differences despite the fact that, for example, two of the populations, Donnington and Condicote, were found in areas where the soil overlies the Cotswold limestone. The reason for the observed differentiation between populations, which in any case is not very great, is thus not known at present.

Now we have dealt with this example at some length because, simple though it is, it is nevertheless capable of yielding most of the information that is required at the initial stage of an investigation of the genetics and ecology of a quantitative character. The other experimental designs shown in Table 2.1 are capable of yielding more precise information because, for example, with these the paternity of a family is known since crosses are made by hand. However, some of these designs might be inappropriate with species that have separate sexes (e.g. designs 5, 6 and 7) and in any case, it is necessary first to collect one's material and to raise it in a laboratory environment in order to use any design other than the first. Thus in nearly all circumstances it is both possible and desirable that one should commence an investigation with natural progenies in view of the information that this design can yield, information which can be used to plain later experiments more efficiently.

GENOTYPE × ENVIRONMENT INTERACTION

As mentioned earlier, one of the problems that confronts us when attempting to investigate the genetics of a quantitative character is the extent to which inferences about variation in the natural habitat that are based on the observation and analysis of material raised in the laboratory are valid. This problem, the third chief problem to be discussed in this paper, does not ordinarily arise with genetical differences that cause large, non-overlapping differences in phenotype. With quantitative characters, on the other hand, the performace of material reared in a laboratory environment may be quite different from that of the same genotype in the natural habitat of the species in question.

Now there is little doubt that plants raised in a laboratory environment are likely to be larger and more uniform than those of a natural population. This effect, however, causes no interpretative problem provided that those genotypes which produce large phenotypes in one environment do so in other environments and *vice versa* (Fig. 2.1a and b). In other words, if the correlation between the phenotypes produced in different environments by the same genotype is large, and positive, variation in the average value of a character over environments may be regarded as unimportant. If, however, this correlation is low because the rank order of genotypes changes over environments (Fig. 2.1c), we would have justifiable doubts about the validity

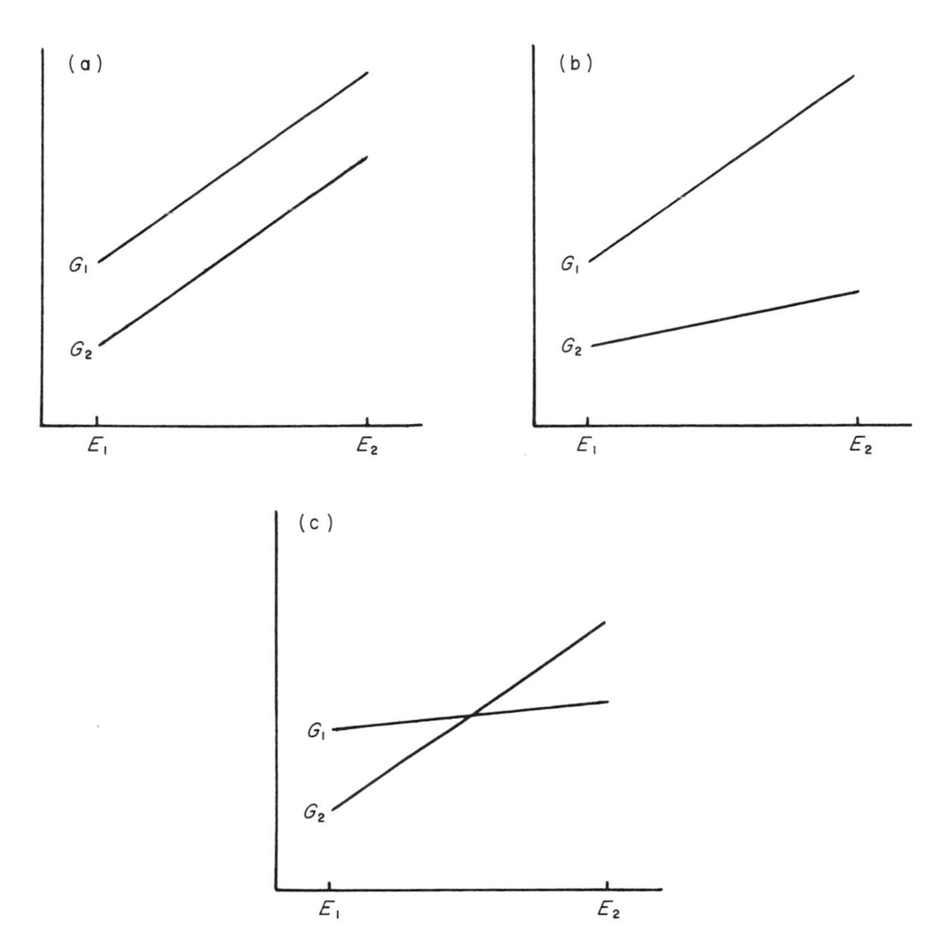

FIG. 2.1. Three possible relationships between the performance of two genotypes (G_1 and G_2) in two environments (E_1 and E_2).

of the connection between the laboratory and natural environment. In short, it is the suspected presence of a particular kind of genotype × environment interaction, not just of genotype × environment interaction in general, which is the cause of our unease in this respect.

One obvious way of investigating this possibility is to carry out an appropriately designed experiment in which the material of interest is raised in a number of different laboratory environments. It is desirable that the environments used in such an experiment should differ for factors known to be important in the natural habitats of the species in question and, if possible, should be greater than the range of the latter. If genotype × environment interaction of the type that weakens the inferential link between the laboratory and nature is not found, our confidence in the validity of this inference will be justified.

Zuberi & Gale (1976) carried out such an experiment, as part of an investigation of variation in *Papaver dubium*, which involved eleven metrical characters measured in each of twenty inbred lines derived from five natural populations. Four sibs from each of these inbred lines were raised in seventeen different environments which differed very greatly in their soil. Thus sixteen of these environments comprised the sixteen factorial combinations of presence or absence of calcium, nitrogen, phosphorous and potassium. Since each of these plots had been maintained in the same way for over 30 years, they provided a very heterogeneous set of soil conditions in which differences in mineral nutrition and pH were certainly much greater than in the natural habitats of the species. The seventeenth environment was provided by the field in which our experiments are usually grown, which thus serves as a control.

Analysis of variance of the data from this experiment showed that the Lines, Environments and, in particular, the Lines × Environments items were highly significant for each of the eleven traits. Product–moment correlation coefficients were then calculated for all pair-wise combinations of the seventeen environments character by character; these are shown in Table 2.5. Since the correlations for *FD*, *SR*, *H10*, *D10* and particularly *FT* are generally high, there is little doubt that performance over environments for these five characters is repeatable. Correlations for three further characters, SHF, HF and FH are much improved if one environment, that receiving nitrogen only, is omitted from the analysis. Since this particular environment was barely capable of supporting *P. dubium* plants, despite the fact that these were raised by sowing seed in compost in the glasshouse, rather than from seed grown directly in the soil, the exclusion of this treatment is justified. For the remaining characters, *LN10*, *BF* and *CN*, however, genotype × en-

TABLE 2.5. *Papaver dubium*. Summary of results of correlation analysis between different environments. Correlations above 0.44 are significant at the 5% level (negative values were never significant). Reproduced by permission of Dr. J.S. Gale and *Heredity* from Zuberi & Gale (1976)

	Number of correlation coefficients within range				
Character	0–0.2 or negative	0.2–0.4	0.4–0.6	0.6–0.8	0.8–1.0
LN10	35	21	21	36	23
H10	0	0	3	38	95
D10	0	0	0	37	99
FT	0	0	0	0	136
FD	1	2	38	64	31
SHF	9	7	18	41	61
HF	12	3	19	43	59
BF	25	38	30	34	9
SR	0	4	28	82	22
CN	47	38	21	27	3
FH	10	23	27	44	32

Key:
*LN*10 = leaf number at 10 weeks of age
*H*10 = height at 10 weeks of age
*D*10 = diameter at 10 weeks of age
FT = flowering time
FD = diameter at flowering time
SHF = stem height at flowering time
HF = height at flowering time
BF = bud number at flowering time
SR = stigmatic ray number
CN = capsule number
FH = final height

vironment interaction is too important a source of variation to be left out of account in assessing how selection might act on these characters. Thus of the eleven characters studied in this experiment only these three gave repeatabilities so low as to suggest that simple inferences about variation in the natural habitats of the species made on the basis of performance in the laboratory are all but worthless. Though this experiment involved the effect of differences in soil on performance, a similar approach could be used with other environmental variables such as time of sowing, crop density, competition with other species and so on.

In certain circumstances a rather more direct approach to the problem of the importance of genotype × environmental interaction with quantitative characters is possible. Thus provided a record is kept of the locations from which seed has been sampled in a natural population of plants, it is possible to compare the performance of plants raised from this seed in the laboratory with those occupying the same locations in the following year. This approach depends on the assumption that the natural population in question is genetically heterogeneous because, for example, it is made up of groups of relatives or because different parts of it contain different genotypes. An experiment of this type was carried out by Snape (1973) on material obtained from a natural population of the predominantly self-pollinating species, *Arabidopsis thaliana*. Though the plants of this population were distributed all over a southward facing slope in the Derbyshire Dales, they were particularly numerous in three more or less distinct locations, A, B and C, which were situated at the top, middle and bottom of the slope respectively. The average flowering times and siliqua numbers of plants raised on agar medium in controlled environment cabinets from seed taken from each of these locations are shown in Table 2.6. Analysis of variance of these data showed that there were significant differences between locations for both characters and that this was due to the material from location A flowering later and bearing more siliquae (pods) than that from B and C. These results suggest that there are genetical differences between plants from location A as compared with those from B and C for both characters. The question here is, Are these differences found in nature? Observation of the performance of plants in the natural habitat suggested that this was indeed the case for flowering time, but not, in any simple way at least, for siliqua number. For technical reasons it was not possible to record the flowering time of individual plants in the population. Instead, the proportion of plants in flower was recorded on each of four occasions in 1972 and on each of five in 1973. In both seasons, the plants of location A came into flower a little later than

TABLE 2.6. *Arabidopsis thaliana*. Average flowering time (*FT*) and siliqua number (*SN*) of plants raised in the laboratory from seed taken from three locations within a single natural population

	Location		
Character	A	B	C
FT	31.5	28.2	27.2
SN	26.1	24.8	25.2

TABLE 2.7. *Arabidopsis thaliana*. Proportion (%) of plants flowering in each of three locations in each of 2 years

Year	Date (day)	Location A	B	C
1972	7.4.72 (7)	0.0	0.8	0.0
	18.4.72 (18)	54.3	80.4	76.8
	4.5.72 (34)	97.3	97.0	98.0
	19.5.72 (49)	100.0	100.0	100.0
1973	11.4.73 (11)	0.0	2.5	0.0
	18.4.73 (18)	17.8	46.0	31.4
	26.4.73 (26)	53.5	79.7	64.5
	4.5.73 (34)	92.4	98.1	93.2
	17.5.73 (47)	100.0	100.0	100.0

those elsewhere in the population (Table 2.7). The siliqua numbers of a random sample of plants measured in each location in each of 2 years is shown in Table 2.8. While the plants in this population produced nearly twice as many siliquae in 1972 as they did in the following year, the rank order is consistent over seasons, the plants at the bottom of the slope (C) producing, on average, more siliquae than those situated higher up (B) which in turn produced more siliquae than those at the top of the slope (A). Comparison of these results with those shown in Table 2.6 reveals, however, that the rank order of locations in this respect has been completely reversed as between the natural habitat and the laboratory. Thus, while inferences about genetical variation for flowering time in the natural habitat that are based on information from laboratory experiments appear to be reasonably sound, comparable inferences about siliqua number are much less so in the absence of a knowledge of the pattern of genotype × environment interaction that this character clearly displays. It will be recalled that we came to very similar conclusions with regard to flowering time and capsule number on the basis

TABLE 2.8. *Arabidopsis thaliana*. Average siliqua numbers of plants growing in each of three locations in each of 2 years

Year	Location A	B	C
1972	11.52	14.13	17.40
1973	6.87	7.65	10.38

of the experiment with poppies that was discussed earlier (Zuberi & Gale 1976).

So far, we have confined our discussion of genotype × environment interaction to its effect on the strength of the inferential link between the performances of material in the laboratory and its performance in the wild. Our discussion of this source of variation should not close, however, without pointing out that the capacity of a genotype to produce different phenotypes in different environments may indicate an adaptive response of the individual to a variable environment. In short, the variability of a quantitative character over different environments may be as suitable a subject for investigation as the variability of the character over different genotypes (e.g. Westerman & Lawrence 1970).

MATHER'S THEORY OF GENETICAL ARCHITECTURE

We turn now to consider the relationship between the type of selection to which a character has been subjected and the genetical architecture of that character, the three chief types of selection being stabilizing, directional and disruptive (Mather 1953).

The theory is based on the notion that the effects of selection are not limited to the more familiar morphological and physiological aspects of the phenotype, but will change also what Darlington has called the nuclear phenotype as well as the properties of the genes themselves. That is to say, selection will change not only the frequencies of genes in populations but also the interrelations of these genes in respect of their effects on the phenotype. In short, the genetical effects that we detect in the genetical analysis of characters are those which have proved successful in the past in that they have given rise to phenotypically adequate individuals.

The next step in the argument is a point made by Fisher (1930) in his theory of the evolution of dominance. He argued that where two alleles, A_1 and A_2, exist in a population, natural selection may be expected to modify the phenotypic expression of heterozygotes, A_1A_2, towards that of the more favoured homozygote, say A_1A_1 ; that is selection, by increasing the frequency of modifying genes in the population, will cause the evolution of dominance of the more advantageous allele, A_1. While there has been some debate about the general validity of Fisher's theory, there is no doubt that genes can alter the dominance relationships of others.

Now where the greater expression of a character in one direction is unconditionally favourable, those alleles acting in that direction will be expected to show dominance over their alternative alleles at all loci; that is, directional selection will be expected to bring about unidirectional dominance in these circumstances. Where, on the other hand, selection is towards a central optimum, as is the case with stabilizing selection, we have no such expectation concerning the direction and magnitude of dominance effects. In short, with stabilizing selection, dominance is expected to be ambidirectional and incomplete, if present.

Much the same argument applies also to the type of non-allelic interaction or epistasis that is expected in these contrasting circumstances. Thus with directional selection, any interaction between genes which minimizes or conceals disadvantageous expression of individual genes will be favoured, so that we expect to find in these circumstances interactions whose effects act in the same direction as dominance; that is, we expect to find interactions of the type described as duplicate epistasis in classical genetics which gives the well-known 15:1 ratio with two such interacting genes of major effect on the phenotype. But as with dominance, we have no such expectation when a character is under stabilizing selection, such non-allelic interaction as may be present being again ambidirectional and weak and hence difficult to detect.

Drawing these arguments together, we expect what Mather (1960, 1966, 1973) has called the genetical architecture of a character to reflect the past history of selection that has been acting on that character. Furthermore, since in many cases, the type of selection that has been acting on a character in the past will be the same as that acting now, we can, with profit, turn the whole argument round. Thus where we can show that a particular trait displays strong unidirectional dominance together with non-allelic interaction of the duplicate type, it is likely that the trait has been and continues to be under directional selection. If, on the other hand, there is either little or no evidence of dominance or, if present, it is ambidirectional and in addition there is little or no evidence of non-allelic interaction, it is likely that the trait is subject to stabilizing selection.

While disruptive selection may be regarded as either the opposite of stabilizing selection or as two-way directional selection, it is less easy to deduce the genetical consequences of this type of selection than that expected from the other types. Thus the expected response will depend not only on the nature and magnitude of the environmental challenge that induces disruptive selection, but also on the gene flow between individuals that inhabit the different habitats. Furthermore, it is by no means clear under what circumstances a population will respond to such a challenge by evolving a polymorphism of two or more forms rather than the necessary degree of developmental flexibility (i.e. adaptive genotype × environment interaction) by which each individual could, in principle, become equally well adapted to any part of the environment. On the other hand, the fact that a trait has been subjected to disruptive selection will often be apparent from the bi- or multimodality of its distribution, so that a knowledge of the genetical architecture of the trait is less important for diagnostic purposes than is the case for the other types of selection.

As mentioned earlier, because the experimental designs required to detect and to estimate the magnitude and direction of non-additive genetical effects (dominance and non-allelic interaction) are, in general, more elaborate than those suitable only for the detection of additive genetical effects, they make greater demands on one's experimental material than the latter. Partly for this reason and because Mather's theory of genetical architecture is perhaps not as widely known as it ought to be, the range of organisms and characters that have been used to test this theory is not yet as great as is desirable. Nevertheless, the experimental evidence that is available, some of which comes from experiments involving one or more of the chief types of selection, is consistent with the theory (Kearsey & Kojima 1967; Mather 1982). Thus components of fitness in *Drosophila*, such as viability, fecundity, egg production and hatchability, which are therefore subjected to directional selection, nearly always display strong directional dominance and, where the experimental design used permits the detection of non-allelic interaction, epistasis of the duplicate type also. Other characters in *Drosophila*, such as wing-length, body weight and chaeta number, which are not so closely related to reproductive success, show incomplete, ambidirectional dominance and little or no non-allelic interaction as expected. Sternopleural chaeta number in *Drosophila melanogaster* is of particular interest because Kearsey & Barnes (1970) have shown that this trait is in fact subject to stabilizing selection as the results from the many experiments with this much studied character suggest.

In general, only two of the experimental designs shown in the appendix allow us both to detect dominance and to determine its direction, the ABIPS design and the early generations of a cross between two inbred lines (designs 6 and 7). Of these, the former is the better, particularly if the parental individuals used to produce the progenies are partly or wholly inbred. Three more designs, NCM2, Selfs and TTC (designs 4, 5 and 8) can also be used provided that it is possible to include both parents and offspring in the same experiment. The simplest designs (1 and 2) are, unfortunately, quite uninformative about dominance.

The genetical architectures of the *P. dubium* characters mentioned earlier were determined in a series of experiments, the earlier of which involved diallel designs and the later, the ABIPS design (Lawrence 1965, 1972; Gale & Arthur 1972; Arthur *et al.* 1973; Gale *et al.* 1976; Thomas & Gale, 1977). The juvenile characters (*LN*10, *H*10 and *D*10) all displayed dominance for rapid development; dominance for capsule number (*CN*) and flower diameter (*FD*) was also in the direction of greater expression for the characters, while that for flowering time (*FT*) was in the early direction. The genetical architecture of this group of characters suggests, therefore, that they are in part under directional selection. Dominance for a second group of characters, stigmatic ray number (*SR*), final height (*FH*) and bud number (*BF*) was weak and ambidirectional which suggests that these characters are subjected to stabilizing selection. As it happens, there is good, direct evidence that two of the juvenile characters, leaf number and diameter, are in fact subjected to natural directional selection for greater expression of the character. Thus in an experiment carried out by MacKay (1981), two groups of nearly 400 seedlings that had been raised from pre-germinated seed on natural soil in an unheated glasshouse were scored for these characters before those of one group were placed outside in January to endure the rigours of winter. In May of the same year, the seedlings of both groups were scored for their survival. The results of this simple experiment are shown in Fig. 2.2; there is little doubt that the most rapidly developing plants, those with the greatest leaf number and diameter at the time of scoring, show the highest percentage survival.

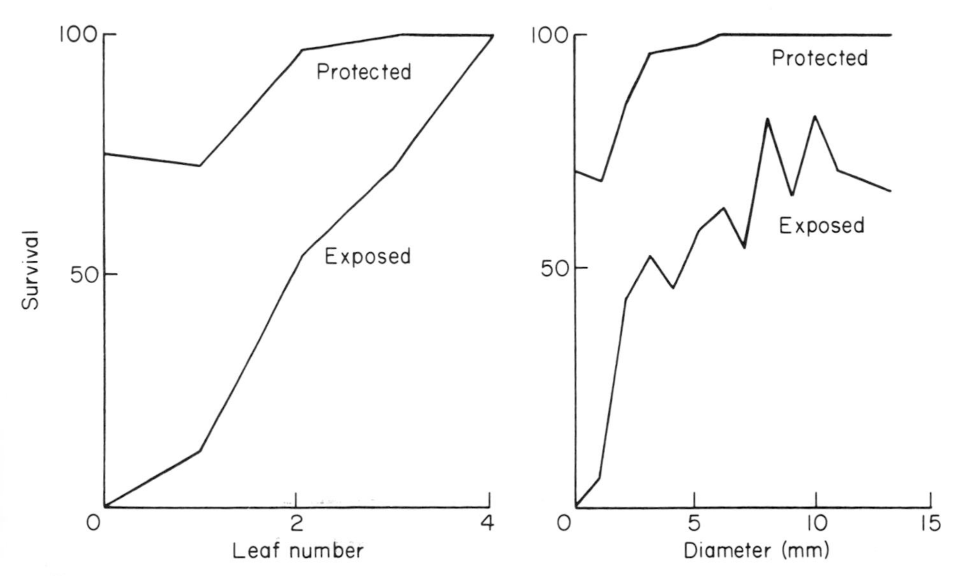

FIG. 2.2. *Papaver dubium*. Survival (%) of exposed and protected seedlings. Leaf number and diameter scored on 13 January and survival on 11 May.

CONCLUSIONS

Biometrical or quantitative genetics has the not wholly justified reputation of being an esoteric, heavily mathematical and statistical branch of genetics. Yet because most of the variation in natural populations is continuous rather than discrete, we can hardly ignore such characters if we wish to understand evolution. The chief purpose of this paper has been to show that quite simple experimental designs and analytical procedures can be employed to answer questions of practical interest by those who wish to investigate the genetics of metrical characters in natural populations. In particular, it is argued that a knowledge of dominance and other non-additive genetical effects can provide a useful guide to the type of natural selection, whether directional, stabilizing or disruptive, to which a trait is subjected. Such information can be used to help with what in nearly all circumstances is the difficult task of detecting directly the type of natural selection that is maintaining variation for quantitative characters in natural populations.

APPENDIX

Genetical expectations of family means and variances

The basic procedure in biometrical genetics is to attempt, first, to detect genetical effects in our material and, second, to estimate these effects either directly or in the form of the heritability of the trait in question. Neither of these tasks can be tackled without first formulating a genetical model which accounts for differences within and between families in terms of one or more of a small number of parameters. The formulation of a suitable model is thus the first step in any analysis of the population genetics of quantitative characters.

The following brief account is intended to give some idea of the way in which such models are formulated. We begin by considering variation between a pair of inbred lines and their descendants in order to introduce some basic notions. Since, however, inbred lines will not ordinarily be available in the early stages of an investigation on a natural population, we consider next the mean and variance of a population of individuals that mate at random in which the gene frequencies are arbitrary.

The assumption of random mating will not be valid with many species, either because some seed is set by self-pollination or because of mating between relatives (e.g. parents and offspring in perennials; or between sibs) or because of both. Models which allow for arbitrary inbreeding and for arbitrary gene frequencies are complex and are therefore unsuitable for present purposes. We can, however, gain some insight into the effect of inbreeding on the detection and estimation of genetical effects by considering the case where all seed is set by self-pollination; that is, the population consists of a lot of different inbred lines. The system of mating in many natural populations is likely to lie somewhere on the continuum between these extremes of random mating and selfing.

Lastly, we consider seven experimental designs that can be used with material of natural origin.

More detailed accounts of biometrical genetics can be found in Jinks (1979, 1981), Lawrence & Jinks (1973) and Mather & Jinks (1977, 1982).

The simple additive–dominance model

Consider a single locus with two alleles, A_1 and A_2. Suppose the effect of substituting A_1 for A_2 is additive in the sense that A_1A_2 individuals are on average d units taller (say) than A_2A_2 plants and that A_1A_1 individuals are on average d units taller than the heterozygotes. The relationship between the mean phenotypes produced by these three genotypes is shown in Fig. 2.3 in diagrammatic form. Since the heterozygote, A_1A_2, is exactly intermediate on

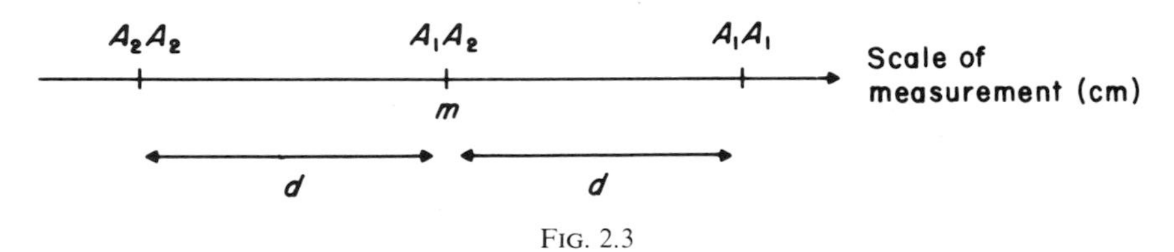

FIG. 2.3

the scale of measurement between the homozygotes, A_2A_2 and A_1A_1, there is no dominance. Call the mid-point between the homozygotes m. Then if A_1A_1 is the genotype of individuals of inbred line P_1 and A_2A_2 that of individuals of inbred line P_2, the genetical expectation of the means of these inbred lines and their F_1 are as follows:

$$\bar{P}_1 = m + d, \qquad \bar{P}_2 = m - d, \qquad \bar{F}_1 = m.$$

Note that P_1 is always taken as the parent of greatest expression, so that d is positive. For many independent genes (i.e. independent in transmission—no linkage; independent in expression—no epistasis) the genetical expectation of these family means are:

$$\bar{P}_1 = m + [d], \qquad \bar{P}_2 = m - [d], \qquad \bar{F}_1 = m,$$

the square brackets allowing for the fact that we shall observe in the phenotype the net effect of allelic substitutions at each of the k loci for which P_1 and P_2 differ.

Suppose, now, that we extend this simple model to allow for dominance. If dominance is present, the score for A_1A_2 is no longer m, though the latter, of course, is still the mid-parent value. Suppose, therefore, that A_1A_2 deviates from m by an amount h. Then in diagramatic form, the model is as shown in Fig. 2.4.

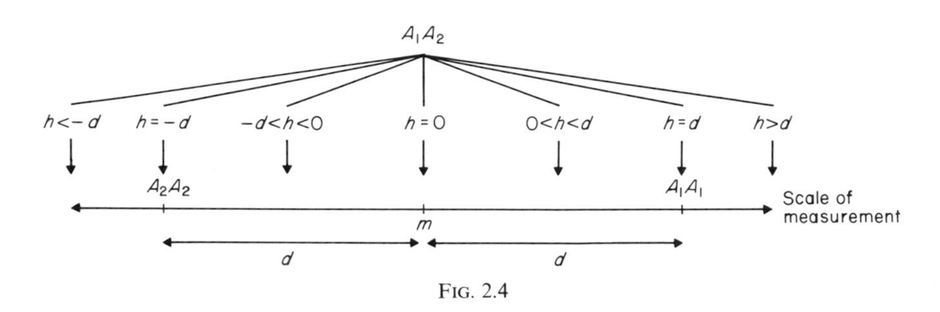

FIG. 2.4

Note that h, unlike d, can take sign to allow for the possibility that A_2 is dominant to A_1 and that $h > d$ or $< -d$ to allow for the possibility of overdominance.

The genetical expectations of the means of the inbred lines and their F_1 now become

$$\bar{P}_1 = m + d, \qquad \bar{P}_2 = m - d, \qquad \bar{F}_1 = m + h.$$

Again, for many independent genes:

$$\bar{P}_1 = m + [d], \qquad \bar{P}_2 = m - [d], \qquad \bar{F}_1 = m + [h].$$

We now have three parameters, m, $[d]$ and $[h]$ to estimate from three equations. Thus if we have numerical values for $\bar{P}_1$, $\bar{P}_2$ and $\bar{F}_1$ we can obtain numerical estimates of m, $[d]$ and $[h]$ since

$$\hat{m} = \tfrac{1}{2}(\bar{P}_1 + \bar{P}_2),$$
$$[\hat{d}] = \tfrac{1}{2}(\bar{P}_1 - \bar{P}_2),$$
$$[\hat{h}] = \bar{F}_1 - \tfrac{1}{2}(\bar{P}_1 + \bar{P}_2).$$

Standard errors of these estimates may be obtained in the usual way, so that we can test whether our estimates $[d]$ and $[h]$ differ significantly from zero; that is, whether there is evidence of additive and dominance effects for the k genes for which P_1 and P_2 differ.

As matters stand at present, however, we have no means of knowing whether this simple additive–dominance model is adequate; to do this, we require more equations. Now by the rules of classical Mendelian genetics, the composition of an F_2 family that is segregating at a single locus is $\frac{1}{4}A_1A_1 + \frac{1}{2}A_1A_2 + \frac{1}{4}A_2A_2$; it follows, therefore, that the mean of this family is:

$$\bar{F}_2 = m + \tfrac{1}{2}h.$$

Similarly, the means of the two backcross families, B_1 and B_2 are

$$\bar{B}_1 = m + \tfrac{1}{2}d + \tfrac{1}{2}h \quad \text{and} \quad \bar{B}_2 = m - \tfrac{1}{2}d + \tfrac{1}{2}h.$$

With these three equations, we now have six in all which, generalizing for many independent genes, are:

$$\begin{aligned} \bar{P}_1 &= m + [d], & \bar{F}_2 &= m + \tfrac{1}{2}[h], \\ \bar{P}_2 &= m - [d], & \bar{B}_1 &= m + \tfrac{1}{2}[d] + \tfrac{1}{2}[h], \\ \bar{F}_1 &= m + [h], & \bar{B}_2 &= m - \tfrac{1}{2}[d] + \tfrac{1}{2}[h]. \end{aligned}$$

Estimates of m, $[d]$ and $[h]$ can be obtained from these six equations by the weighted least squares method and the adequacy of the model tested by the calculation of a residual χ^2. If this χ^2 is not significant and both $[\hat{d}]$ and $[\hat{h}]$ are significantly greater than zero, the simple additive–dominance model adequately accounts for the genetical variation between these family means.

This is the simplest genetical model that we can fit to data. Although this model can be extended without difficulty to include maternal effects, sex linkage, linkage, epistasis and genotype × environment interaction, we shall not pursue these elaborations, because the kinds of experiments which are feasible in the short term with material derived from natural populations will rarely be sufficiently elaborate to permit anything more than the detection and estimation of additive and dominance effects of the genes which determine the trait of interest.

So far we have confined our attention to the genetical expectations of family means. However, it is also possible to derive genetical expectations of family variances, as we shall now see.

Now all variation within the P_1, P_2 and F_1 families must be purely environmental. The F_2 and backcross families, on the other hand, are expected to segregate at each of the k loci for which P_1 and P_2 differ. The variances of these families, therefore, must contain both a genetical and an environmental component.

It can be shown that for a single gene, the genetical components of the variance of these families are:

$$V_{F_2} = \tfrac{1}{2}d^2 + \tfrac{1}{4}h^2, \qquad V_{B_1} = \tfrac{1}{4}(d-h)^2, \qquad V_{B_2} = \tfrac{1}{4}(d+h)^2.$$

In order to avoid the inconvenience (at this stage) of handling terms in

dh, we take the sum of the backcross variances; i.e.

$$V_{B_1} + V_{B_2} = \tfrac{1}{2}d^2 + \tfrac{1}{2}h^2.$$

For many independent genes, the genetic components of these variances are:

$$V_{F_2} = \tfrac{1}{2}\Sigma d^2 + \tfrac{1}{4}\Sigma h^2 \quad \text{and} \quad V_{B_1} + V_{B_2} = \tfrac{1}{2}\Sigma d^2 + \tfrac{1}{2}\Sigma h^2.$$

Defining $D = \Sigma d^2$, $H = \Sigma h^2$ and writing E for the environmental component variation we have

$$V_{P_1} = V_{P_2} = V_{F_1} = E,$$
$$V_{F_2} = \tfrac{1}{2}D + \tfrac{1}{4}H + E,$$
$$V_{B_1} + V_{B_2} = \tfrac{1}{2}D + \tfrac{1}{2}H + 2E.$$

Since we have only three equations from which to estimate the three parameters, D, H and E, we obtain estimates of the latter by perfect-fit solution. An estimate of heritability in the narrow sense, appropriate to an F_2 family is thus

$$ht_N = \frac{\tfrac{1}{2}D}{\tfrac{1}{2}D + \tfrac{1}{4}H + E}.$$

Similarly, for the broad-sense heritability

$$ht_B = \frac{\tfrac{1}{2}D + \tfrac{1}{4}H}{\tfrac{1}{2}D + \tfrac{1}{4}H + E}.$$

Arbitrary gene frequencies

In the model we have been considering, we have not had to allow for arbitrary gene frequencies because the families of interest derive from a cross between two inbred lines. It follows, therefore, that the frequencies of the alleles at all segregating loci must be one half in the F_2 family.

This convenient simplification will not, in general, hold in a natural population; nor shall we have inbred lines at our disposal. We need, therefore, to allow for the possibility that the frequencies of A_1, A_2; B_1, B_2, etc., are not equal. Consider a single locus and suppose that the frequency of A_1 in the population is u and that of A_2 is v, such that $u + v = 1$. Then if individuals in the population mate at random, the frequencies of the three genotypes in the population are:

Genotypes	A_1A_1	A_1A_2	A_2A_2,
Frequencies	u^2	$2uv$	v^2
Score	$m + d$	$m + h$	$m - d$.

The mean of the population with respect to this locus is

$$\bar{P} = m + (u - v)d + 2uvh,$$

and the variance is

$$V_P = 2uv[d + (v - u)h]^2 + 4u^2v^2h^2.$$

For many, independent genes, these expressions become

$$\bar{P}_0 = m + \Sigma(u - v)d + 2\Sigma uvh.$$

and

$$V_{P_0} = 2\Sigma uv[d + (v - u)h]^2 + 4\Sigma u^2v^2h^2.$$

If we now define

$$D_R = 4\Sigma uv[d + (v - u)h]^2$$

and

$$H_R = 16\Sigma u^2v^2h^2,$$

the variance of the population may be recast in the form

$$V_{P_0} = \tfrac{1}{2}D_R + \tfrac{1}{4}H_R + E.$$

Note that the coefficients of D_R and H_R in the equation above are the same as those of D and H for V_{F_2}; this is intentional for in the special case where $u = v = \frac{1}{2}$ at all k loci, the formula above becomes that for V_{F_2}.

Suppose, now, that we inbreed, by self-pollination, the individuals in this population for many generations to the point where all k loci are homozygous. On the one-gene model, in the absence of selection, a proportion u of the plants in this inbred population will be A_1A_1 and a proportion v will be A_2A_2; and the same will be true for each of the remaining $(k - 1)$ loci. Thus for many independent genes

$$\bar{P}_\infty = m + \Sigma(u - v)d$$

and

$$V_{P\infty} = D_1 + E,$$

where

$$D_1 = 4\Sigma uvd^2.$$

We have sketched the outlines of the biometrical genetics we need in order to answer questions of practical interest about variation in natural populations. We can turn, therefore, to consider eight experimental designs that can be used with material of natural origin.

1. *Natural progenies*

Procedure

r sibs are raised from seed taken from each of n randomly sampled wild plants giving a total of nr plants in all.

Analysis of variance

We assume in this and the following designs that the plants are raised in a single, completely randomized block.

Item	df	*MS*	EMS
Between families	$n - 1$	MS_1	$\sigma_w^2 + r\sigma_b^2$
Within families	$n(r - 1)$	MS_2	σ_w^2
Total	$nr - 1$	—	

If MS_1/MS_2 is significantly greater than unity, we can conclude that the trait is heritable (assuming no maternal effects).

Genetical expectations of variance components

(i) If individuals of natural population set seed by self-pollination only, that is, parents are homozygotes:
$\sigma_b^2 = D_1$; the additive genetical component of variation
$\sigma_w^2 = E$, the environmental component of variation
(σ_b^2 is estimated as $(MS_1 - MS_2)/r$)
Hence we can in this case estimate D_1 and E_1 so that

$$ht_N = \frac{D_1}{D_1 + E},$$

where ht_N is the narrow-sense heritability of the trait.

(ii) If individuals in the natural population mate at random and all the members of a family have the *same* male parent, that is, they are related as full-sibs (FS):

$$\sigma_b^2 = \tfrac{1}{4}D_R + \tfrac{1}{16}H_R,$$
$$\sigma_w^2 = \tfrac{1}{4}D_R + \tfrac{3}{16}H_R + E,$$

where D_R and H_R are the additive and dominance components of genetical variation for a population of randomly mating individuals. Note that we cannot estimate the genetical parameters D_R, H_R and E_1 because we have only two equations of estimation. However,

$$t_{FS} = \frac{\sigma_b^2}{\sigma_b^2 + \sigma_w^2} = \frac{\frac{1}{4}D_R + \frac{1}{16}H_R}{\frac{1}{2}D_R + \frac{1}{4}H_R + E} \simeq \frac{1}{2}ht_N,$$

where t is the intraclass correlation coefficient. So that we can, at least, estimate the heritability of the trait.

(iii) If individuals in the population mate at random and each member of a family has a different male parent, that is, they are related as half-sibs (HS):

$$\sigma_b^2 = \tfrac{1}{8}D_R$$
$$\sigma_w^2 = \tfrac{3}{8}D_R + \tfrac{1}{4}H_R + E.$$

Hence

$$t_{HS} = \frac{\sigma_b^2}{\sigma_b^2 + \sigma_w^2} = \frac{\frac{1}{8}D_R}{\frac{1}{2}D_R + \frac{1}{4}H_R + E} = \frac{1}{4}ht_N.$$

For many self-compatible species, the actual mating system is likely to lie between situation (i) and (iii); and for self-incompatible (or dioecious) species, between (ii) and (iii). Hence

Self-compatible species $\quad ht_N \geq t \geq \frac{1}{2}ht_N,$

Self-incompatible species $\quad \frac{1}{2}ht_N \geq t \geq \frac{1}{4}ht_N.$

We cannot be more precise than this unless the mating system is known.

2. *Biparental progenies* (BIPS)

Procedure

(i) Seed is taken from a number (50–100) of plants sampled at random from a natural population. This seed is then bulked and a number of plants raised from this seed in the experimental field. $2n$ plants (100 say) are chosen at random from this material, paired off at random into n pairs and crossed. r sibs are raised from each cross giving a total of nr plants in all.

(ii) Alternatively, the parents of these BIPS could be chosen from the plants of the previous experiment, each natural family, for example, being used to provide one parent.

Analysis of variance

Same as that of the previous design.

Genetical expectations of variance components

(i) Individuals in natural population set seed by self-pollination only, that is, parents are homozygotes:

$$\sigma_b^2 = \tfrac{1}{2}D_R + \tfrac{1}{4}H_R,$$
$$\sigma_w^2 = E.$$

Hence

$$t = \frac{\tfrac{1}{2}D_R + \tfrac{1}{4}H_R}{\tfrac{1}{2}D_R + \tfrac{1}{4}H_R + E} = ht_B,$$

where ht_B is the broad-sense heritability.

(ii) Individuals of natural populations mate at random:

$$\sigma_b^2 = \tfrac{1}{4}D_R + \tfrac{1}{16}H_R,$$
$$\sigma_w^2 = \tfrac{1}{4}D_R + \tfrac{3}{16}H_R + E.$$

Hence

$$t = \frac{\tfrac{1}{4}D_R + \tfrac{1}{16}H_R}{\tfrac{1}{2}D_R + \tfrac{1}{4}H_R + E} \simeq \tfrac{1}{2}ht_N.$$

This is identical to situation (ii) of design 1 because we known with the present design that the members of a family are related as full-sibs.

3. *North Carolina design* 1 (NCM1)

Procedure

A random sample of n_1 plants are obtained from material raised from a bulk sample of seed taken at random from a natural population as in the previous design. These n_1 plants are designated as males. Each of these males is then crossed with n_2 plants designated as females, each male being crossed to a *different set of females*. The total number of parents used in this experiment is thus $n = n_1(1 + n_2)$ and the number of families raised from them, $n_1 n_2$. r sibs are raised from the seed of each cross giving a grand total of $n_1 n_2 r$ plants in all.

Analysis of variance

Item	df	*MS*	EMS
Between males	$n_1 - 1$	MS_1	$\sigma_w^2 + r\sigma_f^2 + n_2 r\sigma_m^2$
Between females within males	$n_1(n_2 - 1)$	MS_2	$\sigma_w^2 + r\sigma_f^2$
Between plants within females and males	$n_1 n_2(r - 1)$	MS_3	σ_w^2
Total	$n_1 n_2 r - 1$	—	

IF MS_2/MS_3 and/or MS_1/MS_2 are significantly greater than unity, we can conclude that trait is heritable.

Genetical expectations of variance components

(i) Individuals in natural population set seed by self-pollination only; that is, parents are homozygotes:

$$\left.\begin{aligned}\sigma_m^2 &= \tfrac{1}{4}D_R,\\ \sigma_f^2 &= \tfrac{1}{4}D_R + \tfrac{1}{4}H_R,\end{aligned}\right\} = \sigma_b^2 \quad \text{of design 2(i).}$$
$$\sigma_w^2 = E,$$

Since we have three equations from which to estimate the three parameters, D_R, H_R and E, we can obtain estimates of the latter by perfect-fit solution as follows:

$$\hat{D}_R = 4\sigma_m^2,$$
$$\hat{H}_R = 4(\sigma_f^2 - \sigma_m^2),$$
$$\hat{E} = \sigma_w^2.$$

Since the total phenotypic variance in this experiment is

$$\sigma_m^2 + \sigma_f^2 + \sigma_w^2 = \tfrac{1}{2}D_R + \tfrac{1}{4}H_R + E,$$
$$ht_N = \frac{\tfrac{1}{2}D_R}{\tfrac{1}{2}D_R + \tfrac{1}{4}D_R + E}$$
$$\left(\text{or } t_m = \frac{\sigma_m^2}{\sigma_m^2 + \sigma_f^2 + \sigma_w^2} = \tfrac{1}{2}ht_N\right).$$

(ii) Individuals in natural population mate at random:

$$\sigma_m^2 = \tfrac{1}{8}D_R,$$
$$\sigma_f^2 = \tfrac{1}{8}D_R + \tfrac{1}{16}H_R,$$

$$\sigma_w^2 = \tfrac{1}{4}D_R + \tfrac{3}{16}H_R + E,$$

$$\hat{D}_R = 8\sigma_m^2,$$

$$\hat{H}_R = 16(\sigma_f^2 - \sigma_m^2),$$

$$\hat{E} = \sigma_w^2 + \sigma_m^2 - 3\sigma_f^2.$$

$$ht_N = \frac{\tfrac{1}{2}D_R}{\tfrac{1}{2}D_R + \tfrac{1}{4}H_R + E}.$$

This can also be estimated from $4t_m$—see (i).

4. *North Carolina design 2* (NCM2)

Procedure

This is similar to the preceding design except that each of the n_1 males is mated to the *same* set of n_2 females. The total number of parents used in this design is thus $n = n_1 + n_2$ and the number of families raised from them, $n_1 n_2$. Assuming as before that r sibs are raised in each family, the total number of plants in the experiment is $n_1 n_2 r$.

Analysis of variance

Item	df	*MS*	EMS
Males	$n_1 - 1$	MS_1	$\sigma_w^2 + r\sigma_{mf}^2 + n_2 r\sigma_m^2$
Females	$n_2 - 1$	MS_2	$\sigma_w^2 + r\sigma_{mf}^2 + n_1 r\sigma_f^2$
Males × females	$(n_1 - 1)(n_2 - 1)$	MS_3	$\sigma_w^2 + r\sigma_{mf}^2$
Within families	$n_1 n_2(r - 1)$	MS_4	σ_w^2
Total	$n_1 n_2 r - 1$	—	

If MS_3/MS_4 is significant, we can conclude that dominance (or other non-additive genetical effects) is present; if either MS_1/MS_3 or MS_2/MS_3 (or both) are significant, we can conclude that the trait is heritable; if MS_2/MS_1 is significant, maternal effects are also present. Note that this is the first design that gives a test for dominance in the analysis of variance.

Genetical expectations of variance components

(i) Individuals in natural population set seed by self-pollination only:

$$\left.\begin{array}{l}\sigma^2_m = \tfrac{1}{4}D_R,\\ \left.\begin{array}{l}\sigma^2_f = \tfrac{1}{4}D_R,\\ \sigma^2_{mf} = \tfrac{1}{4}H_R,\end{array}\right\} = \sigma^2_f \text{ of design 3(i),}\end{array}\right\} = \sigma^2_b \text{ of design 2(i).}$$

$$\sigma^2_w = E,$$

$$\hat{D}_R = 2(\sigma^2_m + \sigma^2_f),$$

$$\hat{H}_R = 4\sigma^2_{mt}$$

$$\hat{E} = \sigma^2_w,$$

from which ht_N can be obtained as before.

(ii) Individuals in natural population mate at random:

$$\left.\begin{array}{l}\sigma^2_m = \tfrac{1}{8}D_R,\\ \left.\begin{array}{l}\sigma^2_f = \tfrac{1}{8}D_R,\\ \sigma^2_{mf} = \tfrac{1}{16}H_R,\end{array}\right\} = \sigma^2_f \text{ of design 3 (ii),}\end{array}\right\} = \sigma^2_b \text{ of design 2(ii).}$$

$$\sigma^2_w = \tfrac{1}{4}D_R + \tfrac{3}{16}H_R + E,$$

$$\hat{D}_R = 4(\sigma^2_m + \sigma^2_f),$$

$$\hat{H}_R = 16\,\sigma^2_{mf},$$

$$\hat{E} = \sigma^2_w - \sigma^2_m - \sigma^2_f - 3\sigma^2_{mf},$$

from which, again, ht_N can be obtained as before,

5. *Self progenies*

Procedure

Each of n plants raised from seed sampled at random from a natural population are self-pollinated. r sibs are then raised in the following season from each of these n parents, giving nr plants in all.

Analysis of variance

Same as for designs 1 and 2.

Genetical expectations of variance components

(i) If individuals of natural population set seed by self-pollination only, then genetical expectations of variance components are the same as for design 1(i).

(ii) Individuals in natural population mate at random:

$$\sigma_b^2 = \tfrac{1}{2}D_1 + \tfrac{1}{8}H_1 - \tfrac{1}{16}H_2 - \tfrac{1}{4}F,$$
$$\sigma_w^2 = \tfrac{1}{4}D_1 + \tfrac{1}{8}H_1 + E.$$

where H_1, H_2 are dominance and F the additive-by-dominance components of genetical variation for a partially inbred population ($D_1 = 4\Sigma uvd^2$; $H_1 = 4\Sigma uvh^2$; $H_2 = 16\Sigma u^2v^2h^2$; and $F = 8\Sigma uv[u-v]dh$). Because the coefficient of D_1 in σ_b^2 is one half, whereas that of D_R is only one quarter in the corresponding equation for design 2(ii), the test for heritable variation is twice as sensitive for self progenies compared with BIPS. In other respects, however, self progenies are less satisfactory because we now have five, rather than three parameters to estimate, a problem which can be resolved only if both parents and offspring of the present generation are included in the experiment. However, it is still possible to obtain a rough estimate of heritability in the present circumstances as follows:

$$t = \frac{\tfrac{1}{2}D_1 + \tfrac{1}{8}H_1 - \tfrac{1}{16}H_2 - \tfrac{1}{4}F}{\tfrac{3}{4}D_1 + \tfrac{1}{4}H_1 - \tfrac{1}{16}H_2 - \tfrac{1}{4}F + E} \simeq \tfrac{2}{3}ht_N.$$

Inbreeding depression and dominance

We saw earlier that the mean of a population of randomly mating individuals is

$$\bar{P}_0 = m + \Sigma(u-v)d + 2\Sigma uvh.$$

The mean of the next generation produced by selfing these individuals is

$$\bar{P}_1 = m + \Sigma(u-v)d + \Sigma uvh,$$

so that $\bar{P}_0 - \bar{P}_1 = \Sigma uvh$.

This makes clear that the sole cause of inbreeding depression is dominance (or strictly, non-additive) effects, a condition which indicates perhaps the most straightforward test for dominance. Thus if seed can be saved from one generation to the next, 100 plants, say, of each generation can be raised in the same randomized experiment and their means compared. If the decline in the performance of the inbred material relative to their non-inbred parents with respect to the trait in question is significant we can conclude that part of the genetical variation for the character is due to dominance effects.

Thus while design 5 does not by itself provide an opportunity of either detecting or estimating dominance, this simple extension yields a very sensitive method for the detection of both the presence and direction of dominance.

6. *Augmented biparental progenies* (ABIPS)

This design combines designs 2 and 5 in one experiment.

Procedure

Each of $2n$ plants raised from seed sampled at random from a natural population are paired off at random, and selfed and crossed reciprocally with the other member of each pair. Each pair of parents thus gives rise to a set of four families, two of which are self and two cross progenies, there being n sets of such progenies. If r sibs are raised in each family, the experiment involves $4nr$ plants in all.

Analysis of variance

Consider *one* set of four families involving $4r$ plants. Then the total SS of these data may be partitioned as follows:

Total SS $(4r-1)$df	Between families SS 3 df	Between self SS	1 df
		Between crosses SS	1 df
		Selfs v. Crosses SS	1 df
	Within families SS $4(r-1)$df	Within selfs SS	$2(r-1)$df
		Within crosses SS	$2(r-1)$df

For genetical purposes it is convenient to consider the three parts of this analysis separately as shown in Table 2.9.

A similar procedure can be followed for each of the remaining $(n-1)$

TABLE 2.9

Analysis of	Item	df	MS	EMS
(a) Self progeny	Between selfs	1	MS_1	$\sigma^2_{ws} + r\sigma^2_{bs}$
	Within selfs	$2(r-1)$	MS_2	σ^2_{ws}
	Total (selfs)	$2r-1$	—	
(b) Cross progeny	Between crosses	1	MS_1	$\sigma^2_{wc} + r\sigma^2_{bc}$
	Within crosses	$2(r-1)$	MS_2	σ^2_{wc}
	Total (crosses)	$2r-1$	—	
(c) Selfs v. crosses	Selfs v. crosses	1	MS_1	$\sigma^2_w + 2r\sigma^2_h$
	Within families	$4(r-1)$	MS_2	σ^2_w
	Total	$4r-3$	—	

TABLE 2.10.

Analysis of	Item	df	MS	EMS
(a) Self progeny	Between selfs	n	MS_1	$\sigma^2_{ws} + r\sigma^2_{bs}$
	Within selfs	$2n(r-1)$	MS_2	σ^2_{ws}
	Total	$n(2r-1)$	—	
(b) Cross progeny	Between crosses	n	MS_1	$\sigma^2_{wc} + r\sigma^2_{bc}$
	Within crosses	$2n(r-1)$	MS_2	σ^2_{wc}
	Total	$n(2r-1)$		
(c) Selfs v. crosses	Selfs v. crosses	n	MS_1	$\sigma^2_w + 2r\sigma^2_h$
	Directional dominance	1	MS_{11}	—
	Dominance variation	$n-1$	MS_{12}	—
	Within families	$4n(r-1)$	MS_2	σ^2_w
	Total	$n(4r-3)$		

sets of families. An overall analysis of differences between (a) selfs (b) crosses and (c) between selfs and crosses can then be obtained by summing the SS's shown in Table 2.9 over all n sets to give the analysis shown in Table 2.10.

Analysis (a) is similar to the analysis of designs 1, 2 and 5; if, therefore, the ratio of MS_1/MS_2 is significantly greater than unity, we can conclude that the trait is heritable. If in analysis (b) MS_1/MS_2 is significant we can conclude that the trait is subject to maternal effects. If in analysis (c) MS_1/MS_2 is significant, then the trait displays dominance. Furthermore, the selfs v. crosses item of this analysis can be partitioned into an overall item for 1 df which gives a measure of directional dominance and another which measures dominance variation over crosses for $(n-1)$ df. The first of these new items is calculated on the grand total of self families and that for cross families and the second is obtained by subtracting the first from the selfs v. crosses SS.

Genetical expectations of variance components

Analysis (a). (i) If individuals of natural populations set seed by self-pollination only then

$$\left.\begin{aligned}\sigma^2_{bs} &= D\\ \sigma^2_{ws} &= E\end{aligned}\right\}\text{ i.e. same as for design 1(i).}$$

(ii) If individuals of natural population mate at random

$$\left.\begin{aligned}\sigma^2_{bs} &= \tfrac{1}{2}D + \tfrac{1}{8}H_1 - \tfrac{1}{16}H_2 - \tfrac{1}{4}F\\ \sigma^2_{ws} &= \tfrac{1}{4}D + \tfrac{1}{8}H_1 + E\end{aligned}\right\}\text{ i.e. same as for design 5(ii).}$$

Analysis (*b*). σ^2_{bc} has, in general, no genetical expectation since it measures maternal effects (however $\sigma_{bc} > 0$ if trait shows sex-linked inheritance in a dioecious species, especially if individuals of the heterogametic sex only are scored in the experiment)

(i) Parents inbred

$$\sigma^2_{wc} = E.$$

(ii) Parents mate at random

$$\sigma^2_{wc} = \tfrac{1}{4}D_R + \tfrac{3}{16}H_R + E \equiv \tfrac{1}{4}D_1 + \tfrac{1}{4}H_1 - \tfrac{1}{16}H_2 - \tfrac{1}{4}F + E.$$

Analysis (*c*). (i) Parents inbred

$$\sigma^2_h = \tfrac{1}{8}H_1.$$

(ii) Parents mate at random

$$\sigma^2_h = \tfrac{1}{64}H_1 + \tfrac{3}{128}H_2.$$

Since σ^2_w is simply the average of σ^2_{ws} and σ^2_{wc}, it is of little interest.

7. *Generations of the selfing series*

All of the designs that we have discussed so far, involve one round of experimental crossing only. In certain circumstances, such as where we are dealing with a species that is known to set seed predominantly by self-pollination or where we are interested in the genetics of a very large phenotypic difference (e.g. heavy metal tolerance v. non-tolerance), it could be worth while to carry out two successive rounds of crossing between the descendents of two parents, P_1 and P_2 to give the standard set of early generation of the selfing series (e.g. P_1, P, F_1, F_2, B_1, B_2).

We have discussed the basic information that can be obtained from this set of families in the introduction to this appendix. The minimum experiment involves these six families. However, if all possible reciprocal crosses are made, this number can be increased to sixteen. The number of plants to be raised in each family depends on a number of considerations, but should be larger for the segregating families than for the non-segregating ones. The data need be dealt with by the analysis of variance only in circumstances where crosses are replicated over reciprocals and/or blocks in order to establish homogeneity within families. If the simple m, [d], [h] model turns out to be inadequate, the model may be expanded to account for additional genetical effects (note that this is the first design that we have considered which provides an opportunity of testing assumptions made in formulating genetical models). However, though this design is simple in principle, the interpretation of the results of analysis may not be and advice should be sought (and the literature studied thoroughly) before embarking on an investigation which uses this design.

8. *Triple test cross* (TTC)

This is the best design currently available because it provides independent tests for the presence of additive, dominance and epistatic effects and yields independent estimates of the additive (D) and dominance (H) components of genetical variation that are equally precise. It was originally designed for the analysis of an F_2 family (as shown below), but *in appropriate circumstances* it can be used with other kinds of segregating families or populations provided that individuals of extreme phenotype (hence diverse genotype) can be extracted from the population; these individuals can then be used as testers.

Procedure

Each of nF_2 individuals ($n = 20 - 30$ say) are crossed to each of their grandparental inbreds, P_1 and P_2, and also to their reciprocal F_1 parents, $F_1(1 \times 2)$ and $F_1(2 \times 1)$, which serve as testers (the design was called a triple test cross because in the original version only one F_1 tester was used). Each F_2 individual, therefore, gives rise to four progenies. For convenience we may label these progenies as follows: $F_2 \times P_1 = L_1$, $F_2 \times P_2 = L_2$, $F_2 \times F_1(1 \times 2) = L_3$, $F_2 \times F_1(2 \times 1) = L_4$. In all, there will be $4n$ such progenies. If r sibs are raised in each family ($r = 5 - 10$ say), the experiment will involve a grand total of $4nr$ individuals.

Analysis of variance

Although the analysis of a triple test cross experiment can take a conventional form, for genetical purposes it is convenient to follow a procedure analogous to that used with the ABIPS design 6.

For each set of four families with the same F_2 parent in common, we calculate the following four comparisons and their associated SS's as follows:

$$C_1 = L_1 + L_2 + L_3 + L_4 \quad \text{and } C_1^2/4r$$
$$C_2 = L_1 - L_2 \quad \text{and } C_2^2/2r$$
$$C_3 = L_1 + L_2 - L_3 - L_4 \quad \text{and } C_3^2/4r$$
$$C_4 = L_3 - L_4 \quad \text{and } C_4^2/2r$$

where L_1, L_2, L_3 and L_4 are the totals of each of the four kinds of family obtained by summing over the r scores in each. These are the between-family comparisons. We need also to calculate four within-family *SS*'s in the usual way, each of which will have $(r-1)$df, assuming no individuals are missing or have been otherwise excluded from the data. Call these within-family *SS*'s, SS_{L_1}, SS_{L_2}, SS_{L_3} and SS_{L_4}. Since we are concerned with n such sets of families, we will have n sets of four between-family *SS*'s and n sets of within-family *SS*'s when we have completed these calculations.

We next obtain the following sums of these *SS*'s and comparisons:

$$\begin{array}{lll} \Sigma C_1 & \Sigma(C_1^2/4r) & \Sigma SS_{L_1} \\ \Sigma C_2 & \Sigma(C_2^2/2r) & \Sigma SS_{L_2} \\ & \Sigma(C_3^2/4r) & \Sigma SS_{L_3} \\ & \Sigma(C_4^2/2r) & \Sigma SS_{L_4} \end{array}$$

where $\Sigma C_1 = C_1$ for first set $+ C_1$ for second set and so on through to the nth set; and similarly for the other sums. The *SS*'s required for the analysis of variance of these data are then calculated from these sums as follows:

(i) Additive $SS = \Sigma(C_1^2/4r) - (\Sigma C_1)^2/4nr$ with $(n-1)$df.
(ii) Dominance $SS = \Sigma(C_2^2/2r) - (\Sigma C_2)^2/2nr$ with $(n-1)$df.
(iii) Epistasis $SS = \Sigma(C_3^2/4r)$ with n df.
(iv) Reciprocals $SS = \Sigma(C_4^2/2r)$ with n df.
(v) Error *SS* for Additive and Epistasis items $= \Sigma SS_{L_1} + \Sigma SS_{L_2} + \Sigma SS_{L_3} + \Sigma SS_{L_4}$ with $4n(r-1)$df.
(vi) Error *SS* for Dominance item $= \Sigma SS_{L_1} + \Sigma SS_{L_2}$ with $2n(r-1)$df.
(vii) Error *SS* for Reciprocals item $= \Sigma SS_{L_3} + \Sigma SS_{L_4}$ with $2n(r-1)$df.

The results of this analysis can then be set out as shown in Table 2.11.

Tests for the presence of additive, dominance, epistatic and reciprocal

TABLE 2.11.

	Item	df	*MS*	*EMS*
(a) Additive effects	Additive (1)	$n-1$	MS_1	$\sigma_w^2 + 4r\sigma_a^2$
	Error (5)			
	Error (5)	$4n(r-1)$	MS_2	σ_w^2
(b) Dominance effects	Dominance (2)	$n-1$	MS_1	$\sigma_{wd}^2 + 2r\sigma_d^2$
	Error (6)	$2n(r-1)$	MS_2	σ_{wd}^2
(c) Epistatic effects	Epistasis (3)	n	MS_1	$\sigma_w^2 + 4r\sigma_e^2$
	Error (5)	$4n(r-1)$	MS_2	σ_w^2
(d) Reciprocal effects	Reciprocals (4)	n	MS_1	$\sigma_{wr}^2 + 2r\sigma_r^2$
	Error (7)	$2n(r-1)$	MS_2	σ_{wr}^2

effects are made by calculating the ratio of MS_1/MS_2 for analyses (a), (b), (c) and (d) respectively. If this ratio is significant in analysis (c) *seek advice and ignore what follows*.

Genetical expectations of variance components

It can be shown that *in the absence of epistasis*:

$$\sigma_a^2 = \tfrac{1}{8}D,$$
$$\sigma_d^2 = \tfrac{1}{8}H$$
$$\sigma_{wd}^2 = \tfrac{1}{8}D + \tfrac{1}{8}H + E.$$

from which estimates of *D*, *H* and *E* may be obtained by perfect-fit solution; and estimates of heritability that are appropriate to an F_2 family may be obtained in the usual way.

The analytical power of this design can be increased considerably by including P_1, P_2 and F_1 individuals (i.e. the testers) and also F_3 families that have been produced by selfing the F_2 individuals that have been crossed to the testers; or families that have been produced by mating these F_2 individuals at random. Consult the literature for further details.

Though this design is fairly straightforward in principle, a number of interpretative problems may be encountered in practice and advice should be sought (and the literature studied) before embarking on an investigation which employs this design.

REFERENCES

Arthur A.E., Rana M.S., Gale J.S., Humphreys M.O. & Lawrence M.J. (1973) Variation in wild populations of *Papaver dubium*. VI. Dominance relationships for genes controlling metrical characters. *Heredity*, **30**, 177–187.

Fisher R.A. (1930) *The Genetical Theory of Natural Selection*. Clarendon, Oxford (2nd ed., Dover, New York, 1958).

Gale J.S. & Arthur A.E. (1972) Variation in wild populations of *Papaver dubium*. IV. A survey of variation. *Heredity*, **28**, 91–100.

Gale J.S. Solomon R., Thomas W.T.B. & Zuberi M.I. (1976) Variation in wild populations of *Papaver dubium*. XI. Further studies on direction of dominance. *Heredity*, **36**, 417–422.

Humphreys M.O. & Gale J.S. (1974) Variation in wild population of *Papaver dubium*. VIII. The mating system. *Heredity*, **33**, 33–42.

Jinks J.L. (1979) The biometrical approach to quantitative variation. *Quantitative Genetic Variation* (Ed. by J.N. Thompson jr, & J.M. Thoday), pp. 81–109. Academic Press, New York.

Jinks J.L. (1981) The genetic framework of plant breeding. *Philosophical Transactions of the Royal Society of London*, **292**, 407–419.

Kearsey M.J. & Barnes B.W. (1970) Variation for metrical characters in *Drosophila* populations. II. Natural selection. *Heredity*, **25**, 11–21.

Kearsey M.J. & Kojima K. (1967) The genetical architecture of body weight and egg hatchability in *Drosophila melanogaster*. *Genetics*, **56**, 23–37.

Lawrence M.J. (1965) Variation in wild populations of *Papaver dubium*. I. Variation within populations, diallel crosses. *Heredity*, **20**, 183–204.

Lawrence M.J. (1972) Variation in wild populations of *Papaver dubium*. III. The genetics of stigmatic ray number, height and capsule number. *Heredity*, **28**, 71–90.

Lawrence M.J. & Jinks J.L. (1973) Quantitative Genetics. *Practical Genetics*. (Ed. by P.M. Sheppard) pp. 86–129. Blackwell, Oxford.

Mackay I.J. (1981) *Population Genetics of Papaver dubium*. Ph.D. thesis, University of Birmingham.

Mather K. (1953) The genetical structure of populations. *Symposium of the Society for Experimental Biology*, **7**, 66–95.

Mather K. (1960) Evolution in polygenic systems. *Evoluzione e Genetica*, pp. 131–152. Academia Nazionale dei Lincei, Rome.

Mather K. (1966) Variability and selection. *Proceedings Royal Society London B*. **164**, 328–340.

Mather K. (1973) *Genetical Structure of Populations*. Chapman and Hall, London.

Mather K. (1982) Response to Selection. *The Genetics and Biology of Drosophila*, vol. 3c. (Ed. by M. Ashburner & J.N. Thompson jr). Academic Press, London.

Mather K. & Jinks J.L. (1977) *An Introduction to Biometrical Genetics*. Chapman and Hall, London.

Mather K. & Jinks J.L. (1982) *Biometrical Genetics* 3rd ed. Chapman and Hall, London.

Ooi S.C. (1970) *Variation in Wild Populations of Papaver rhoeas. L*. Ph.D. thesis, University of Birmingham.

Snape J.W. (1973) *Population and Biometrical Genetics of Arabidopsis thaliana: A Study of a Single Population*. Ph.D. thesis, University of Birmingham.

Snedecor G.W. & Cochran W.G. (1967) *Statistical Methods*, 6th ed. Iowa State University Press, Ames.

Thomas W.T.B. & Gale J.S. (1977) Variation in wild populations of *Papaver dubium*. XII. Direction of dominance during development. *Heredity*, **39**, 305–312.

Westerman J.M. & Lawrence M.J. (1970) Genotype-environment interactions and developmental regulation in *Arabidopsis thaliana*. I. Inbred lines; Description. *Heredity*, **25**, 609–627.

Zuberi M.I. & Gale J.S. (1976) Variation in wild populations of *Papaver dubium*. X. Genotype-environment interaction associated with differences in soil. *Heredity*, **36**, 359–368.

3. THE DEFINITION AND MEASUREMENT OF FITNESS

F.B. CHRISTIANSEN
Department of Ecology and Genetics, University of Aarhus, Ny Munkegade, DK-8000 Aarhus C, Denmark

INTRODUCTION

Fitness is the character of the individual on which natural selection acts. This definition is embedded in Darwin's definition of natural selection: 'This preservation of favourable individual differences and variations, and the destruction of those which are injurious, I have called Natural Selection, or the Survival of the Fittest' (Darwin 1872). Thus, fitness is the theoretical quantification of the attributes of an individual in relation to its 'struggle for life'. The role of the fitness concept in Neo-Darwinian evolutionary theory has been heavily influenced by Fisher's Fundamental Theorem of Natural Selection which states that: 'The rate of increase in fitness of any organism at any time is equal to its genetic variance in fitness at that time' (Fisher 1930). Fisher defined fitness as the Malthusian parameter calculated in terms of the instantaneous individual birth and death rates. The fundamental theorem introduced a great simplification of evolutionary reasoning in conjunction with the Fisherian view of Darwinian evolution, namely, the accumulation over a long period of time of changes at many loci influencing a character subject to natural selection. On this view, the process of evolution of a character may be sufficiently described by specifying the fitness of the various phenotypes and the heritable fraction of the fitness variation attributed to the character. Further, the description of the fitnesses of an array of phenotypes would indicate the phenotype that is most favoured by natural selection and hence the Optimal Adaptive Stragegy which would be the goal of evolution, assuming sufficient genetic variation for the character. This highly deterministic view of evolution has been and is still much debated (Wright 1931; Haldane 1932). However, Fisher's fundamental theorem reconciled Darwinian evolution with genetics in that it linked the evolutionary improvement of the species with individual fitness and the genetic variation of fitness within the species and, furthermore, permitted evolutionary reasoning without immediate reference to genetics. Thus, fitness became established as a strong intuitive concept.

Fisher's fundamental theorem rests on two important assumptions:

(a) the process of natural selection may be described through the concept of individual fitness, and (b) the process of genetic evolution may be described for each gene separately. Both these assumptions are assumptions of independence, that is, neither individuals nor genes interact. The first assumption, which is our major concern here, breaks down for sexually reproducing organisms in which reproduction usually requires the interaction of two or more individuals (for asexually reproducing organisms the theorem is almost tautological). Ever since Darwin, fitness has been viewed as a composite of two aspects or components, namely, survival and fecundity. However, while the probability of survival can easily be visualized as an individual character, reproduction and number of offspring produced typically are a property of parental pairs. Indeed, an important part of the fundamental theorem, the mean fitness principle, which states that the mean fitness of the population increases between generations (Scheuer & Mandel 1959; Mulholland & Smith 1959; Kingman 1961), has been shown to hold in populations with discrete non-overlapping generations for fitness variation only due to differential viability. Fitness variation due to fecundity differences proved to be genetically rather more complex (Bodmer 1965), and mean fitness does not necessarily increase between generations (Pollak 1978). In age-structured populations the situation becomes even more complex and the description of evolution by natural selection of a simple character like individual fitness, and the fundamental theorem of natural selection, is therefore only applicable if interactions between individuals can be neglected. This is a reasonable first approximation if the per-locus variation in survival and fecundity is very small (Pollak & Kempthorne 1970; Charlesworth 1974, 1976, 1980); that is, only if the Fisherian view of evolution is applicable.

The second assumption of Fisher's fundamental theorem, the assumption of independent evolution among genes affecting the character, is crucial even if variation in fecundity is neglected. The genotypic frequencies at each genetic locus need to be in Hardy–Weinberg proportions and dominance requires a modification of the theorem (Kimura 1958; Turner 1970). For two loci the theorem fails (Kojima & Kelleher, 1961; Moran 1964; Turner 1981), but nevertheless it seems that the mean viability of the population usually increases in most cases (Karlin & Carmelli 1975; Ewens 1980). In general the only part of the fundamental theorem that generalizes to multiple loci is that the directly heritable part of the fitness variance (Fisher's genetic variance) vanishes when the population reaches a stable genetic equilibrium (Ewens & Thomson, 1977). For multiple loci, qualitative as well as quantitative aspects of the evolution of a character are influenced by the way in which the genes segregate and by the genetic history of the population (Thomson 1977). This modern realization of indeterminism puts a new emphasis to Darwin's summary of his view of evolution: 'There seems to be no more design in the variability of organic begins and in the action of natural selection, than in the course which the wind blows' (Darwin, 1876).

FITNESS AND EVOLUTION

Prediction of the evolution by natural selection acting on a varying character therefore requires a detailed description of both the mode of action of selection and of the genetic basis of the character. From this description the induced selection acting on the underlying genotypic variation can be inferred (Fig. 3.1). However, the genes contributing to the inheritance of a given character may well have pleiotropic effects which are also subject to natural selection, so the genotypic selection depends on a wide range of characters. The simplest summary of this situation is that phenotypic evolution is brought about by selection acting on the total genotypic variation which is itself a product of natural selection acting on the total phenotypic variation (Lewontin 1974). Thus, predicting phenotypic evolution becomes an insurmountable task unless we have guiding principles like Fisher's funda-

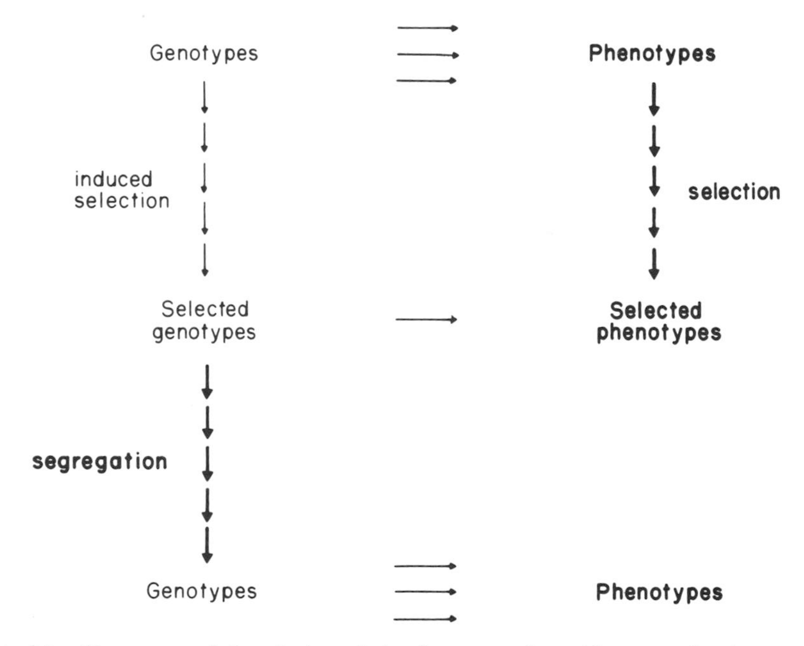

FIG. 3.1. The process of phenotypic evolution in an organism with non-overlapping generations. This figure has been simplified by assuming that natural selection only works through differential mortality of the individuals.

mental theorem or the mean fitness principle, which, as we have seen, both break down for selection concerning fecundity and in other situations involving interactions between individuals.

To advance in evolutionary theory from this despair we can choose to follow either of two directions: (a) the quantitative genetic approach, that is, the use of genetically and experimentally based phenotypic inheritance models, or (b) the population genetic approach, that is, the study of the induced selection process on simple genetic systems. The first alternative considers the direct inheritance of characters with Mendelian preservation of variance through breeding, and this approach is good for arguments on short term evolution. However, simple observations show both the strength and the weakness of this approach. For instance, in *Drosophila melanogaster* egg production is a character which is obviously related to fitness and shows a much reduced heritability compared to other characters, but the directly heritable part of the variance is by no means negligible (Scharloo *et al.* 1977). Correlated effects on other aspects of fitness may explain this, but a direct causal link is difficult. The second approach is natural when the variation of interest is discrete. However, even for characters related to an obvious 'major gene' as, for example, in melanic forms (Kettlewell 1973), the rest of the genome cannot be neglected. The evolutionary interaction of *carbonaria* and *insularia* in *Biston betularia* being an obvious illustration (Kettlewell 1958). In general, the study of the evolution of simple genetic systems should be viewed as a means of learning about the genetic consequences of natural selection. The two approaches compliment each other as modern biometric inheritance models derive from generalizations of the properties of simple genetic systems (Fisher 1918). Further, evolution progresses by allelic substitutions of single genes, and theories of genetical evolution must take account of this.

The description and measurement of fitness may involve either the quantitative genetic approach or the population genetic approach. The first requires the investigation of natural selection on the phenotype, the second on a simple genotype. The second approach is particularly appealing to a geneticist, so in the following I will be concerned with this, rather than the first approach. However, the basic structure of fitness description does not differ between the approaches. In both cases the action of natural selection is specified in its various aspects to obtain a description of the complicated and multifaceted conceptualization of fitness.

COMPONENTS OF FITNESS

Natural selection may act at any stage in the life-cycle of an organism and the object of selection may differ over these stages. In a diploid, sexually reproducing organism selection may act on gametes, individuals or combinations of individuals, each with different evolutionary consequences. The simplest situation concerns organisms with discrete non-overlapping generations where fitness is specified in terms of four selection components: gametic selection, zygotic selection, sexual selection and fecundity selection (Bundgaard & Christiansen 1972; Christiansen & Frydenberg 1973). Of these selection components, only zygotic selection can be a true individual fitness component, as it describes the survival of the individual from zygote to sexually mature adult. Gametic selection describes the survival of gametes from meiosis to fertilization, but usually deviations from Mendelian segregation are included as an aspect of gametic selection. Sexual selection describes the transition from sexually mature adults to parents, and finally, fecundity selection describes the zygotic yield of parental pairs. This last component clearly includes interactions among individuals, but if sexual selection was to be viewed on an individual basis it could also depend on the population of available mates. Gametic selection seems to be the haploid counterpart of zygotic selection, but in many organisms the survival of gametes may depend on the parent and its mate.

The selection components describe the change in genotypic and gene frequencies from the zygotic stage through the life-cycle until the stage of the

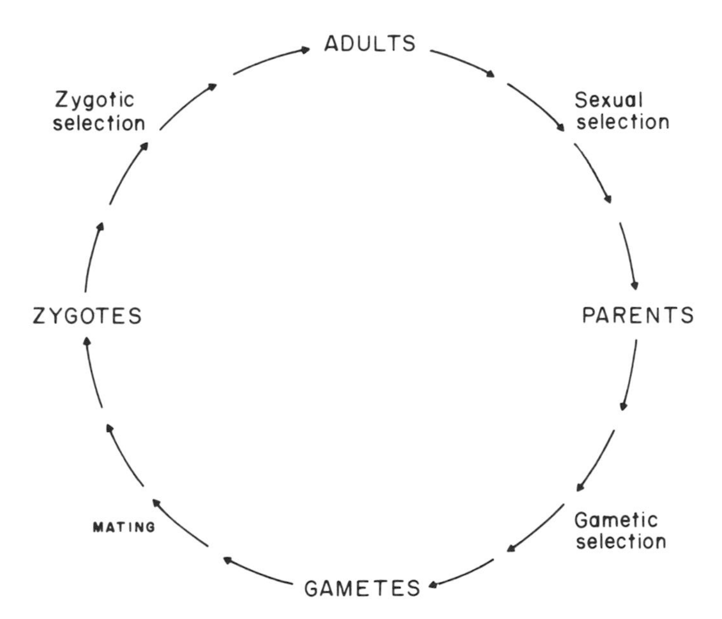

FIG. 3.2. Life stages in an organism with non-overlapping generations. The three selection components zygotic, sexual and gametic selection are acting between the life stages shown. The mating among parents in the population determines the rule for union of gametes into zygotes.

gametes that unite to form the next zygotic stage (Fig. 3.2). To complete the description of the full life-cycle we need the rule for union among gametes or formation of zygotes. This rule of combination reflects the mating of individuals, and it is not to be considered as a selection component (Christiansen & Frydenberg 1973), as it can include, for example, the effect of inbreeding in the population. However, the description of mating is usually intimately connected to the description of sexual selection (O'Donald 1980) and the rules for gametic union may have a strong influence on the evolutionary consequences of other selection components. In any case, description of mating is required for a comprehensible description of selection.

The concept of selection components may immediately be extended to organisms with overlapping generations and discrete breeding times. The components of selection concerned with breeding are specified for each age group in the population and the survival between breeding times is treated as a new selection component with relevance to breeding; that is, adult zygotic selection is recognized as being different from juvenile zygotic selection (Christiansen & Frydenberg 1976). The discreteness of breeding times in this model allows us to consider the population at important stages in the life cycle, as in the case of non-overlapping generations; the description of selection components is thus a description of transitions between these stages.

The specification of fitness in terms of selection components immediately suggests a way to measure fitness directly. For convenience let us consider only the case of non-overlapping generations. Here complete information about selection may be obtained from a sample of the zygote population; a sample of the mature adult population; a sample of the breeding population, that is, the population of parents, recording the breeding pairs; determination of the number of zygotes produced by each parental pair; and finally samples of the offspring population of zygotes (Christiansen & Frydenberg 1973; Lewontin 1974). This ideal measuring system is, unfortunately, impractical with many organisms. Organisms with brood protection, however, provide a special opportunity to design an experiment for mating observations which approach the ideal measuring system through the application of mother–offspring measuring systems (Christiansen 1980).

DATA ON MOTHER–OFFSPRING COMBINATIONS

The data from a simple mother–offspring system applied to the investigation of a two-allele autosomal polymorphism take the form shown in Table 3.1. The data represent a population sample of the marine teleost, *Zoarces viviparus*, taken while the females are in late pregnancy, and the genotypic

TABLE 3.1. Observed genotypic distribution in the *EstIII* polymorphism in *Zoarces viviparus* (L.) in Kalø Cove (north of Aarhus, Denmark). Data from Christiansen, Frydenberg & Simonsen (1973)

Adult genotype*	Offspring genotype*				NPF[†]	Adults		
	11	12	22	Sum		Females	Males	Sum
11	41	70		111	8	119	54	173
12	65	173	119	357	32	389	200	589
22		127	187	314	29	343	177	520
Sum	106	370	306	782	69	851	431	1282

*The alleles $EstIII^1$ and $EstIII^2$ are indicated by the numbers 1 and 2.
[†]Non-pregnant females.

classification refers to the locus *EstIII* (Christiansen, Frydenberg & Simonsen 1973). A sample of adults is obtained and sorted according to sex, the females are separated into mothers and non-pregnant females and one foetus is randomly picked form each mother. The genotype at the esterase locus of all adults and the chosen foeti are then determined, providing the data of Table 3.1. The total number of foeti carried by each mother is also scored. These data immediately provide the genotypic proportions among adults and among breeding females. In addition, the mother–offspring genotypic combinations provide information on the segregation of alleles in the females and on the fertilizing male gametes; that is, the population of breeding males is represented in the sample through the gametes they produce. Thus, the data of Table 3.1 provide nearly all the information required by the ideal measuring system, the main deficiency being that the population of breeding males is only represented by the fertilizing gametes. Male sexual selection and male gametic selection are therefore confounded and the genotypes of mated pairs are only represented by pairs of female genotype and male gamete type, so that fecundity selection cannot be fully analysed in terms of genotypic effects of the male parent. Female gametic selection, female sexual selection and the contribution of the female parent to fecundity selection, however, may all be fully analysed. Zygotic selection is evaluated by a comparison between the population of zygotes and the adult population, and the zygotic population can be estimated from the mother–offspring genotypic combinations and the fecundity of these. A full analysis of zygotic selection therefore requires data covering two generations.

The statistical procedure for analysis of simple mother–offspring data on a one-locus, two-allele polymorphism is discussed by Christiansen & Frydenberg (1973; Christansen *et al.* 1973), and the extension to multiple alleles and multiple loci is described by Østergaard & Christiansen (1981). The organism, *Zoarces viviparus*, we have used for illustration has discrete overlapping generations, so more detailed data are actually required. The extended analysis of selection components of the *EstIII* polymorphism is presented by Christiansen, Frydenberg & Simonsen (1977). Similar analyses of plant populations are given by Allard, Kahler & Clegg (1977) and Clegg, Kahler & Allard (1978).

This selection component analysis assumes that the ideal life stages are the same as the observed stages even though the correspondence is rarely perfect. For instance, in the *Zoarces* data the foeti are considered as *zygotes*. In fact this assumption creates *ad hoc* selection components, because, zygotic selection is partitioned into two components; a pre-observational component (foetal zygotic selection) and a post-observational component (juvenile zygotic selection). Only juvenile zygotic selection is adequately determined, whereas determination of foetal zygotic selection requires knowledge of an unknown stage, the true zygotic stage. This transfers effects of foetal zygotic selection to the breeding components of selection among the parents, and although some effects of foetal zygotic selection may be detected they can never be adequately evaluated (Østergaard & Christiansen 1981). The discrepancy between ideal and observational selection components may give unwanted, but impressive biases (Prout 1965; Christiansen, Bundgaard & Barker, 1977), and any measuring system should be evaluated on its power to resolve the ideal selection components.

For monogamous species more complete information on fitness may be gained by investigating more than one offspring per mother (Cooper 1968; Nadeau 1978; Nadeau, Dietz & Tamarin 1981; Nadeau & Baccus 1981; Christiansen 1980). However, for polyandrous species like *Zoarces viviparus* further offspring provide very limited information. With singly sired broods, male sexual and gametic selection may be partitioned if the genotype of three offspring per female is determined (Christiansen 1980); the information per additional offspring thereafter rapidly diminishes. This more complete mother–offspring measuring system, therefore, approaches the resolving power of the ideal measuring system.

The discreteness of breeding times is crucial to the simple description of fitness used above. This discreteness allows selection to be described between stages and the measurement of fitness to be made on populations of individuals. If the breeding of the individual occurs repeatedly over a period of time, the observational definition of a breeding population becomes difficult. Here, the breeding components of selection have to be specified in relation to the individuals, such as the number of broods an individual participates in, and the fecundity per brood. Further, the survival of breeding individuals through the breeding period is important for the interaction among individuals in sexual and fecundity selection, apart from being a determinant in the number of broods. These problems of fitness description occur whenever interactions among individuals occur through an extended period of life and inference on fitness require demographic information on individuals (Bodmer 1968). In most organisms, except man, this kind of information on individual survival and reproduction is very difficult to obtain in the quantities required for a description of genotypic selection. In organisms with non-discrete breeding an impression of fitness variation is usually obtained by indirect methods, as in *Drosophila melanogaster*, for example, where the life-cycle may be discrete by experimental manipulation (Prout 1971a, 1971b; Bundgaard & Christiansen, 1972; Clark, Feldman & Christiansen 1981; Clark & Feldman 1981).

FITNESS IN EVOLUTIONARY ECOLOGY

The measurement of selection components in a natural population depends on the presence of detectable genotypic variation and currently the most applicable method of investigating such genotypic variation is protein electrophoresis. Description of fitness variation related to a genetic polymorphisms allows detailed predictions about the genetic evolution of the considered loci. However, it is rarely possible to relate the genotypic variation disclosed by electrophoresis to any phenotypic variation other than the biologically trivial variation in electrophoretic mobility. Thus, detailed predictions of genetic evolution do not necessarily say much about phenotypic evolution. On the other hand, if fitness variation related to the variation of a given character is well described, then the predictions on phenotypic evolution will depend on the genetic basis of the observed phenotypic variation. However, in evolutionary ecology the main interest lies in predictions of phenotypic evolution as a response to the environmental circumstances of the population. The prediction of phenotypic evolution requires modelling of the genetic basis of phenotypic variation and the specification of evolution in relation to the environment requires modelling of the relation between the character and the environment. In genetics, natural selection is an agent that shapes the genotype through evolution, and in ecology, fitness is a property of the organism in relation to its environment. Together the two fields can describe the adaptation of an organism to its environment through evolution by natural selection and this description rests evenly on genetic and ecological conceptions.

In relation to the fitness description of population genetics probably the most important contribution of ecology is the description of the biotic environment of the individual. If, for simplicity, we restrict attention to juvenile zygotic selection, then consideration of the biotic environment allows for mutualistic or competitive interactions among individuals in the population. Thus, the mean fitness principle is expected to breakdown even when differential survival is the only source of fitness variation. Indeed, investigation of simple models of intraspecific competition has proved this (Anderson 1971; Roughgarden 1971; Charlesworth 1971; Clarke 1972). If, however, the intraspecific competition can be formulated as density-dependent individual survival, then a version of the mean fitness principle for an autosomal locus holds, in that the population size is predicted to increase during evolution (Roughgarden 1976; Prout 1980). A similar result for symmetric intraspecific competition given by Matessi & Jayakar (1976) states that the average intraspecific competition experienced by an individual will decrease during evolution. These results, however, seem to depend on formulation in terms of individual fitness as they do not apply in models where survival is determined as a function of interactions between individuals present in the population at any time (Poulsen 1979; Iwasa & Teramoto 1980). The measurement of fitness in cases of intraspecific competition meets the same problems as in the case of extended breeding periods, in that demographic data describing the fate of the individual is required.

Evolution of intraspecific mutualistic interactions through kin selection (Hamilton 1964; Maynard Smith 1964) has been described through the use of individual fitness by the concept of inclusive fitness. As in the case of competition or other individual interactions, however, individual fitness is an insufficient determinant of evolution by natural selection (Cavalli-Sforza & Feldman 1978; Uyenoyama & Feldman 1981). Again, the description of natural selection involves a description of the fate of the individual as a function of its interaction with other individuals in the population, that is, as a function of its biotic environment.

The importance of genetic assumptions in the prediction of evolution of phenotypic characters may be illustrated in some simple models of intraspecific exploitative competition (Christiansen & Loeschcke 1980a, 1980b). The models employ the niche concept of MacArthur & Levins (1967) and Levins (1968) as modified by Christiansen & Fenchel (1977). The three genotypes A_1A_1, A_1A_2 and A_2A_2 at an autosomal locus contribute differently to a character involved in the expolitation of a resource continuum. Their genotypic values are given as $2d_1$ and $2d_2$ for the two homozygotes and $d_1(1-a)+d_2(1+a)$ for the heterozygote, where it is assumed that $d_1 > |d_2|$ and $d_1 > 0$. The resource availability is described by a Gaussian resource spectrum with zero mean (arbitrary normalization), and the resource utilization by the genotypes is described by a Gaussian utilization function with mean equal to the character value and variance equal among genotypes but less than the resource variance. The model of the population growth is closely related to the model of Anderson (1971) and Roughgarden (1971).

For the model of no dominance, that is $a = 0$ (Christiansen & Loeschcke 1980a), the locus will be maintained at a unique polymorphic equilibrium when $d_2/d_1 < (\kappa^2 - 1)/(\kappa^2 + 7)$, where κ^2 is the ratio of the resource variance to the variance of the utilization function ($\kappa^2 > 1$). When $(\kappa^2 - 1)/(\kappa^2 + 7) < d_2/d_1$ (<1 as assumed), then allele A_2 will replace A_1 no matter what the initial conditions are. For $a = 1$ or $a = -1$ (full dominance) then the results are qualitatively the same; A_2 will replace A_1 when $(\kappa^2 - 1)/(\kappa^2 + 3) < d_2/d_1$ or a globally stable polymorphism exists (assuming $d_2/d_1 < 1$). For general intermediate dominance, $-1, < a < 1$ a globally stable polymorphism exists when $d_2/d_1 < (1 - a)(\kappa^2 - 1)/[(\kappa^2 + 7) - a(\kappa^2 - 1)]$ or A_2 will replace A_1 except for the case where $(\kappa^2 + 3)/(3\kappa^2 + 1) < a < 1$ when an interval of d_2/d_1 values exists in which A_2 only replaces A_1 if its initial frequency is sufficiently high; for low initial frequencies of A_2 a stable polymorphism will result.

Thus, for the classical cases of no dominance and full dominance the behaviour of the model is consistent, and the qualitative outcome of selection can be predicted from arguments on initial evolution, namely, whether the frequency of an allele can increase in a population initially monomorphic for the alternative allele. However, slightly different assumptions give a much more complicated picture, where the final outcome of evolution at this locus depends on the initial conditions. Moreover, this complication is more likely to occur when the range of the genotypes at the locus, $2(d_1 - d_2)$, is small. Thus quantitative genetic reasoning on trophic characters related to resource exploitation is expected to be influenced by the particular assumptions about dominance.

Finally, ecological reasoning may contribute to our understanding and formulation of correlations between components of selection, the trade-off pleiotropic effects of Prout (1980). The various fitness components cannot be independent and even the formulation of seemingly obvious correlations can give surprising results. For instance, the recognition of an inverse relationship between fecundity and offspring size in simple organisms unveils an evolutionary instability giving rise to evolution for either small or large offspring, only on the assumption that survival of the offspring generally increases with size (Vance 1973; Christiansen & Fenchel 1979).

CONCLUSION

The classical concept of individual fitness is insufficient to account for the action of natural selection. Instead we have to use a much more detailed description of fitness which recognizes various selection components and interactions among individuals. It is important to know the genetic basis of the phenotypic variation of interest in order to infer the induced genotypic selection as a prerequisite of evolutionary predictions. The main contribution of ecology to evolutionary theory is the description of connections between fitness and the environment of the organism. In particular, the description of the biotic environment of the organism is important for the investigation of intraspecific competition and mutualism. Hybridization of ecological and genetical knowledge will advance the study of individual interactions in fitness, interactions which constitute one of the basic forces in Darwin's Theory of Evolution by Natural Selection.

ACKNOWLEDGEMENTS

Comments and criticism of the manuscript by Andrew Clark, Gerdien de Jong and Volker Loeschcke are gratefully acknowledged.

REFERENCES

Allard R.W., Kahler A.L. & Clegg M.T. (1977) Estimation of mating cycle components of selection in plants. *Measuring Selection in Natural Population* (Ed. by F.B. Christiansen & T.M. Fenchel). *Lecture Notes in Biomathematics*, vol. 19, p. 1–19. Springer, Heidelberg.

Anderson W.W. (1971) Genetic equilibrium and population growth under density-regulated selection. *American Naturalist*, **105**, 489–498.

Bodmer W.F. (1965) Differential fertility in population genetics models. *Genetics*, **51**, 411–424.

Bodmer W.F. (1968) Demographic approaches to the measurement of differential selection in human populations. *Proceedings of the National Academy of Sciences, U.S.A.*, **59**, 690–699.

Bundgaard J. & Christiansen F.B. (1972) Dynamics of polymorphisms: I. Selection components in an experimental population of *Drosophila melanogaster*. *Genetics*, **71**, 439–460.

Cavalli-Sforza L.L. & Feldman M.W. (1978) Darwinian selection and 'altruism'. *Theoretical Population Biology*, **14**, 268–280.

Charlesworth B. (1971) Selection in density-regulated populations. *Ecology*, **52**, 469–474.

Charlesworth B. (1973) Selection in populations with overlapping generations. V. Natural selection and life histories. *American Naturalist*, **107**, 303–311.

Charlesworth B. (1974) The Hardy-Weinberg law with overlapping generations. *Advances in Applied Probability*, **6**, 4–6.

Charlesworth B. (1976) Natural selection in age-structured populations. *Lectures on Mathematics in the Life Sciences* (Ed. by S. Levin), vol. 8, pp. 69–87. American Mathematical Society, Providence.

Charlesworth B. (1980) Selection in Age-Structured Populations. *Cambridge Studies in Mathe-Matical Biology*, vol. 1. Cambridge University Press, Cambridge.

Christiansen F.B. (1980) Studies on selection components in natural populations using population samples of mother–offspring combinations. *Hereditas*, **92**, 199–203.

Christiansen F.B., Bundgaard J. & Barker J.S.F. (1977) On the structure of fitness estimates under post-observational selection. *Evolution*, **31**, 843–853.

Christiansen F.B. & Fenchel T.M. (1977) Theories of Populations in Biological Communities. *Ecological Studies*, vol. 20. Springer. Berlin.

Christiansen F.B. & Fenchel T.M. (1979) Evolution of marine invertebrate reproductive patterns. *Theoretical Population Biology*, **16**, 267–282.

Christiansen F.B. & Frydenberg O. (1973) Selection component analysis of natural polymorphisms using population samples including mother–offspring combinations. *Theoretical Population Biology*, **4**, 425–445.

Christiansen F.B. & Frydenberg O. (1976) Selection component analysis of natural polymorphisms using mother–offspring samples of successive cohorts. *Population Genetics and Ecology* (Ed. by S. Karlin & E. Nevo), pp. 277–301. Academic Press, New York.

Christiansen F.B., Frydenberg O. & Simonsen V. (1973) Genetics of *Zoarces populations*, IV Selection component analysis of an esterase polymorphism using population samples including mother–offspring combinations. *Hereditas*, **73**, 291–304.

Christiansen F.B. Frydenberg O. & Simonsen V. (1977) Genetics of *Zoarces* populations. X. Selection component analysis of the *EstIII* polymorphism using samples of successive cohorts. *Hereditas*, **87**, 129–150.

Christiansen F.B. & Loeschcke V. (1980a) Evolution and intraspecific exploitative competition I. One-locus theory for small additive gene effects. *Theoretical Population Biology*, **18**, 297–313.

Christiansen F.B. & Loeschcke V. (1980b) Intraspecific competition and evolution. *Vito Volterra Symposium on Mathematical Models in Biology* (Ed. by Claudio Barigozzi), *Lecture Notes in Biomathematics*, vol. 39, pp. 151–170. Springer, Berlin.

Clark A.G. & Feldman M.W. (1981) The estimation of epistasis in components of fitness in experimental populations of *Drosophila melanogaster* II. Assessment of meiotic drive, viability, fecundity and sexual selection. *Heredity*, **46**, 347–377.

Clark A.G., Feldman M.W. & Christiansen F.B. (1981) The estimation of epistasis in components of fitness in experimental populations of *Drosophila melanogaster* I. A two-stage maximum likelihood model. *Heredity*, **46**, 321–346.

Clarke B. (1972) Density-dependent selection. *American Naturalist*, **106**, 1–13.

Clegg M.T., Kahler A.L. & Allard R.W. (1978) Estimation of life cycle components of selection in an experimental plant population. *Genetics*, **89**, 765–792.

Cooper D.W. (1968) The use of incomplete family data in the study of selection and population structure in marsupials and domestic animals. *Genetics*, **60**, 147–156.

Darwin C. (1872) *The Origin of Species.* 6th ed.

Darwin C. (1876) *The autobiography of Charles Darwin.* W.W. Norton & Co., London (republished 1969).

Ewens W.J. (1980) *Mathematical Population Genetics.* Springer, Berlin.

Ewens W.J. & Thomson G. (1977) Properties of equilibria in multi-locus genetic systems. *Genetics*, **87**, 807–819.

Fisher R.A. (1918) The correlation between relatives on the supposition of Mendelian inheritance. *Transactions of the Royal Society of Edinburgh*, **52**, 399–433.

Fisher R.A. (1930) *The Genetical Theory of Natural Selection.* Clarendon Press, Oxford (republished 1958, Dover Publications, New York).

Haldane J.B.S. (1932) *The Causes of Evolution.* Longmans, Green, London (republished 1966, Cornell University Press, Ithaca, New York).

Haldane J.B.S. (1962) Natural selection in a population with annual breeding but overlapping generations. *Journal of Genetics*, **58**, 122–124.

Hamilton W.D. (1964) The genetical evolution of social behavior. I. *Journal of Theoretical Biology*, **7**, 1–16.

Hamilton W.D. (1964) The genetical evolution of social behavior. II. *Journal of Theoretical Biology*, **7**, 17–52.

Iwasa Y. & Teramoto E. (1980) A criterion of life history evolution based on density-dependent selection. *Journal of Theoretical Biology*, **84**, 545–566.

Karlin S. & Carmelli D. (1975) Numerical studies on two-loci selection models with general viabilities. *Theoretical Population Biology*, **7**, 399–421.

Kettlewell B. (1958) A survey of the frequencies of *Biston betularia* L. (Lep.) and its melanic forms in Britain. *Heredity*, **12**, 51–72.

Kettlewell B. (1973) *The Evolution of Melanism*. Clarendon Press, Oxford.

Kimura M. (1958) On the change of population fitness by natural selection. *Heredity*, **12**, 145–167.

Kingman J.F.C. (1961) A mathematical problem in population genetics. *Proceedings of the Cambridge Philosophical Society*, **57**, 574–582.

Kojima K. & Kelleher T.M. (1961) Changes of mean fitness in random mating populations when epistasis and linkage are present. *Genetics*, **46**, 527–540.

Levins R. (1968) Toward an evolutionary theory of the niche. *Evolution and Environment* (Ed. by E.T. Drake), pp. 325–340, Yale University Press, New Haven.

Lewontin R.C. (1974) *The Genetic Basis of Evolutionary Change*. Columbia University Press, New York.

MacArthur R.H. & Levins R. (1967) The limiting similarity, convergence and divergence of coexisting species. *American Naturalist*, **101**, 377–385.

Matessi C. & Jayakar S.D. (1976) Models of density-frequency dependent selection for exploitation of resources. *Population Genetics and Ecology* (Ed. by S. Karlin & E. Nevo), pp. 707–721. Academic press, New York.

Maynard Smith J. (1964) Kin selection and group selection. *Nature*, **201**, 1145–1147.

Moran P.A.P. (1964) On the nonexistence of adaptive topographies. *Annals of Human Genetics*, **27**, 383–393.

Mulholland H.P. & Smith C.A.B. (1959) An inequality arising in genetical theory. *American Mathematical Monthly*, **66**, 673–683.

Nadeau J.H. (1978) *The Measurement of Selection in Natural Populations*. PhD. thesis, Boston University, Boston, Massachusetts.

Nadeau J.H. & Baccus R. (1981) Selection components of four allozymes in natural populations of *Peromyscus maniculatus*. *Evolution*, **35**, 11–20.

Nadeau J.H., Dietz K. & Tamarin R. (1981) Gametic selection and the selection component analysis. *Genetical Research, Cambridge*, **37**, 275–284.

O'Donald P. (1980) *Genetic Models of Sexual Selection*. Cambridge University Press, Cambridge.

Østergaard H. & Christiansen F.B. (1981) Selection component analysis of natural polymorphisms using population samples including mother–offspring combinations, II. *Theoretical Population Biology*, **19**, 378–419.

Pollak E. (1978) With selection for fecundity the mean fitness does not necessarily increase. *Genetics*, **90**, 383–389.

Pollak E. & Kempthorne O. (1970) Malthusian parameters in genetic populations. I. Haploid and selfing models. *Theoretical Population Biology*, **1**, 315–345.

Pollak E. & Kempthorne O. (1971) Malthusian parameters in genetic populations. II. Random mating populations in infinite habitats. *Theoretical Population Biology*, **2**, 357–390.

Poulsen E.T. (1979) A model for population regulation with density- and frequency-dependent selection. *Journal of Mathematical Biology*, **8**, 325–343.

Prout T. (1965) The estimation of fitnesses from genotypic frequencies. *Evolution*, **19**, 546–551.

Prout T. (1971a) The relation between fitness components and population prediction in *Drosophila*. I. The estimation of fitness components. *Genetics*, **68**, 127–149.

Prout T. (1971b) The relation between fitness components and population prediction in Drosophila. II. Population prediction. *Genetics*, **68**, 151–167.

Prout T. (1980) Some relationships between density independent selection and density dependent growth. *Evolutionary Biology*, **13**, 1–96.

Roughgarden J. (1971) Density-dependent natural selection. *Ecology*, **52**, 453–468.

Roughgarden J. (1976) Resource partitioning among competing species—a coevolutionary approach. *Theoretical Population Biology*, **9**, 388–424.

Scheuer P.A.G. & Mandel S.P.H. (1959) An inequality in population genetics, *Heredity*, **13**, 519–524.

Scharloo W., van Dijken F.R., Hoorn A.J.W., de Jong G. & Thörig G.E.W. (1977) Functional aspects of genetic variation. *Measuring Selection in Natural Populations* (Ed. by F.B. Christiansen & T.M. Fenchel), pp. 131–147, Springer, Heidelberg.

Thomson G. (1977) The effect of a selected locus on linked neutral loci. *Genetics*, **85**, 753–788.

Turner J.R.G. (1970) Changes in mean fitness under natural selection. *Mathematical Topics in Population Genetics* (Ed. by K. Kojima), pp. 32–78, Springer, Berlin.

Turner J.R.G. (1981) 'Fundamental theorem' for two loci. *Genetics*, **99**, 365–369.

Uyenoyama M.K. & Feldman M.W. (1981) On relatedness and adaptive topography in kin selection. *Theoretical Population Biology*, **19**, 87–123.

Vance R.R. (1973) On reproductive strategies in marine benthic invertebrates. *American Naturalist*, **107**, 339–352.

Wright S. (1931) Evolution in Mendelian populations. *Genetics*, **16**, 97–159.

4. ECONOMICS OF ONTOGENY—ADAPTATIONAL ASPECTS

P. CALOW

Department of Zoology, University of Glasgow, Glasgow G12 8QQ

INTRODUCTION

Even organisms surrounded by unlimited resources have limited resources available for their metabolism. This is because their resource-acquiring mechanisms are restricted and limiting. Hence, resource utilization is likely to have been influenced importantly by natural selection according to ecological pressures. Moreover, the time-varying investment of resources in the processes and structures of organisms is the basis of their ontogeny. Van Valen (1980) thinks of evolution as the control of ontogeny by ecology and here I view this as happening through the metabolic allocation of limited resources between often-conflicting processes and structures. Hence the aims of this paper are: (a) to make the relationship between ontogeny and resource utilization more explicit; (b) to begin to formulate the principles according to which resource-utilization patterns have evolved (cf. Townsend & Calow 1981); (c) to emphasize the ecological basis of these principles. This approach views the organism as a crucial unit of selection and such an orientation will be justified more explicitly in a later section. Clearly the emphasis is on organisms functioning in ecosystems rather than on the functioning of ecosystems themselves (cf. Calow 1977a). For convenience, most of the discussion will relate to animals, but the approach is applicable to all organisms.

CURRENCY

In analysing the economics of ontogeny it is important that the appropriate currency should be used. A large variety of resources are acquired and utilized by functioning organisms, each is involved in a multitude of processes, and these contribute to the formation of a variety of products, not all related in an obvious way to neo-Darwinian fitness. Irrespective of what is being formed, however, the most crucial resource, from an investment point of view, ought to be that which: (a) is required most generally and (b) is most limiting. In principle, (b) might apply to any resource and, indeed, the supply of all resources by the feeding or photosynthetic processes is limited (see above). In practice, however, there is only one resource which is required in all metabolic processes and in the building of all structures. This is energy which is also easily measured, and used in models. For plants, Gutschick (1981) considers energetics to be at the heart of nitrogen budgeting but Abrahamson & Caswell (1982) found poor correlations between biomass and mineral allocation patterns. For animals, energy is likely to be a good general currency in situations where food and feeders are of similar composition (carnivores) but less good in other situations (e.g. detrivores and herbivores). In this chapter, currency is defined generally as 'resource', and in units of energy where more specific terminology is needed.

RATIONALE OF APPROACH

The relative spread of genes through a population depends upon how the traits they determine influence the survival, developmental rate and reproduction of the organisms which carry them. Genes associated with traits that maximize survival and fecundity and minimize time to breeding will spread at the expense of others, or resist being displaced by new mutations. This is the *Core Neo-Darwinian Hypothesis*. (Often fitness is expressed by a single summarizing term, such as r, the Malthusian parameter, which incorporates these demographic components; see Charlesworth in this volume, and below). Hence, to judge the success of particular gene-determined, resource allocation traits, the physiological processes should be mapped directly onto their consequences for the demographic parameters. Yet it is usually difficult to define how short-term physiological processes influence long-term demographic effects; for example, to define how an extra watt of energy spent in a moment of metabolism might influence long-term ability to leave offspring. More immediate, if less direct criteria of fitness are required, and it is the thinking out and evaluation of these criteria which is often the main objective of this adaptationist programme. Biologists involved in this kind of work are not usually interested in trying to refute the Core Neo-Darwinian Hypothesis. Rather, they assume that this hypothesis is approximately correct and then proceed to discover what it means for ontogeny, by the development of a number *auxiliary hypotheses* which redefine the core in more immediately measurable and observable criteria.

For example, in some relatively early work of this kind Cohn (1954, 1955; cf. Rashevsky 1965) considered how, given the neo-Darwinian core, the vertebrate blood system would be expected to develop? In particular, how wide should individual vessels be? Now it would be extremely difficult to map such a property onto whole-organism survivorship and fecundity but auxiliary hypotheses can easily be formulated. Cohn, for example, thought that selection should favour economization in metabolic expenditure since savings made in the ontogeny of one system could be used to promote survivorship (e.g. in predator escape) or alternatively to produce more gametes. Figure 4.1 shows how Cohn used this auxiliary hypothesis to generate testable predictions and these turned out to be reasonably accurate (Rashevsky 1965).

Are there any other general auxiliary hypotheses, relating to resource

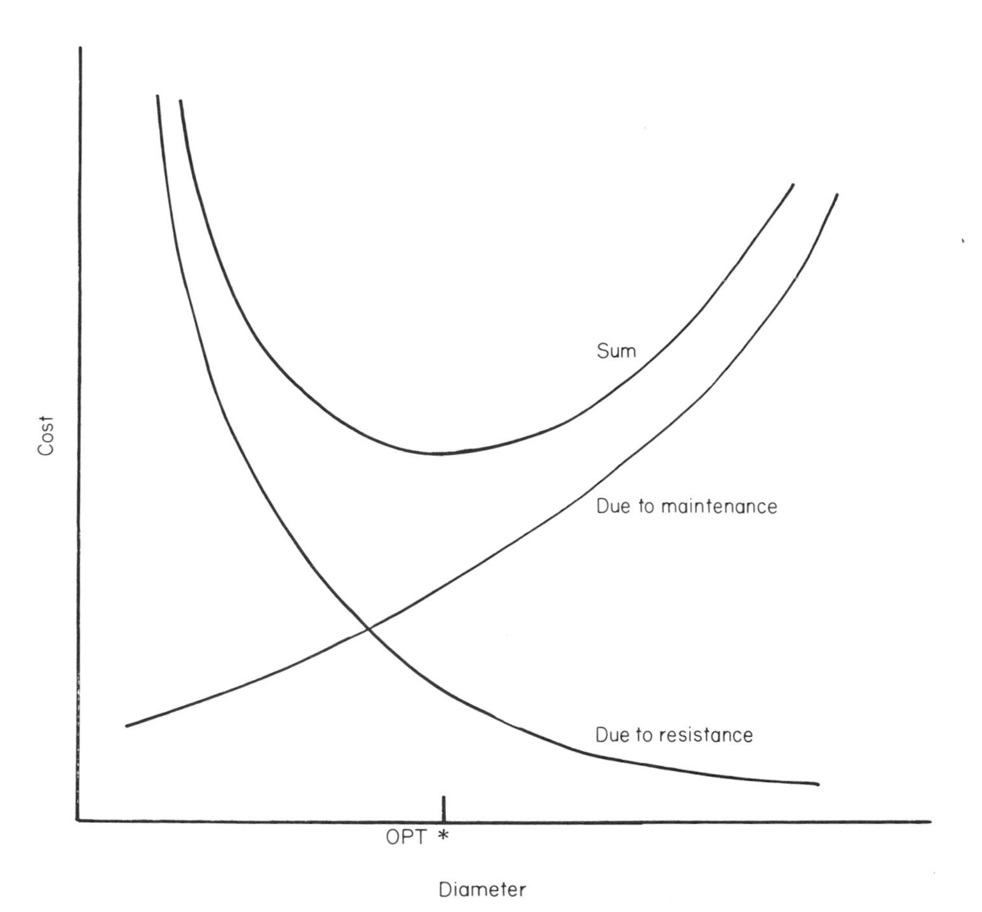

FIG. 4.1. Cohn's model. The auxiliary hypothesis is that the most economic solution should be favoured (i.e. is optimum). Two cost functions are defined—one which increases with diameter (costs of building and maintenance of the system) and one which reduces (costs of pumping blood increase with finer bore vessels due to increased friction). Total costs are the sum of these two components. The optimum diameter (*) is the one with minimum total costs.

utilization, which might be used in the same way? A widely canvassed one is the so-called maximization principle which postulates that all traits that cause biomass production to be maximized should be favoured since this minimizes time from birth to reproduction and ultimately maximizes reproductive output. The economization principle, noted above, can be considered as a consequence of this, for organisms should economize on all aspects of metabolism in order to maximize growth rates and, ultimately, reproductive output (see p. 89 below). On the other hand, maxima might not be achieved because of various constraints (i.e. inexorable limits imposed by physico-chemical, genetic, developmental, morphological, and physiological factors) and trade-offs (i.e. limits which can be shifted by natural selection but only at the expense of other things). A particularly apposite class of trade-offs in the present context is what I call the costs of living; i.e. the costs needed to maintain biomass against disrepair, disease and predation—these are expensive in resources, but bring gains in individual survival. Clearly, natural selection is likely to have maximized production within the limits imposed by the constraints and according to the various trade-offs, so it is more appropriate to consider optimization rather than maximization models and to view natural selection as an optimizing process.

In practice, however, it is often useful to begin by assuming that organisms behave *as if* they conform to simple maximization principles and then to use this to discover what complicating constraints and costs are involved. (As Dawkins [1982] noted this is a restricted version of what Sibly & McFarland [1976] have described as the reverse optimality programme). Generally, what costs there are and how much is invested in them will depend upon ecological factors and are likely to be exposed by ecological analysis. This paper emphasizes this ecological dimension. The approach is global; pointing to the kinds of models, costs and constraints that ought to be considered rather than to the construction of detailed models of particular systems or situations.

BASIC PATTERN OF RESOURCE ALLOCATION

A simplified version of the basic pattern of resource allocation in animals is illustrated in Fig. 4.2. The bolder pathways represent the ones that should be emphasized according to the simple maximization principle. However, the broken lines indicate that benefits are to be obtained from the alternative investments and imply an optimization process. It is convenient to divide ontogeny into two parts, one emphasizing somatic and the other reproductive processes, and this is what will be done below. However, it is also important to realize that ontogeny is a continuous process and that there are likely to

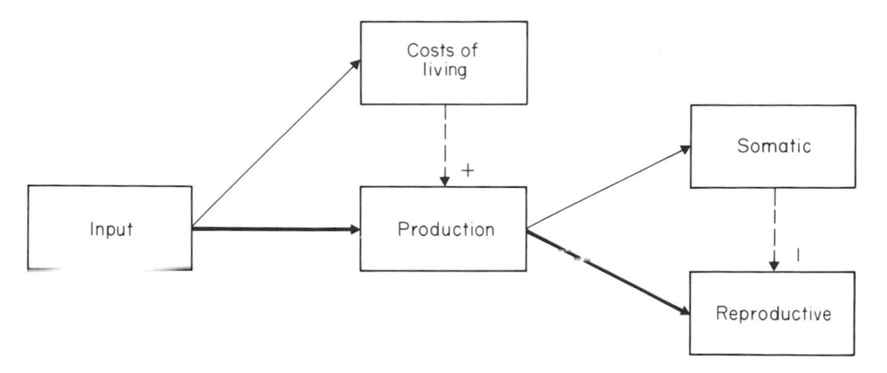

FIG. 4.2. Basic pattern of resource allocation. Bolder lines indicate pathways favoured according to the *Maximization Principle*. Broken lines are non-energetic interactions.

be interactions between somatic and reproductive processes. These, in fact, are crucial to the understanding of life-cycle phenomena and will be considered in the section on reproduction (p. 91).

RESOURCE ALLOCATION DURING THE GROWTH PHASE OF ONTOGENY

Trade-offs between metabolic speed and effciency

At the outset it is important to realize that biomass accumulation (production) is the product of a rate and an efficiency:

$$\text{Biomass accumulation} = \text{rate of supply of resources} \times \text{efficiency of conversion.}$$

Trade-offs are possible between the speed and efficiency of conversion (see review in Calow 1977a; cf. Schiemer 1983) but these should always be in the direction of increased biomass accumulation. For example, homeotherms have lower conversion efficiencies than poikilotherms because of their endothermic generation of heat (Calow 1977b; Schroeder 1981). Yet this enables their metabolism to occur at higher, more constant levels and the growth rates (somatic biomass accumulation) of homeothermic vertebrates are an order of magnitude higher than most poikilothermic vertebrates and invertebrates (Calow & Townsend 1981). The population measure of growth, r, is also correlated positively with metabolic rate and is higher in homeotherms than poikilotherms (Fenchel 1974; McNab 1980; Henneman 1983).

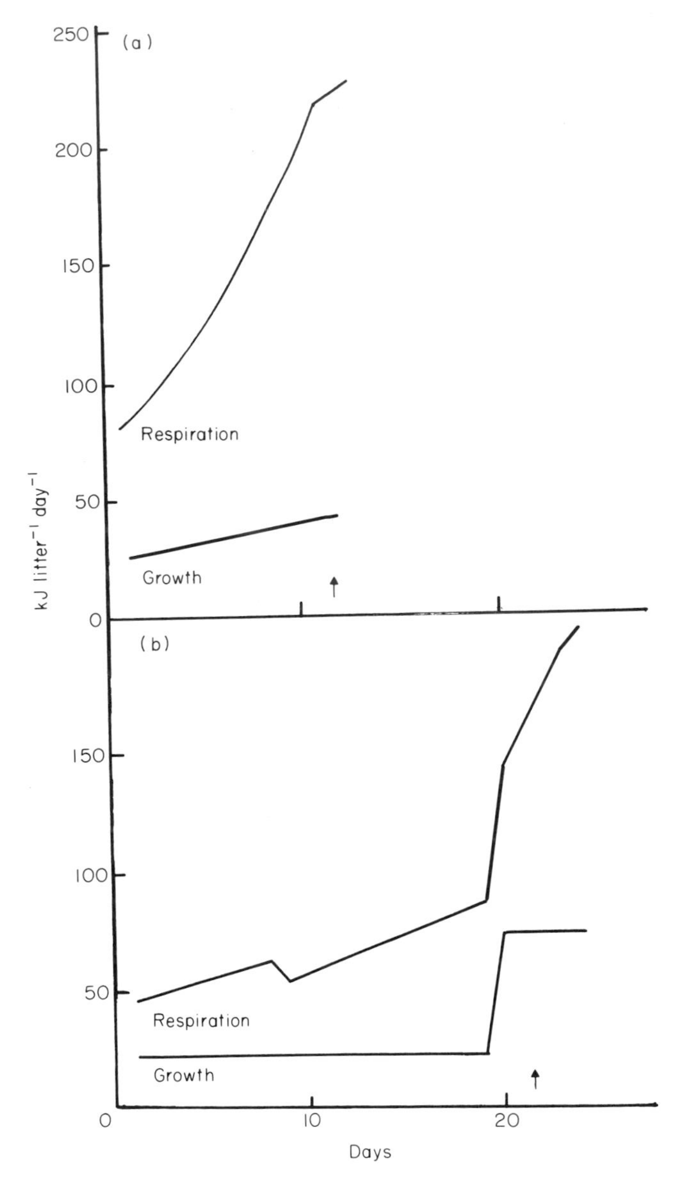

FIG. 4.3. Energetics of juvenile rodents: (a) *Sigmodon hispidus*; (b) *Neotoma floridana*. The arrows show the times of weaning. The total amounts of energy required for growth and maintenance before weaning are similar in both species (approx. 2000 kJ) but more of this energy goes to production in *N. floridana*. Nevertheless, *S. hispidus* reaches a size at which it can be weaned earlier. (After McClure & Randolph 1980).

Moreover, there are interesting differences between homeotherms in the developmental timing of the onset of homeothermy. This has consequences for the speed and efficiency of ontogeny and these are related, at least in part, to ecological pressures. For example, the small rodent, *Sigmodon hispidus*, becomes homeothermic at a smaller size and earlier age than a similar species, *Neotoma floridana*, and therefore expends more energy in respiratory metabolism and has a lower growth efficiency and a more rapid development (Fig. 4.3) (McClure & Randolph 1980). The metabolic strategy of *S. hispidus* has evolved under conditions of continuous food abundance or greatly fluctuating rations, whereas that of *N. floridana* has evolved under chronic food shortage. (Body size constraints are probably also important in the ontogeny of homeothermy—McClure & Randolph 1980).

In general it is possible that metabolic speed is favoured when resources are non-limiting, and efficiency when resources are more limiting (Roughgarden 1971) and this fits neatly into the *r*- and *K*-selection dichotomy (Calow 1977a; McNab, 1980).

Costs of maintenance

Organisms living in stressful conditions often have high metabolic rates and low rates of production. This has been observed in aquatic invertebrates living in exposed and fast-flowing habitats (e.g. Fox & Simmonds 1933), estuarine fishes suffering osmotic stress (Stearns 1980), and invertebrates in polluted environments (Bayne 1976). In organisms as widely different as micro-organisms, vertebrates and waterweeds (Price 1972; Goldberg & Dice 1974; Cooke, Oliver & Davies 1979) stress, particularly of a chemical kind, is known to stimulate molecular turnover in tissues (presumably required to replace and repair damaged structures) and paying the costs of this may account for 10–25% of basal metabolism (Waterlow 1980). These are extreme manifestations of more general maintenance costs which arise because biological steady states are some distance from thermodynamic equilibrium (Calow 1978a). Just how much maintenance is required depends upon the environment and how much is invested in it on demographic factors. Kirkwood (1981) has suggested, for example, that the extent to which organisms pay these costs of living might determine their rates of ageing since the vitality of an organism is likely to depend on the density of damage in its biomass and the latter depends, in turn, on investment in repair, viz.:

$$\text{Density of damage in tissues} = \text{rate of generation of damage} - \text{rate of repair}$$

Two testable predictions follow. First, that intrinsically long-lived species should invest more in repair and turnover than short-lived ones, and there is some (though not very precise) support for this (Calow 1978a; Kirkwood 1981). Second, that organisms invest only enough in somatic repair to ensure the retention of vigour through the normal expectation of life in the wild (economization principle). Hence the greater the level of extrinsic mortality (from predation etc.) the less should be invested in repair and the shorter should be the life-span when these sources of mortality are excluded. Primates, birds and tortoises, all with long lives in protected situations (Comfort 1979), do appear to be relatively predator-free in nature, but more precise evidence relating to this prediction has yet to be found. A recent paper by Gibbons & Semlitsch (1982) suggests, for example, that the turtle *Pseudemys scripta* can be expected to attain very old ages in protected situations and, though having a shorter life-span in natural populations than promoted by popular conception, it is nevertheless more long-lived than other species for which complete life-table data are available (see Fig. 1 of Gibbons & Semlitsch 1982).

Costs of activity

Physical activity is metabolically costly, yet an investment in it can bring benefits in terms of survivorship (e.g. predator escape) and resource returns. The former cannot be evaluated in energy units and here submaximal growth rates may be optimum. Alternatively, physical activity can enhance net energy returns (from the food) and hence growth and gamete production. For example, the actively foraging lizard, *Cnemidophorus tigris*, respires about twice as much as the more sedentary, but ecologically related *Calisaurus draconoides* and yet has a significantly greater net foraging return of metabolizeable energy per energy invested in activity (Anderson & Karasov 1981). At a fixed food concentration, Ware (1975) has shown that food returns increase at a reducing rate and metabolic costs increase at an increasing rate with the swimming speed of fishes. From this, he was able to calculate the optimum speeds for maximizing growth rate, growth efficiencies and growth efficiency per unit food intake, all of which have been proposed as metabolic measures of neo-Darwinian fitness in the literature on fishes. Using data obtained by Ivler (1960) on bleak, Ware (1975) demonstrates convincingly that in this species, fishes swim at speeds that maximize growth rates—a result that is expected on the basis of the discussion on p. 84.

There might also be an intimate relationship between feeding mode, body form and metabolic performance (see also p. 89). This has been well-illustrated for predatory, aquatic notonectids by Giller (1982). Here *Notonecta obliqua*, a sit and wait predator, was found to have relatively poor streamlining (high drag) and hence high metabolic rates under sustained swimming, but a high mechanical advantage (ratio of lever arm length to total limb length) in the swimming legs and therefore the capacity for short rapid bursts of activity. Alternatively, *Notonecta maculata* and *N. glauca*, more seek-out feeders, are better streamlined, have lower metabolic rates under sustained swimming but have lower mechanical advantages and are therefore less capable of rapid bursts of activity. Similarly, in fishes there appears to have been general selection for cost reduction by streamlining (i.e. economization) but this is particularly true of active, pelagic species (Ware 1982).

In mammals, no more than 15% of daily energy expenditure is invested in transport costs—this proportion, known as the ecological cost of transport, is greatest in largest animals (Garland 1983). Hence, locomotory specialization should be more prominent in big mammals—but specialization might also be related to non-economic factors such as speed and endurance. These possibilities are reviewed in Garland (1983).

Theoretical studies have suggested that the energy invested in searching activities should depend upon the density of the prey—increasing as prey density increases (Norberg 1977). But the mobility of the prey is also important, since feeders on less mobile foods have to search for these resources and often become more active as food density reduces (Calow 1981). Alternatively, feeders on mobile food are able to sit and wait for it when it is not immediately available.

Ecological restraints

As well as the physiological trade-offs described above there might also be ecological restraints on growth. These have been discussed elsewhere (Calow 1982) and include the costs in parent survivorship of feeding fast-growing offspring and the gains in individual survivorship from tracking a submaximal growth rate as a result of predator avoidance, food-gathering and responding to disturbances in the physico-chemical environment.

Costs of form

Structures as well as processes are costly; i.e. they require resources for their building and maintenance and have implications for the mechanics and hence metabolic efficiency of the organism, as discussed above for the notonectids and fishes. The maximization principle implies economization in these costs (p. 83–84) such that maximum benefits are gained from a structure for minimum metabolic costs—what Rashevsky (1965) called the principle of adequate design. Evidence that economization can be, and has been important in the evolution of form is as follows:

1 Epp & Lewis (1980) have claimed that the change in whole-body form from a nauplius to copepodid stage, mid-way through the development of copepods, is for economic reasons. Nauplii have lower metabolic rates than similar-sized copepodids would have, but have metabolic rates which increase more rapidly with mass than in the copepodid form (Fig. 4.4). Hence, there is a body mass at which the copepodid form becomes more economic and metamorphosis occurs at approximately that size (Fig. 4.4).
2 Economization provides a reasonable explanation of the evolutionary loss of useless characters, for example in cavernicolous and parasitic animals (e.g. Barr 1968). There are also other possible explanations (Regal 1977).
3 A good experimental demonstration of the economization principle at the

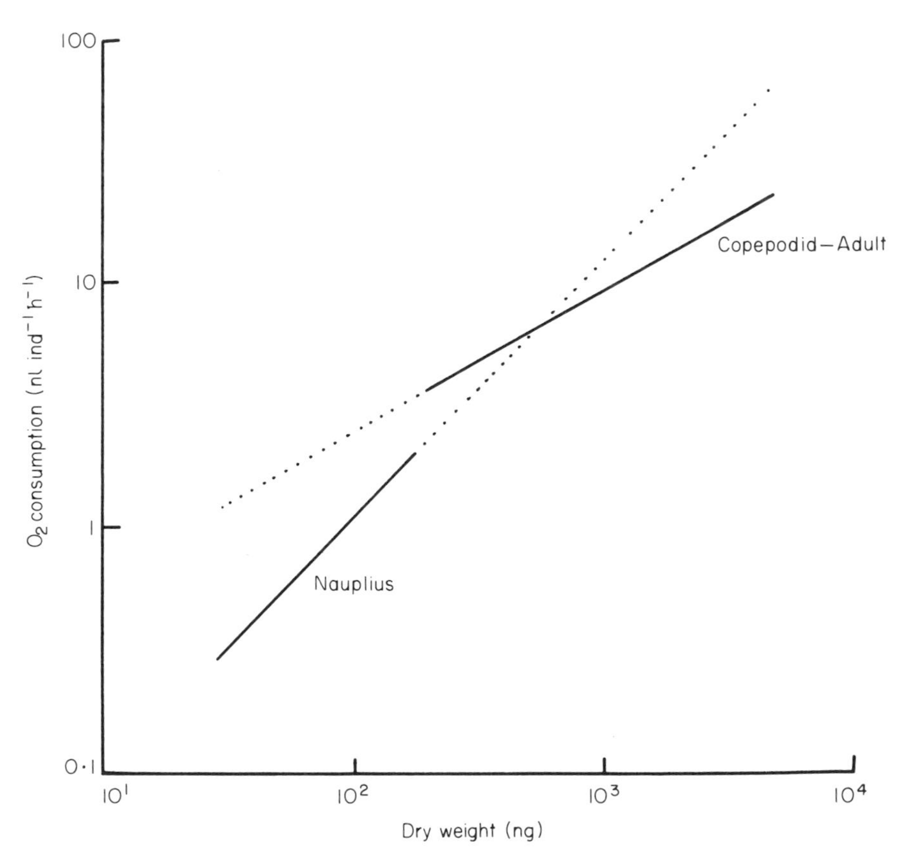

FIG. 4.4. Size dependency of metabolic rate in life-cycle stages of copepods. Broken lines are extrapolated. Based on *Mesocyclops biasilianus*. (After Epp & Lewis 1980).

biochemical level is that of Zamenhof & Eichorn (1967). They showed that mutant microbes that lacked the capacity to synthesize the amino acid tyrosine were competitively superior to wild types when both were cultured in media containing this compound.

4 Big guts are more effective than small ones in providing the space and surface for digestion and absorption. Yet big guts are costly to build, maintain and carry around. The fact that in some animals gut size is plastic and that it increases on poor rations (Sibly 1981) is consistent with the economization principle.

Of course the important costs and benefits need not always be metabolic ones. The functioning of the nervous system and brain, for example, cannot readily be reduced to a metabolic basis (but see Martin 1981). 'Alba' polymorphism in *Colias* butterflies illustrates the interplay of metabolic and non-metabolic costs in the evolution of form—in this case in wing pigmentation. Alba (non-pigmented) females redirect resources usually used in producing pigments to alternative metabolic ends. As a result, alba females mature earlier, retain more larva-derived resources in their fat bodies for somatic maintenance and for reproduction, and in some conditions mature their eggs faster than females with pigmented wings (Scott, Watt & Lawrence 1980). This gives alba genes an advantage in cold and resource-limited environments where they occur at higher frequencies (Hovanitz 1944). Alternatively, alba females are less attractive to males than pigmented females and may be more vulnerable to predators (Scott *et al.* 1980). Hence, alba does not necessarily spread at the expense of pigmentation and polymorphism for alba and non-alba genes is common.

In analysing the selective basis of form, all costs and benefits should be quantitatively assessed, and then combined into a single cost function which represents the precise way in which the various costs and benefits trade-off against each other. This is what Cohn did in analysing blood vessel diameter (p. 82) and in the models employed there all terms could be related to metabolic energy costs (cf. Milsum & Roberge 1973; McFarland 1976). In analysing the form of skeletal systems, Alexander (1981, 1982) uses a cost function which includes the metabolic costs of building, maintaining and using the structure as well as a term which expresses the mortality risks of failure of a given-sized structure. The extent to which metabolic criteria are important in particular models will depend upon both the structures in question (i.e. they are more likely to be important for guts and respiratory surfaces than brains) and the ecological circumstances in which they are required to operate (as in the *Colias* butterflies, where alba is favoured in metabolically stringent circumstances).

RESOURCE ALLOCATION DURING REPRODUCTION

Somatic processes are, in a sense, in continuous competition with reproductive ones for limited resources. Why produce more somatic tissues? When should resources be switched from somatic to reproductive production? How should this be done? There are a number of possible ways of trying to answer these questions, but from an economics point of view it is important to note that in most organisms the size dependency of the energy input and dissipation processes (see Fig. 4.5a) is such that the energy available for production follows the peaked distribution illustrated in Fig. 4.5a and b. A simple hypothesis is, therefore, that reproductive output is maximized if production is switched from growth to reproduction at the peak (as in

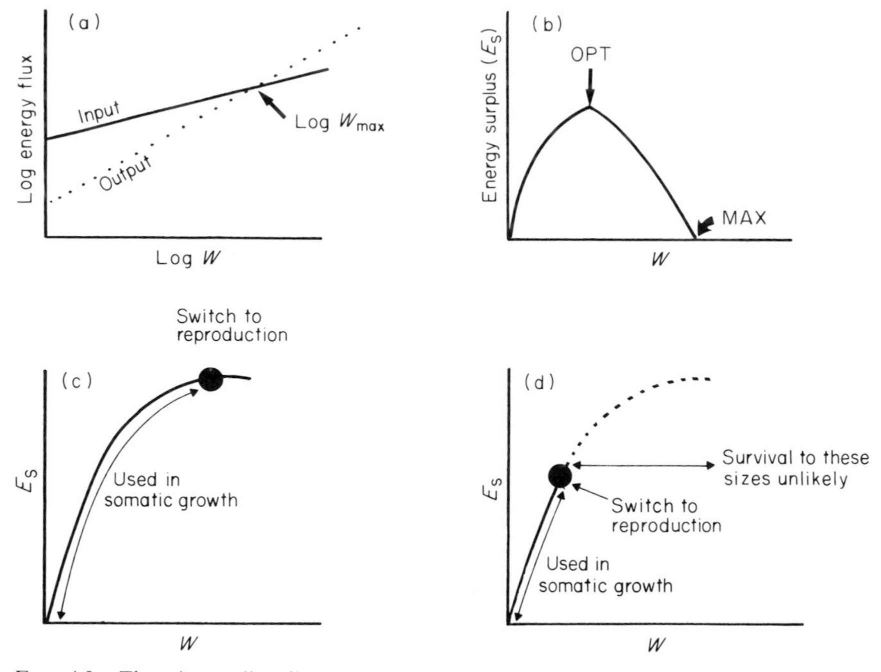

FIG. 4.5. There is usually a linear relationship of the form illustrated in (a) between the logarithm of energy fluxes (input from feeding and output from respiration) and the logarithm of organismic mass (W). The difference between these two lines is the energy surplus for production which, because of the logarithmic form of (a), takes on the peaked form in (b). MAX is the maximum possible size. OPT is the size, according to maximization criteria, at which energy should be switched from somatic to reproductive production. Patterns of investment in reproduction are illustrated in (c)—breed when productive potential is maximum—and (d)—breed before productive potential is minimum because survival reduces with size.

Fig. 4.5c; see Sebens 1979, 1982). A similar model has been used to explain sexual dimorphism in size (Reiss 1982). However, this is too simple, for the probability of survival to this optimum size must be taken into account (Lynch 1980) as must the time it takes to reach the peak. That is to say, organisms with little chance of surviving to the size which maximizes production should breed before the optimum (cf. Fig. 4.5c and d). Thus Lynch (1980) has shown that Cladocera that live in habitats with vertebrate predators (that feed on the adults perferentially breed before the maximum production potential is reached (e.g. as in Fig. 4.5d) because survival chances reduce with increasing size, whereas those which live in habitats where there are only invertebrate predators (that concentrate on young Cladocera) breed when the production potential is maximum (e.g. as in Fig. 4.5c) because survival chances increase with increasing size. Using simulation techniques, Lewontin (1965) has also demonstrated that in colonizing species fitness is more sensitive to generation time (increasing as the latter reduces) than it is to reproductive productivity.

Some theory exists on whether the switch from growth to reproduction should be sharp (the so-called bang-bang strategy) or graded (Cohen 1971; Oster & Wilson 1978; Alexander 1982) and most models predict the former rather than the latter as an optimum solution. Within the plant and animal kingdoms, however, there are numerous examples of both bang-bang and graded strategies and so further scrutiny of the theory is required here.

In the case where reproduction occurs to the exclusion of somatic production it is to be expected, on the basis of the maximization principle, that reproductive production be maximized at the time it is initiated. However, both reproductive conversion efficiencies (i.e. the efficiency of conversion of input resources to gametes—sometimes referred to as reproductive effort) and absolute rates of gamete production show considerable variation even within taxonomically related species (e.g. Calow 1978b) i.e. there are no consistent maxima. One of the most obvious reasons for this is that maximizing reproductive output minimizes the energy available for maintaining and moving the soma. With less energy available for maintaining the vitality of the soma, for supporting key somatic physiological processes, for powering movement to escape predation or to capture prey, then the mortality risks of the parent increase and this is well documented (Calow 1978b). The fact that reducing rations can interact with reproduction to reduce parent survival (Calow 1977a; Browne 1982) lends support to this hypothesis. Hence there is a trade-off between reproductive output and parent survival and the varation noted should be explicable in these terms.

The optimization models here are well-known (Stearns 1976; Law 1979; Bell 1980; Sibly & Calow 1983) but are still not very rigorously tested. The

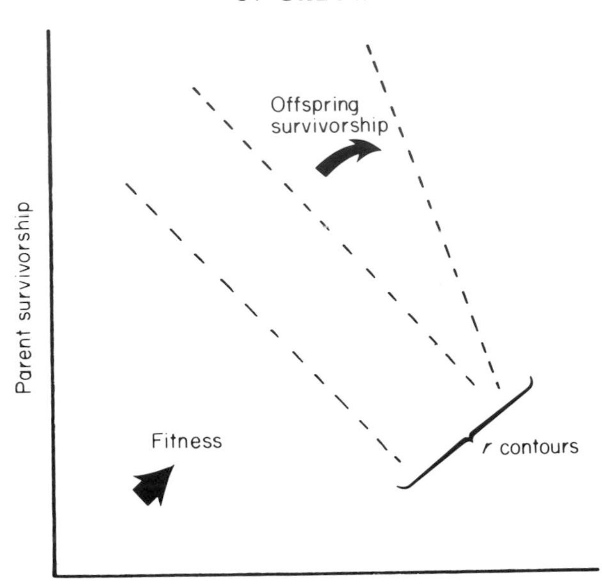

FIG. 4.6. Relationship between fitness, investment in reproduction, parent survivorship and offspring survivorship. Fitness is expressed in terms of the Malthusian parameter, r, defined classically as $1 = \sum_{t}^{\infty} e^{-rt} l_t n_t$ (where l and n are respectively survivorship and fecundity, e = base of natural logarithms, t = time). If reproduction occurs once per year and it is assumed that the survival and fecundity of adults are independent of age, this simplifies (Sibly & Calow 1982; Calow & Sibly 1983) as follows: $1 = e^{-r} l_a + e^{-r} l_j n$ (where subscripts a and j refer to parents and offspring respectively and n = number of offspring). Thus, $r = \log_e = (l_a + l_j n)$ so assuming n is directly proportional to investment in reproduction (i.e. gamete size is independent of investment in reproduction) contours of equal r appear as straight lines on the above graph increasing in value as they move from the origin and having a slope of $-l_j$.

rationale behind them is sketched out in Fig. 4.6. Fitness increases with increasing investment in reproduction and increasing parent survivorship. Assuming that the size of gametes remains constant (such that there is a linear relationship between resources invested in reproduction and the numbers of gametes produced; which is not universal, Calow 1978b) contours of equal fitness are likely to be linear with a negative slope which is sensitive to offspring survival chances (i.e. increasing as offspring survival increases) (explained in Fig. 4.6). With no trade-off between parent survivorship and reproduction, selection would drive the reproductive processes to their physiological limit. However, there are trade-offs, and their form is particularly crucial in defining optimum compromises (Law 1979). It is unfortunate, therefore, that we still have little precise information on the functional relationship between reproductive investment and parent survivorship. The reason for this is that these relationships are almost impossible to define since selection (by definition) is likely to have restricted the operation of organisms in particular populations to optimal segments of the trade-off curve. Caswell (1982) proposes an interesting inverse approach in which cost functions are derived from observations on life-history parameters by assuming these to have been optimized by natural selection.

A convenient and plausible starting point, however, is the convex up relationship depicted in Fig. 4.7a and b (Calow & Sibly 1983). This might operate, for example, if investment in reproduction was based upon the following physiological priorities: first few gametes formed from resources in excess of somatic requirements (little or no effect on adult survivorship); next few gametes formed from resources needed by but not essential to the soma (survivorship reduced); next few gametes formed by resources essential to the soma (survival rapidly reduced to zero). In this case it is possible to specify optimal compromises on the trade-off curve as those which make contact with the highest fitness contours. Hence, the optimal solution is unique and is dependent on the slope of the fitness contours and therefore on offspring survivorship. With low offspring survivorship (fitness contours have shallow slope) the investment in reproduction should be low (Fig. 4.7a); a condition which I have referred to as reproductive restraint (Calow 1978b) and which allows repeated breeding or iteroparity. With higher offspring survivorship (fitness contours steeper) the investment in reproduction should be increased as shown in Fig. 4.7b. Here the organism operates at a part of the curve where increased investment in reproduction causes massive increases in

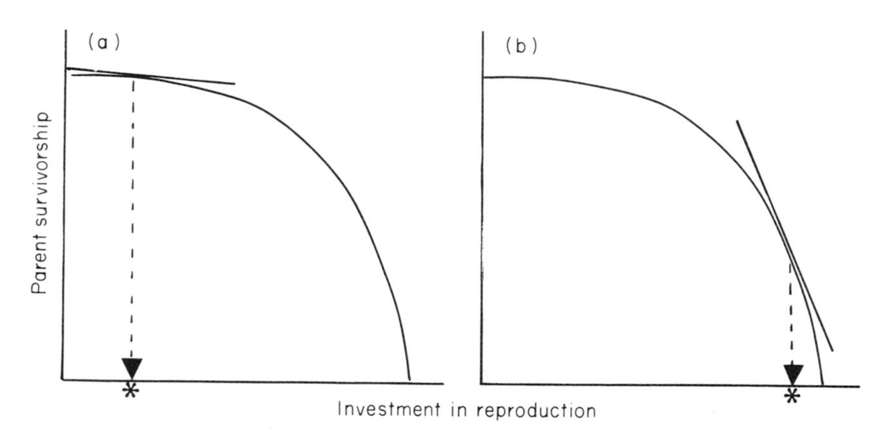

FIG. 4.7. Optimum solutions with a simple trade-off between investment of resources in reproduction and parent survivorship; (a) for poor offspring survival, (b) for good offspring survival. An * denotes optimal investment. (a) predicts iteroparity and (b) predicts semelparity.

mortality; a condition I have referred to as reproductive recklessness and which leads to single-episode, big-bang reproduction or semelparity. Adult mortality, from extrinsic causes (such as accident, disease and predators), is also likely to have an effect on optimal reproductive investment and it can be shown (Schaffer 1974; Calow 1983), *ceteris paribus*, that increased extrinsic adult mortality should lead to higher levels of reproductive investment.

There is still surprisingly little precise information on the relationship between age-specific survivorship and life-cycle pattern. However, the following findings are in agreement with the above predictions:

1 The mussel *Anodonta piscinalis* invests more in reproduction in populations where offspring survivorship is highest (Haukioja & Hakala 1978).

2 Iteroparous millipedes exploit patchy habitats and their offspring suffer high mortality in migrating between patches whereas semelparous ones live in more continuous habitats (Blower 1969).

3 Marsupial mice are semelparous in predictable, seasonal environments where juvenile survivorship is high, but iteroparous in ecological circumstances where offspring survivorship is more precarious, e.g. in temperate grasslands and the arid central Australia (Braithwaite & Lee 1979).

It should also be noted that, in principle, more complex relationships between reproductive investment and survivorship are possible than that indicated in Fig. 4.7 even on the basis of the physiological priorities discussed above (Calow in press). These might lead to non-unique optima (Fig. 4.8) (Schaffer & Rosenzweig 1977). Moreover, the possibility cannot be discounted that the relationship between the investment of resources in reproduction and parent survivorship and the level of juvenile survival have stochastic components and in this case the predictions become more 'fuzzy'. Occam's

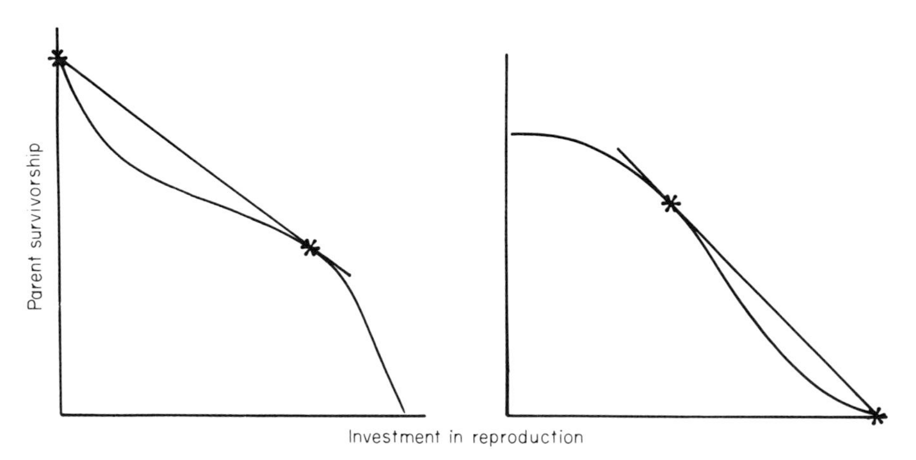

FIG. 4.8. Two forms of the trade-off curve in Fig. 4.7 which lead to multiple optima.

Razor, would probably argue for emphasis on the simple curve (Fig. 4.7) as an initial working hypothesis but, clearly, more experimental information is urgently required in this area.

In the case where somatic and reproductive production occur together, the form of the models can be similar to those observed above (e.g. Schaffer 1974; Law 1979) but as well as involving post-reproductive survival, the trade-off with reproduction also involves subsequent growth, and hence the adult size and fecundity at the subsequent breeding periods. The optimum solutions therefore depend on the form of the trade-off between somatic and reproductive production and on the functional relationship between adult size and fecundity—on which there is surprisingly little precise information (Calow & Sibly 1983).

ORGANISMS AS INTEGRATED WHOLES

One pressing conclusion which emerges from the economics approach to ontogeny is that organisms operate as integrated wholes and it is in this context that the genes they contain are subjected to natural selection. There are two reasons for this: (a) the finite nature of the resources available for ontogeny means that an investment in one aspect of metabolism has consequences for all others; (b) by influencing survival, developmental rate and fecundity the resource investment at one time has implications for optimum investment strategies at other times. Hence, metabolic integration and co-adaptation are likely outcomes of natural selection though there may have been some genetic constraints on this (Dawkins 1982).

Some examples of metabolic co-adaptation have been reviewed by Pianka (1981) and Solbrig (1981) for animals and plants respectively. Another example has emerged from work on lake-dwelling triclads (Calow, Davidson & Woollhead 1981). Here it is well established that populations are food-limited and that this condition intensifies during the breeding season. With respect to feeding strategies there are two main groups of species—one which feeds on active prey and one which feeds on sluggish and even immobile prey. The former are less active than the latter and adopt sit-and-wait rather than seek-out feeding strategies (see also p. 89). In consequence, the sit-and-wait feeders have lower metabolic rates, higher conversion efficiencies, and higher growth rates than the other species and are less susceptible (in a survival sense) to reduced food supply (Calow & Woollhead 1977). The survival chances of offspring of the sit-and-wait feeders are therefore invariably better and do not reduce as markedly with reducing rations as those of seek-out feeders. Hence the sit-and-wait feeders are more reckless (usually semelparous) than the seek-out feeders (usually

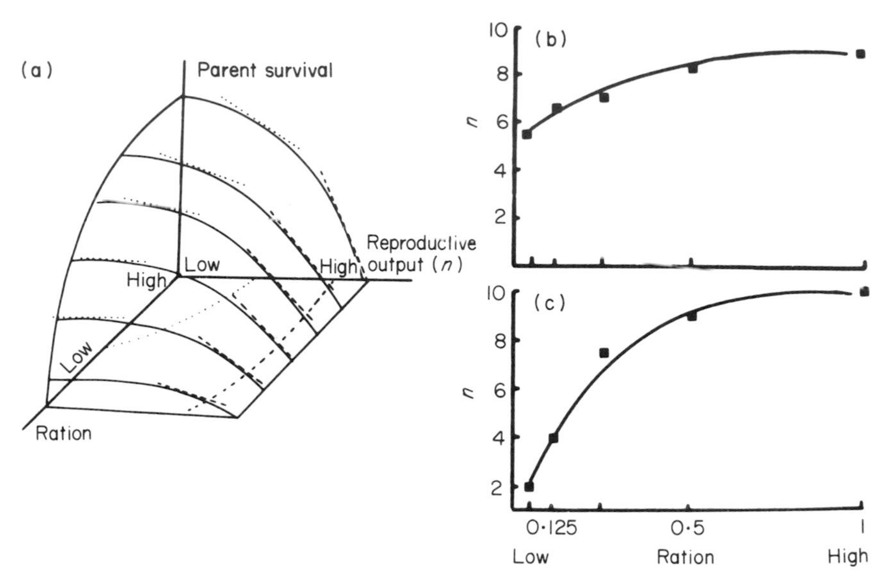

FIG. 4.9. Model and data on reproductive strategies of triclads. (a) Laboratory observations suggest that adult survivorship reduces at an increasing rate with both reducing ration and increasing reproductive output (Calow & Woollhead 1977). Broken lines are hypothetical r contours for a semelparous species, *Dendrocoelum lacteum*, (offspring survival good and relatively insensitive to ration) dotted lines are for an iteroparous species, *Polycelis tenuis*, (offspring survival poor and more sensitive to ration). *Dendrocoelum lacteum* is not made to invest in reproduction to such an extent that survival probability is reduced to zero because there is a suggestion in *some* populations that a few individuals do survive into a second year after reproduction. The relationships between juvenile survival and ration have implications for relationships in the ration/reproductive output plane—insensitivity of juvenile survival to ration (*D. lacteum*) predicts insensitivity in reproductive investment and *vice versa* (*P. tenuis*). (b) Actual relationship between reproductive output (n) and ration for *D. lacteum*—shows insensitivity. (c) Same as (b) but for *P. tenuis*—shows sensitivity. n = number of cocoons produced per observational period. Cocoons contain several to many offspring, but the actual number is independent of ration. The results in (b) and (c) conform to expectations in (a). After Calow & Sibly (1983).

iteroparous), investing more in reproduction and being less sensitive in this respect to reducing ration levels. A model showing a plausible relationship between reproductive investment, parent and offspring survivorship and ration for triclads is given in Fig. 4.9, together with graphs showing the relationship between investment in reproduction and ration for *Dendrocoelum lacteum* (sit-and-wait feeder, semelparous) and *Polycelis tenuis* (seek-out feeder, iteroparous). The model accounts adequately for the observations.

Clearly the feeding strategies of triclads have implications for energy allocation in growing organisms, which in turn influences age-specific developmental rates and survival chances. The latter, finally, has an impact on energy allocation and adult survivorship during the reproductive phase of the life cycle.

To appreciate co-adaptations of this kind, in-depth whole-life-cycle studies involving physiological, ecological and genetic analyses are essential. I have referred to this kind of adaptational programme as functional biology (Calow 1981) since it is concerned not just with the way traits assist in the operation of the whole organism, but with what *function* they play in the fitness of the organism. This, I would argue, focuses on the economics of metabolism because it is through that, that we can best appreciate the impact of a particular property (e.g. the size of an organ) or process (e.g. a particular mode and pattern of locomotion) on the developmental rate, to some extent the survivorship and, most definitely, on the reproductive output of the organism.

There is sometimes debate about the relative importance of genotypic and phenotypic correlations as a basis for trade-offs and constraints in evolution. Genetic correlations must, of course, always be more important than phenotypic ones (cf. Charlesworth in this volume). However, the main concern in this chapter has been with the physical basis (finite resources for utilization) for trade-offs and constraints in the evolution of organismic metabolism and life cycles. This will influence both the genetic and phenotypic interactions. Just as not all phenotypic traits have a genetic basis, not all genetic correlations are grounded in fundamental physical constraints (e.g. linkage phenomena are based on chromosomal organization) and insofar as they are not, they can, in principle, themselves be subject to natural selection. So in developing a *general theory* of the evolution of organismic, functional biology it is the fundamental factors which are important and information of genetic correlations is necessary but not sufficient for this theory.

HOMEOSTASIS AND FITNESS

Another important conclusion from this economics analysis of ontogeny is that physiological constraints and trade-offs and ecological restraints often impinge on the maximization principle. This could have been anticipated from the observation that endocrine systems have evolved to control both growth and reproduction, i.e. experimental manipulations of hormones can accelerate the growth and increase the reproductive output of vertebrates

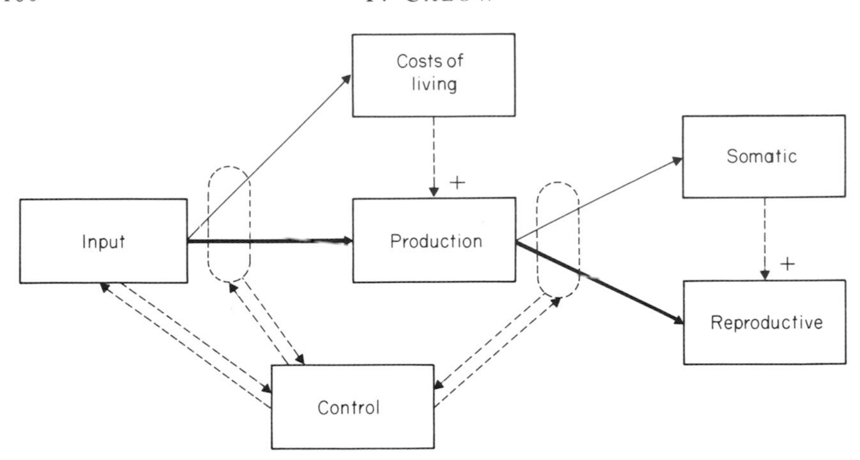

FIG. 4.10. As in Fig. 4.2 but with control introduced. This emanates from a 'black box' (representing endocrine and neural controls). It operates on the basis of information received from the system and by passing information to it.

above normal, suggesting that 'normal' is held below the physiological capacity of the system by active, endocrine control (Calow 1982). The metabolic models must, therefore, be modulated by active control systems (e.g. Fig. 4.10) and this is likely to have implications for our understanding not only of growth and reproduction but also feeding strategies (Calow 1982). A major problem here will be in specifying submaximum targets because if these are defined from observations on feeding, growth and reproduction they will inevitably constrain the models to track observations and so introduce an element of circularity. Clearly targets should be defined by independent criteria, e.g. with respect to the requirements of getting food, reducing mortality risks, and raising of offspring. And again this will require in-depth studies of particular systems from a physiological, ecological and evolutionary perspective.

The principles of metabolic integration and homeostasis have become joined in the concept of plastic *Developmental Trajectories* (Alberch *et al.* 1979; Stearns & Crandall, in press). This assumes that the track along which size and other developmental parameters change under some naturally experienced stress is predictable since it is the whole integrated trajectory that has been subjected to selection. Epigenetics will undoubtedly come to play a more and more important part in the adaptationist programme but its implications are at present only dimly perceived. Stearns & Crandall (in press) have, nevertheless, used plastic developmental trajectory theory to advantage in the analysis of age and size at maturity of fishes.

REFERENCES

Abrahamson W.G. & Caswell H. (1982) On the comparative allocation of biomass, energy and nutrients in plants. *Ecology*, **63**, 982–991.

Alexander R.McN. (1981) Factors of safety in the structure of animals. *Science Progress*, **67**, 109–130.

Alexander R.McN. (1982) *Optima for Animals*. Edward Arnold, London.

Alberch P., Gould S.J., Oster G.F. & Wake D.B. (1979) Size and shape in ontogeny and phylogeny. *Paleobiology*, **5**, 296–317.

Anderson R.A. & Karasov W.H. (1981) Contrasts in energy intake and expenditure in sit-and-wait and widely foraging lizards. *Oecologia*, **49**, 67–72.

Barr T.C. (1968) Cave ecology and the evolution of troglobites. *Evolutionary Biology* (Ed. by T. Dobzhansky, M.K. Hecht & W.C. Steere), vol. 2, pp. 35–102. Appleton-Century, Crofts, New York.

Bayne B.L. (1976) *Marine Mussels: Their Ecology and Physiology*. Cambridge University Press, Cambridge.

Bell G. (1980) The costs of reproduction and their consequences. *American Naturalist*, **116**, 45–76.

Blower G. (1969) Age-structure of millipede populations in relation to activity and dispersion. *The Soil Ecosystem* (Ed. by J.G. Sheal), pp. 209–216. Systematics Association, Publication No. 8 London.

Braithwaite R.W. & Lee A.R. (1979) A mammalian example of semelparity. *American Naturilist*, **113**, 151–155.

Browne R.A. (1982) The costs of reproduction in Brine Shrimp. *Ecology*, **63**, 43–47.

Calow P. (1977a) Ecology, evolution and energetics: a study in metabolic adaptation. *Advances in Ecological Research* (Ed. by A. Macfadyen), vol. 10 pp. 1–62. Academic Press, London.

Calow P. (1977b) Conversion efficiencies in heterotrophic organisms. *Biological Reviews*, **52**, 385–409.

Calow P. (1978a) *Life Cycles*. Chapman Hall, London.

Calow P. (1978b) The cost of reproduction—a physiological approach. *Biological Reviews*, **54**, 23–40

Calow P. (1981) *Invertebrate Biology: A functional Approach*. Croom Helm Publishers, London.

Calow P. (1982) Homeostasis and fitness. *American Naturalist*, **120**, 416–419.

Calow P. (in press) Exploring the adaptive landscapes of invertebrate life cycles. *Advances in Invertebrate Reproduction*, **3**.

Calow P. (1983) Pattern and paradox in parasite reproduction. *Parasitology*, **86**, 197–207.

Calow P., Davidson A.F. & Woollhead A.S. (1981) Life-cycle and feeding strategies of freshwater triclads—a synthesis. *Journal of Zoology*, **193**, 215–237.

Calow P. & Sibly R.M. (1983) Physiological trade-offs and the evolution of life cycles. *Science Progress*, **68**, 177–188.

Calow P. & Townsend C.R. (1981) Resource utilization in growth. *Physiological Ecology. An Evolutionary Approach to Resource Use*. (Ed. by C.R. Townsend & P. Calow). pp. 220–244. Blackwell Scientific Publications, Oxford.

Calow P. & Woollhead A.S. (1977) The relation between ration, reproductive effort and age-specific mortality in the evolution of life-history strategies—some observations on freshwater triclads. *Journal of Animal Ecology*, **46**, 765–781.

Caswell H. (1982) Optimal life histories and age-specific costs of reproduction. *Journal of Theoretical Biology*, **98**, 519–529.

Cohen D. (1971) Maximizing final yield when growth is limited by time or by limiting resources. *Journal of Theoretical Biology*, **33**, 299–307.

Cohn, D.L. (1954) Optimal systems: I. The vascular system. *Bulletin of Mathematics and Biophysics*, **16**, 59–74.

Cohn D.L. (1955) Optimal systems: II. The vascular system. *Bulletin of Mathematics and Biophysics*, **17**, 219–227.

Comfort A. (1979) *The Biology of Senescence*, 3rd ed. Churchill Livingstone, Edinburgh.

Cooke R.J., Oliver J. & Davies D. (1979) Stress and protein turnover in *Lemna minor*. *Plant Physiology* (Bethesda), **64**, 1109–1113.

Dawkins R. (1982) *The Extended Phenotype*. W.H. Freeman & Co., Oxford and San Francisco.

Epp R.W. & Lewis W.M. (1980) The nature and ecological significance of metabolic changes during the life history of copepods. *Ecology*, **61**, 259–264.

Fenchel T. (1974) Intrinsic rate of natural increase: the relationship with body size. *Oecologia*, **14**, 317–376.

Fox H.M. & Simmonds B.G. (1933) Metabolic rates of aquatic arthropods from different habitats. *Journal of Experimental Biology*, **10**, 67–74.

Garland T. (1983) Scaling the ecological cost of transport to body mass in terrestrial mammals. *American Naturalist*, **121**, 571–587.

Gibbons J.W. & Semlitsch R.D. (1982) Survivorship and longevity of a long-lived vertebrate species: how long do turtles live? *Journal of Animal Ecology*, **51**, 523–527.

Giller P.W. (1982) Locomotory efficiency in the predation strategies of the British *Notonecta* (Hemiptera, Heteroptera). *Oecologia*, **52**, 273–277.

Goldberg A.L. & Dice J.F. (1974) Intracellular protein degradation in mammalian and bacterial cells. *Annual Review of Biochemistry*, **43**, **835–869**.

Gutschick, V.P. (1981) Evolved strategies in nitrogen acquisition by plants. *American Naturalist*, **118**, 607–637.

Haukioja E. & Hakala T. (1978) Life-history evolution in *Anodonta piscinalis* (Mollusca, pelecypoda). Correlation of parameters. *Oecologia*, **35**, 253–266.

Henneman W.W. (1983) Relationship among body mass, metabolic rate and intrinsic rate of natural increase in mammals. *Oecologia*, **56**, 104–110.

Hovanitz W. (1944) The distribution of gene frequencies in wild population of *Colias*. *Genetics*, **29**, 31–60.

Ivler V.S. (1960) On the utilization of food by planktophage fishes. *Bulletin of Mathematical Biophysics*, **22**, 371–389.

Kirkwood T.B.L. (1981) Repair and its evolution: survival *versus* reproduction. *Physiological Ecology. Evolutionary Aspects of Resource Use*. (Ed. by C.R. Townsend & P. Calow), pp. 165–189. Blackwell Scientific Publications, Oxford.

Law R. (1979) Ecological determinants in the evolution of life histories. *Population Dynamics*, (Ed. by R.M. Anderson, B.D. Turner & L.R. Taylor), B.E.S. Symposium 20, pp. 81–103. Blackwell Scientific Publications, Oxford.

Lewontin R.C. (1965) Selection for colonizing ability. *The Genetics of Colonizing Species* (Ed. by H.G. Baker & G.L. Stebbins) pp. 79–94. Academic Press, New York.

Lynch M. (1980) The evolution of cladoceran life histories. *Quarterly Review of Biology*, **55**, 23–42.

Martin R.D. (1981) Relative brain size and basal metabolic rate. *Nature*, **293**, 57.

McClure P.A. & Randolph J.C. (1980) Relative allocation of energy to growth and development of homeothermy in the eastern wood rat (*Neotoma floridiana*) and hispid cotton rat (*Sigmodon hispidus*). *Ecological Monographs*, **50**, 199–219.

McFarland D.J. (1976) Form and function in the temporal organisation of behaviour. *Growing Points in Ethology* (Ed. by P.P.G. Bateson & R.A. Hinde) pp. 55–93. Cambridge University Press, Cambridge.

McNab B.K. (1980) Food habits, energetics and the population biology of mammals. *American Naturalist*, **116**, 106–114.

Milsum J.H. & Roberge F.A. (1973) Physiological regulation and control. *Foundations of Mathematical Biology, 3,* (Ed. by R. Rosen) pp. 1–95. Academic Press, London and New York.

Norberg R.A. (1977) An ecological theory on foraging time and energetics and choice of optimal food searching method. *Journal of Animal Ecology*, **46**, 511–529.

Oster G.F. & Wilson E.O. (1978) *Caste and Ecology in the Social Insects*. Princeton University Press, Princeton.

Pianka E.R. (1981) Resource acquisition and allocation among animals. *Physiological Ecology. An Evolutionary Approach to Resource Use.* (Ed. by C.R. Townsend & P. Calow) pp. 300–314. Blackwell Scientific Publications, Oxford.

Price M.J. (1972) Turnover of intracellular proteins. *Annual Review of Microbiology*, **26**, 103–126.

Rashevsky N. (1965) Models and mathematical principles in biology. *Theoretical and Mathematical Biology* (Ed. by T.H. Waterman & H.J. Morowitz) pp. 36–53. Blaisdell Publ. Co., New York, Toronto and London.

Regal P.J. (1977) Evolutionary loss of useless features: is it molecular noise suppression. *American Naturalist*, **111**, 123–133.

Reiss M. (1982) Males bigger, females biggest. *New Scientist*, **96**, 226–229.

Roughgarden J. (1971) Density dependent natural selection. *Ecology*, **52**, 453–468.

Schaffer W.M. (1974) Selection for optimal life histories: the effects of age structure. *Ecology*, **55**, 291–303.

Schaffer W.M. & Rosenzweig M.L. (1977) Selection for optimal life histories. II. Multiple equilibria and the evolution of alternative reproductive strategies. *Ecology*, **58**, 60–72.

Schiemer F. (1983) Comparative aspects of food dependence and energy use of free-living nematodes. *Oikos*, **41**, 32–49.

Schroeder L.A. (1981) Consumer growth efficiencies: their limits and relationships to ecological energetics. *Journal of Theoretical Biology*, **93**, 805–828.

Scott M.G., Watt W.B. & Lawrence F.G. (1980) Metabolic resource allocation vs. mating attractiveness: Adaptive pressures on the 'alba' polymorphism of *Colias* butterflies. *Proceedings of the National Academy of Sciences, USA*, **77**, 3615–3619.

Sebens K.P. (1979) The energetics of asexual reproduction and colony formation in benthic marine invertebrates. *American Zoologists*, **19**, 683–697.

Sebens K.P. (1982) The limits of indeterminate growth: an optimal size model applied to passive suspension feeders. *Ecology*, **63**, 209–227.

Sibly R.M. (1981) Strategies of digestion and defecation. *Physiological Ecology. An evolutionary approach to resource use.* (Ed. by C.R. Townsend & P. Calow) pp. 109–139. Blackwell Scientific Publications, Oxford.

Sibly R.M. & McFarland D.J. (1976) On the fitness of behaviour sequences. *American Naturalist*, **110**, 601–617.

Sibly R.M. & Calow P. (1983) An integrated approach to life-cycle evolution using selective landscapes. *Journal of Theoretical Biology*, **102**, 527–547.

Solbrig O.T. (1981) Energy information and plant evolution. *Physiological Ecology. An Evolutionary Approach to Resource Use.* (Ed. by C.R. Townsend & P. Calow) pp. 274–299. Blackwell Scientific Publications, Oxford.

Stearns S.C. (1976) Life-history tactics: a review of the ideas. *Quarterly Review of Biology*, **51**, 3–47.

Stearns S.C. (1980) A new view of life-history evolution. *Oikos*, **35**, 266–281.

Stearns S.C. & Crandall R.E. (1982) Plasticity for age and size at maturity: a life-history response to unavoidable stress. *Fisheries Society of the British Isles Symposium*, July 1982.

Townsend C.R. & Calow P. (1981) *Physiological Ecology. An Evolutionary Approach to Resource Use.* Blackwell Scientific Publications, Oxford.

Van Valen L.M. (1980) Evolution as a zero-sum game for energy. *Evolutionary Theory*, **4**, 289–300.

Ware D.M. (1975) Growth, metabolism and optimal swimming speed of pelagic fish. *Journal of the Fisheries Board of Canada*, **32**, 33–41.

Ware D.M. (1982) Power and evolutionary fitness of teleosts. *Canadian Journal of Fisheries and Aquatic Sciences*, **39**, 3–13.

Waterlow J.C. (1980) Protein turnover in the whole animal. *Investigative and Cell Pathology*, **3**, 107–119.

Zamenhof S. & Eichorn H. (1967) Study of microbial evolution through loss of biosynthetic functions: establishment of 'defective' mutants. *Nature*, **216**, 456–458.

5. FITNESS VARIATION AMONG BRANCHES WITHIN TREES

DOUGLAS E. GILL AND TIMOTHY G. HALVERSON
Department of Zoology, University of Maryland, College Park, Maryland 20742 *USA*

INTRODUCTION

The thesis in this Chapter concerns a speculation that organisms with extensively branched architecture, such as large, tree-like plants, are not genetically uniform but are actually colonies of many heritable genotypes. The genetic diversity within their structure is generated by the accumulation of 'somatic mutations' that arise spontaneously in the meristems of the constitutive modular parts. The amount of genetic variation is expected to be a stochastic variable that grows as a product of the logistic growth in numbers of apical meristems (units of mutability) and the rates of mutation per meristem. Although such a deliquescent organism as a tree may function as a single ecological organism by virtue of its branches all sharing a common root system, we propose that they also function as a multitude of evolutionary individuals because of the developmental and genetic independence of their parts.

The term 'somatic mutation' as applied here may be misleading to ecologists from zoological backgrounds. In animals, mutations that occur in somatic tissues and exclusive of the germ line have no heritability and are irrelevant in an evolutionary context. In plants mutations that arise in meristems during development have the dual potentiality of being expressed phenotypically in the somatoplasm and reproductively in the gametophytes. That such mutations persist vegetatively after grafting and may be transmitted through pollen and seed show that they are heritable both asexually and sexually. Hence somatic mutations can have evolutionary importance in plants.

One essential feature of the hypothesis is that phenotypic variation generated among branches by developmental mutations is manifest as fitness differentials. Dependent upon its genotype, a modular part may be short- or long-lived and a branch may produce many or few seeds. In this way natural selection operates among the genotypically diverse 'population' of branches just as though they were genuine units of selection. We view deliquescent organisms as having a source of genetic variation perhaps orders of magnitude greater than conventional populations. Genetic change in such species is greatly accelerated by the natural selection of reproductive modules within each ecological organism. The hypothesis also has profound implications about the evolution of plant–animal interactions and the appropriate time scale on which rates of evolution in plants are measured (Whitham 1981).

In order to validate the claim that variation found among the branches of trees is relevant to natural selection and has evolutionary significance, it is necessary to show that (a) significant phenotypic variation among branches exists, (b) differential survival or reproduction of modular parts occurs, and (c) traits that confer the differential fitness are inherited. In this chapter we present original data that illustrate the first two criteria, namely significant phenotypic variation and differential fitness among branches within trees. In the first data set we show the significant autocorrelated variability in infestation among ramets of a temperate shrub by a galling aphid. In the second data set we apply a nested variance analysis to the amount of reproductive variation among branches. Experiments testing for the genetic basis of the variation, the third criterion, are in progress. In another paper (Gill, unpublished) the literature that supports all three points, especially the evidence for genetic variation within trees, is reviewed. That literature review covers over 500 references and is too large to include here.

The concept of an important evolutionary role played by genetic mosaicism within plants, with particular attention to the evolutionary interaction of herbivorous insects and trees, has been developed independently by Whitham & Slobodchikoff (1981) and Whitham (1983).

PATTERNS OF LEAF GALLS IN *HAMAMELIS VIRGINIANA* L.

The hypothesis that trees are genetically mosaic requires that the genetic variation is expressed as phenotypic variation among branches. One expectation is that there should exist morphological and phytochemical variation that renders branches variously attractive or repulsive to herbivorous insect pests. If such phenotypic variation does have a genetic basis, then the pattern of variation expressed on a branch-by-branch basis should be autocorrelated between successive years. Indeed, the hypothesis of significant genetic variation among branches within trees would be seriously refuted by a failure to demonstrate significant autocorrelated variation in the pattern of infestation between years.

If a deliquescent plant were genetically uniform, some patterning in levels of infestation among plant parts is expected from random forces alone. However, a null hypothesis of no significant genetic mosaicism within a tree predicts that the patterning between years should be independent and uncorrelated. That null hypothesis is explicitly testable.

Direct measurements of morphological or phytochemical variation may reveal the existence of among-branch variation but not its significance in the context of selective propagation. On the other hand the pattern of infestation by the insect pests themselves can serve as an assay of the variation within the plant as it relates to potentially strong selective agents. Hence branches that are genetically susceptible to attack should experience heavy infestation every year while branches that are genetically resistant should remain relatively free from infestation for a series of years. Repeatedly damaged branches have reduced growth and fruit production or eventually die, while undamaged branches flourish and are fruitful.

In order to test the prediction of autocorrelated variation in levels of infestation among branches between successive years, we studied the infestation of witch-hazel, *Hamamelis virginiana* L., a common understory shrub of the hardwood forests in the Shenanoah Mountains in western Virginia, USA, by the gall-aphid, *Hormaphis hamamelidis* (Homoptera: Hormaphidae) in 1980 and 1981. In both years the dorsal surfaces of the leaves of witch-hazel were conspiciously covered with the characteristic cone-shaped galls of *H. hamamelidis.* The variation among individual plants (genet) was very obvious, but there was also conspicuous variation in the levels of infestation among shoots (ramets) within each plant.

The life cycle of *H. hamamelidis* is complex (Pergande 1901) but favourable for a study of its interaction with witch-hazel. Aphid stem-mothers hatch from sexually produced eggs that overwinter in tiny crevices in witch-hazel bark. At leaf bud burst the newly hatched stem-mothers settle on leaves and mechanically induce galls to grow and envelop them. They produce up to 100 first-generation progeny in the gall. The alate second generation is produced in May and disperses to the alternative host plant, black birch, *Betula lenta* L. Additional asexual generations are produced on birch and culminate in an alate sixth generation. These alates disperse back to witch-hazel in early autumn and produce the sexual (seventh) generation, which in turn lays five to ten winter diapausing eggs. We assume that this life cycle, described by Pergande from the Potomac River Valley near Washington D.C. pertains to our study in the Shenandoah Mountains, but field studies confirming it are in progress.

It is the dispersal of the alate aphids from birch to the deciduous witch-hazel every autumn that validates the null hypothesis that the pattern of infestation on witch-hazel leaves each growing season is independent of the pattern the previous year. Moreover, the pattern of galls on leaves within ramets each year is expected to be be distributed as a Poisson variate if the settling of the seventh generation on a genet is random. Deviations from a Poisson distribution could arise only as a consequence of oviposition preferences for particular branches in the autumn, behavioural interactions among stem mothers in the spring, or discrimination of leaves by stem mothers for gall formation. Evidence for territorial defence (Whitham 1979) and leaf selection by stem mothers in response to leaf phenolics (Whitham 1978; Zucker 1982) has been described for the cottonwood galling aphid, *Pemphigus betae*, Doane. All three possibilities depend on a functional response of aphids to the phenotypic variation of the leaves in the host plant. A positive correlation of the patterns of infestation between successive years would imply that the proximal cues among the parts of a genet to which the aphids respond are carried as traits in stem or bud tissue over the winter and re-expressed in the new leaves in the new year.

The method employed in this study was to permanently mark with tags the repeated modular parts of several individual witch-hazels and count the number of galls that were maturing on every leaf. The subdivision of

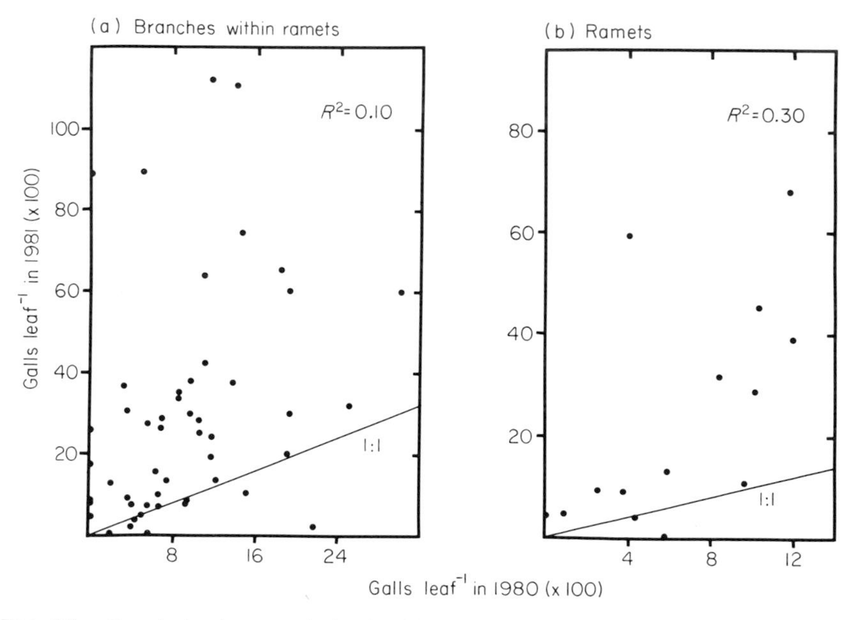

FIG. 5.1. Correlation between the levels of infestation by a gall-forming aphid, *Hormaphis hamamelidis*, on branches within ramets (a) and on the same ramets as a whole (b) of witch hazel genet *HI*, *Hamamelis virginiana* L. in 1980 and 1981. The straight line indicates a hypothetical 1 : 1 relationship between years. Locality is the George Washington National Forest, Rockingham County, Virginia.

each genet was rather coarse in 1980, typically involving only ramets and occasionally extending to branches within ramets. In 1981 a finer level of subdivision in the plant architecture was accomplished, but close attention to the previous years units was paid so that direct comparison between years could be made.

The numbers of ramets and branches within ramets in only two plants, *HI* and *DRI*, were large enough for statistical analysis. When the number of galls on the fifty-one marked branches within ramets on genet *HI* were compared between 1980 and 1981, the correlation between years was found to be significant at the 5% level (Fig. 5.1a). The level of infestation in 1981 was, on average, three times greater than it was in 1980. The correlation between years accounted for only 10% of the observed variation, suggesting a low level of asexual heritability of the traits that generated the variation in infestation.

Examining the same bush at one architectural division coarser, namely the entire ramet, the correlation of numbers of galls between years was significant and accounted for 30% of the observed variation. Thus, ramets that were heavily infested in 1980 tended to be heavily infested in 1981; those that were lightly attacked in 1980 tended to be lightly attacked in 1981 also.

The twenty-nine branches composing seven ramets in bush (genet) *DRI* were also three times more heavily infested in 1981 than they were in 1980 (Fig. 5.2). The correlation in infestation between years on a branch-to-branch basis was very conspicuous, accounting for over 61% of the observed variation. This correlation was very highly significant ($P < 0.001$).

When all ramets from all plants were pooled together, so that genets with few ramets could be included in the analysis, the resulting correlations were enormously significant no matter what test was used. The reason for such huge correlations was that the pooling confounded within-genet and among-genet variation. The autocorrelation of genets with respect to aphid infestation was overwhelming: entire bushes that were very heavily infested in 1980 were very heavily infested in 1981, and genets that were not attacked in 1980 escaped in 1981 also. The variation among genets extended the total range of variation by a factor of 25 greater than the variation within genets. This contrast is evident by comparing the scales of the infestations in *HI* and *DRI*.

The results revealed significant correlations in levels of infestation by aphids making galls on the blades of leaves of witch-hazel between 2 successive years when examined on a branch-by-branch and ramet-by-ramet basis. The consistent positive correlations are taken as preliminary evidence consistent with the hypothesis that repeated modular parts of plants con-

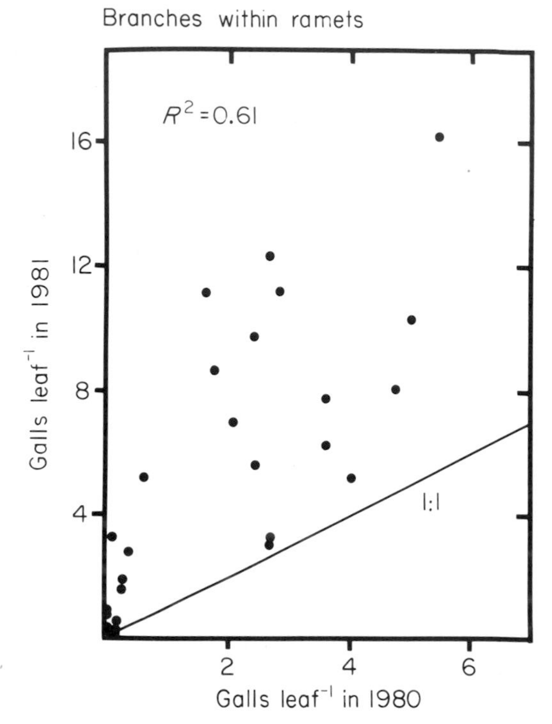

FIG. 5.2. Correlation between the levels of infestation by a gall-forming aphid, *Hormaphis hamamelidis*, on branches within ramets of witch-hazel genet *DRI*, *Hamamelis virginiana* L. in 1980 and 1981. The straight line indicates a hypothetical 1 : 1 relationship between years. Locality same as Fig. 5.1.

tained genetic variation rendering some parts susceptible and others resistant to herbivorous insect attack. The fact that witch-hazel in deciduous is viewed as important because it imposes an independence of infestability in the foliage between years. The data, of course, do not prove the existence of within-plant genetic variation; proof requires the satisfaction of all three criteria set out earlier.

If stem mothers of *Hormaphis hamamelidis* behave similarly to *Pemphigus betae*, then it is likely that the clumped pattern of galls per leaf was established by their settling activities in the spring. The proximal mechanism of attraction to the few, preferred leaves is probably suitable chemical quality (Zucker 1982). Immediate discrimination of the quality of plant tissue is a well-documented behaviour in aphids (Dixon 1970a, b; Kennedy & Booth 1951, Van Emden & Bashford 1969).

Assuming that the proximal cue was chemical in nature, then the significant between-year correlations of infestation patterning implies that the chemical characteristics of the ramets and branches within ramets were retained over the winter within stems. One mechanism of retention would be genetic. It is also possible that another cause of the autocorrelation was repeated microclimate differences around each plant part that induced the observed variation. We discount this possibility because the branches themselves were so intermingled each year that leaves from separate ramets were often touching one another. These observations argue against environmentally induced traits associated with variation in microclimate. Only reciprocal grafting experiments can unambiguously rule on this alternative to the genetic variation hypothesis. Presently, the most parsimonious explanation is that stems and branches posses inherent qualities of attractiveness and repellance that are re-expressed every year in the foliage.

The results of this study also revealed a highly significant difference among genets that was one to two orders of magnitude greater than the within-genet variation. This was the kind of variation among individuals most commonly expected by traditional concepts of Darwinian evolution in plants. The contrast in the among-genet v. the within-genet variation is consistent with the expectation of logistic expansion in the number of apical meristems, considering the small number of meristems in these modest-sized bushes. The small value of the within-bush variation found in *Hamamelis virginiana* contrasts with the variation in fruit production and seed set found in caper trees (presented below) largely because of the difference in size between the two species.

The most direct test of sexual heritability of plant characteristics that result in the autocorrelated clumped patterns of galls is to germinate seed from controlled pollinations. Reciprocal crosses between the flowers of heavily infested and non-infested branches would control for male and female parentage. Subjecting the potted seedlings (or more mature stems if necessary) to natural infestations in the field, or controlled infestations in the laboratory, should rigorously test whether the observed variation is heritable through pollen or seed. Such experiments are feasible with the gall-aphid witch-hazel system and are planned.

FRUIT PRODUCTION AND SEED SET IN *CAPPARIS ODORATISSIMA* JACQ.

Although there exists abundant literature confirming that phenotypic variation within trees is widesperead (Gill, unpublished), few data were found that quantitatively compare the amount of variation within trees with the amount among trees. Chemical variation in samples of the same tissue types from different parts of the same tree is claimed to be as great as that found among individual trees (Dyson & Herbin 1970; Parker 1977; Von Rudloff 1969, 1972). Rather precise quantitative estimates of relative amounts of phenotypic variation are obtainable by an experimental design that is directed toward a hierarchical analysis of variance. In such a sampling programme, samples of plant parts would be collected from portions of several plants among several locations, from several geographic areas, and so on.

In an effort to obtain direct measures of the relative amounts of reproductive variation within and among trees, and to illustrate the level of statistical significance in fitness differentials in reproduction among branches, one of us (Gill) co-ordinated a study of the fruit production and seed set in a tropical forest tree, *Capparis odoratissima* Jacq. under natural conditions. The species was chosen arbitrarily from among many because of the convenience of sampling.

Trees in a grove of caper trees at the edge of the flooded marshland on the Rio Tempisque at Palo Verde, Costa Rica were selected for study. They were chosen because their bole divided into three equivalent limbs which penetrated to a fully exposed crown. Their architecture was exceptionally

TABLE 5.1. Mean ($\bar{X}$) and standard deviation (SD) of the fruit production (fruits leaf^{-1}) and seed set (seeds fruit^{-1}) per branchlet in three major branches of caper trees, *Capparis odoratissima* Jacq., at Palo Verde, Costa Rica. Five branchlets per major branch were sampled from two trees on 14 July 1979 and four branchlets were sampled from three major branches in three trees on 13 July 1981. Analysis of variance are in Tables 5.2 and 5.3

	Fruit production				Seed set			
	1979		1981		1979		1981	
	$\bar{X}$	SD	$\bar{X}$	SD	$\bar{X}$	SD	$\bar{X}$	SD
Tree 1								
Branch A	0.391	0.093	0.092	0.066	2.272	0.540	1.683	0.251
Branch B	0.085	0.079	0.059	0.026	1.516	0.746	1.883	0.684
Branch C	0.345	0.149	0.070	0.012	1.944	0.198	1.243	0.303
Total	0.274	0.173	0.073	0.040	1.911	0.597	1.602	0.494
Tree 2								
Branch D	0.172	0.046	0.052	0.034	1.392	0.336	1.418	0.347
Branch E	0.210	0.060	0.024	0.011	1.280	0.311	0.795	0.701
Branch F	0.223	0.089	0.037	0.022	2.098	0.420	1.213	0.487
Total	0.202	0.066	0.038	0.025	1.590	0.501	1.142	0.552
Tree 3								
Branch G	—		0.129	0.013	—		1.253	0.122
Branch H	—		0.153	0.069	—		2.185	0.853
Branch J	—		0.260	0.077	—		2.163	0.124
Total	—		0.181	0.081	—		1.867	0.642

favourable for an analysis of variance of reproductive variation between and within trees. On 14 July 1979, all the leaves, all the fruit pods, and all the maturing seeds in each fruit were counted from five equivalent-sized branchlets clipped from each of the three major branches of two trees. The data (Table 5.1) were expressed as the number of fruits per leaf on each branchlet and the number of viable seeds per fruit in order to standardize sample sizes. Two years later the procedures were repeated on the same two trees and a third tree, but only four branchlets were collected per major branch. An effort was made to collect branches of similar size and exposure but without conscious bias with respect to fruit load. Natural logarithms of the fruit production were taken prior to analysis to assure homogeneity of variance, but no transformations of the seed set data were necessary. The variation among branchlets was nested within major branches, which in turn were nested within trees.

There was significant variation in number of fruits produced among major branches but not between the two trees in 1979 (Table 5.2). The addition of a third tree in 1981 produced a significant between-tree variance, but the variation among major branches in the original two trees persisted over the 2-year period (Table 5.3). With respect to the set of viable seeds, the variation among major branches was always significant, but there was no

TABLE 5.2. Nested ANOVA of fruit production (ln fruits leaf^{-1}) and seed set (seeds fruit^{-1}) in caper trees, *Capparis odoratissima* Jacq. within years at Palo Verde, Costa Rica. Data from Table 5.1

	Year	Source of variation	df	Mean squares	F ratio	significance (*P*)
(a) Fruit production	1979	Trees	1	0.071	0.039	NS
		Among major branches	4	1.823	2.780	< 0.05
		Among branchlets	24	0.656		
		Total	29	0.797		
	1981	Trees	2	0.368	21.441	< 0.001
		Among major branches	6	0.390	1.637	NS
		Among branchlets	27	0.238		
		Total	35	0.729		
(b) Seed set	1979	Trees	1	0.771	0.906	NS
		Among major branches	4	0.851	4.013	< 0.05
		Among branchlets	24	0.212		
		Total	29	0.319		
	1981	Trees	2	1.616	2.468	NS
		Among major branches	6	0.654	2.661	< 0.05
		Among branchlets	27	0.246		
		Total	35	0.394		

TABLE 5.3. ANOVA on fruit production (ln fruits leaf^{-1}) and seed set (seeds fruit^{-1}) in two caper trees, *Capparis odoratissima* Jacq. at Palo Verde, Costa Rica. Samples from the same two trees were collected on 14 July 1979 and 13 July 1981

	Source of variation	df	Mean squares	F ratio	Significance (*P*)
(a) Fruit production	Trees	1	2.059	1.924	NS
	Years	1	29.471	27.549	< 0.001
	Tree–year interaction	1	1.450	1.355	NS
	Among major branches	8	1.070	2.127	< 0.05
	Among branchlets	42	0.503	—	
	Total	53	1.182	—	
(b) Seed set	Trees	1	1.980	3.126	NS
	Years	1	1.908	3.012	NS
	Tree–year interaction	1	0.065	0.103	NS
	Among major branches	8	0.633	2.812	< 0.05
	Among branchlets	42	0.225	—	
	Total	53	0.349	—	

significant differences among the trees as wholes (Tables 5.2 and 5.3).

The results reveal that the variation in reproductive characteristics among major branches within a tropical forest tree was greater than the differences between the trees as wholes. There were highly significant differences in fruit production in the two trees between the 2 years. The fact that there was no significant tree-year interaction indicates that the sharp reductions in fruit production from 1979 to 1981 was equivalent in the two trees. Clearly the trees responded to environmental conditions as whole entities, but the results are also consistent with the hypothesis that the modules shared a common change in environment. Of course, the data in no way prove that there is a genetic basis to the among-branch variation. They simply demonstrate that within-tree variation can be greater than between-tree variation in reproductive output.

The sample sizes of trees were minimal and did not represent the average tree in this dry, deciduous forest. Most of the trees in the grove had no visible fruit, and did not lend themselves to a study of potential differentials in reproductive output from parts of trees. Thus, the trees chosen for study were exceptionally fruit laden (productive) compared to many conspecifics in the same grove. It is also the case that our samples of two and three trees did not provide meaningful estimates of among-tree variation in a forest because the addition of a third tree to the previous sample of two drastically altered the statistical patterns. These meager data indicate that the variation within individual trees with respect to reproductive characters was demonstrably as large or larger than the variation among trees as wholes.

CONCLUSIONS

A hypothesis that the genetic variation among the modular parts of highly branched plants is generated through somatic mutation is postulated. It is suggested that phenotypic variation in morphology and phytochemistry is expected in turn, and that the interaction of herbivorous insects and other plant pests with foliage that is variously attractive and repellant should produce fitness differentials among plant parts that take on evolutionary importance. Proof of the existence of evolutionary significant quantities of genetic variation within deliquescent plants requires the fulfillment of three criteria: (a) demonstration of phenotypic variation, (b) differential fitness among plant parts, and (c) heritability of the variation.

Original data are presented that satisfy the first two criteria. The infestation of witch-hazel, *Hamamelis virginiana*, by a galling aphid, *Hormaphis hamamelidis*, was variable within genets as well as among genets. The pattern of infestation among plant parts (ramets and branches within ramets) within genets was autocorrelated between successive years in a manner consistent with the hypothesis. A second data set presented a quantitative estimate of the within-tree variation relative to among-tree variation. In that study, the fruit production and set of viable seeds in caper trees in Costa Rica were found to be at least as variable within trees as between trees. If genetically based, the variation observed would not be negligible in an evolutionary context. Demonstration of criterion (c) awaits further research.

ACKNOWLEDGEMENTS

The idea for genetic mosaicism in trees was originally suggested by E. Bird during a quantitative field ecology course, Zoology 674, at the University of Maryland in Spring 1976. During the 6 years of library work, innumerable people have made valuable suggestions of sources of information and implications to consider. Most stimulating have been the discussions with J.D. Allan, J. Antonovics, K. Bawa, K. Berven, M. Clegg, J. Coddington, R. Denno, R. Fritz, D.J. Futuyma, F.B. Gill, W. Haber, R. Foster, J.C. Hickman, W.J. Libby, J. Meers, B. Mock, D.H. Morse, N. Nadkarni, B.S. Rathcke, D. Rhoades, W. Rice, R.P. Seifert, W. Stearns, G.L. Stebbins, R.N. Stewart, A.H. Thompson, G.J. Vermeij, J. White, T. Whitham, D.H. Wise, and the Evolutionary Ecology Group at The University of Maryland. The participants of the 79–3 and 81–3 courses in Tropical Biology of the Organization for Tropical Studies, assisted with data collection from the caper trees. Conceptual and factual errors are of course mine alone. In 1979 T. Witham and D.E.G. discovered they had developed the same idea independently. The work has been supported by NSF grants DEB 76–20326, 77–04817, 78–10832 and 80–05080.

REFERENCES

Dixon A.F.G. (1970a) Quality and availability of food for a Sycamore aphid population. *Animal Populations in Relation to their Food Resources* (Ed. by A. Watson), pp. 271–287. British Ecological Society Symposium No. 10, Blackwell Scientific Publications, Oxford.

Dixon A.F.G. (1970b) Stabilization of aphid populations by an aphid-induced plant factor. *Nature*, **227**, 1368–1369.

Dyson W.G. & Herbin G.A. (1970) Variation in leaf wax alkanes in cypress trees growth in Kenya. *Phytochemistry*, **9**, 585–589.

Kennedy J.S. & Booth C.O. (1951) Host alternations in *Aphis fabae* Scop. I. Feeding preferences and fecundity in relations to the age and kind of leaves. *Annals of Applied Biology*, **38**, 25–64.

Parker J. (1977) Phenolics in black oak bark and leaves. *Journal of Chemical Ecology*, **3** (5), 489–496.

Pergande T. (1901) *The Life History of Two Species of Plant-lice.* Technical Series No. 9: 7–44. Department of Agriculture, Government Printing Office, Washington D.C.

Van Emden H.F. & Bashford M.A. (1969) A comparison of the reproduction of Brevicoryne brassicae and the Myzus persicae in relation to soluble nitrogen concentration and leaf age (leaf position) in the Brussels Sprout plant. *Eutomologie experimental et Applié*, **12,** 351–364.

Von Rudloff E. (1969) Scope and limitations of gas chromatography of terpenes in chemosystematic studies. *Recent Advances in Phytochemistry*, **2**, 127–159.

Von Rudloff E. (1972) Seasonal variation in the composition of the volatile oil of the leaves, buds, and twigs of white spruce (*Picea glauca*). *Canadian Journal of Botany*, **50**, 1595–1603.

Whitham T.G. (1978) Habitat selection by *Pemphigus* aphids in response to resource limitation and competition. *Ecology*, **59**, 1164–1176.

Whitham T.G. (1979) Territorial behavior of *Pemphigus* gall aphids. *Nature* (*London*), **279**, 324–325.

Whitham T.G. (1981) Individual trees as heterogeneous environments: adaptation to herbivory or epigenetic noise? *Insect Life History Pastterns: Habitat and Geographic Variation* (Ed. by R.F. Denno & H. Dingle), pp. 9–27. Springer, New York.

Whitham T.G. (1983) Host manipulation of parasites: within-plant variation as a defense against rapidly evolving pests. *Import of Variable Host Quality on Herbivorous Insects* (Ed. by R.F. Denno & S. McClure). Academic Press, New York.

Whitham T.G. & Slobodchikoff C.N. (1981) Evolution by individuals, plant-herbivore interactions, and mosaics of enetic variability: the adaptive significance of somatic mutations in plants. *Oecologia*, **49**, 287–292.

Zucker W.V. (1982) How aphids choose leaves: the roles of phenolics in host selection by a galling aphid. *Ecology* **63** (4), 972–981.

6. THE EVOLUTIONARY GENETICS OF LIFE HISTORIES

BRIAN CHARLESWORTH
School of Biological Sciences, University of Sussex, Brighton BN1 9QG

INTRODUCTION

Evolutionary ecologists have devoted a good deal of attention to life-history phenomena (cf. reviews by Stearns 1976, 1977; Law 1979; Charlesworth 1980; and Calow in this volume). The fundamental problem with which they are concerned is that of how natural selection shapes patterns of survivorship and reproduction as functions of age. This breaks down into a set of related questions, such as: Why be iteroparous rather than semelparous? Why is senescence an apparently universal phenomenon in multicellular organisms? How should resources be allocated between reproduction, growth, and survival? It has commonly been assumed by theorists that natural selection acts in such a way as to maximize the intrinsic rate of increase of a population (r) in a density-independent environment, or to maximize carrying capacity (K) in a density-dependent environment (MacArthur & Wilson 1967). Theories of life-history evolution thus frequently reduce to the construction of models in which r or K are maximized according to some plausible set of constraints. The most popular of these has been the reproductive effort model (Williams 1966; Gadgil & Bossert 1970; Schaffer 1974a). This assumes that the proportion of available energy at a given age that is devoted to reproduction, $E(x)$, controls survivorship to age $x + 1$, $P(x)$, and fecundity at age x, $m(x)$, such that $P(x)$ and $m(x)$ are decreasing and increasing functions of $E(x)$ respectively. Given the forms of these functions, it is not difficult to find sets of $E(x)$ values for each age that correspond to a local maximum of r or K, and to determine the consequences of changes in the external environment for the form of the optimal life-history (e.g. Gadgil & Bossert 1970, Michod 1979).

Many useful insights have been obtained with such models, and there often seems to be good qualitative agreement between the predictions of the theory and the nature of the correlations between ecological factors and observed life-histories (Law 1979; and Calow in this volume). I do not propose in this chapter to discuss these models and their detailed predictions. Instead, I shall consider two areas in which ideas and data from genetics are relevant to life-history evolution. The first concerns the validity of the assumption that selection maximizes r or K. This has been rather taken for granted by ecologists, following the example of Fisher (1930). But in dealing with life-history evolution we are usually concerned with selection in sexually reproducing populations with age structure, where it is not at all obvious how fitness is to be measured. The second area concerns the nature and extent of genetic variation for life-history characteristics. If natural selection is to be effective in moulding life histories, appropriate genetic variability must be available, and it is therefore of importance to the evolutionary ecologist to know whether or not it is legitimate to assume the existence of such variability. Furthermore, it is likely that the nature of this variation may place constraints on the course of life-history evolution. For example, positive genetic correlations between survivorship and fecundity will have quite different consequences from negative correlations for the response of a life history to selection. Knowledge of such constraints is clearly of considerable importance for predicting the direction of life-history evolution.

SINGLE LOCUS MODELS OF LIFE-HISTORY EVOLUTION

A necessary condition for the validity of the usual assumptions about the use of r and K is that gene frequency change at a single locus under selection can be correctly predicted from allelic effects on r and K. Single-locus theories of selection in age-structured populations were first constructed by Haldane (1927, 1962) and Norton (1928). Interest in this topic revived during the 1970s, and I have recently reviewed the major developments (Charlesworth 1980). The following account summarizes some of the main results.

Assumptions of the model of selection

Consider a diploid, sexual population segregating for alleles $A_1, A_2, \ldots, A_n$ at an autosomal locus, and which reproduces according to a standard Leslie model with discrete age-classes. Each genotype has its characteristic age-specific survival and fecundity schedule, such that $P_{ij}(x, t)$ is the probability of survival of an A_iA_j female aged x at time t to age $x + 1$ at time $t + 1$, and $m_{ij}(x, t)$ is the total number of daughters she produces at time t, counted as newly formed zygotes. If age 0 is used to denote new zygotes, age 1 for individuals who have survived one time-interval from conception, and so on, the net probability of survival from conception to age x at time t is

$$l_{ij}(x, t) = P_{ij}(O, t - x)P_{ij}(1, t + 1 - x) \ldots P_{ij}(x - 1, t - 1). \qquad (1)$$

TABLE 6.1. Assumptions of the model of selection

1.	The fecundity of a mating between a given pair of individuals is dependent only on the age and genotype of the female.
2.	Mating is at random with respect to age and genotype.
3.	The primary sex-ratio is constant and independent of parental age and genotype.
4.	The age-specific survival probabilities of males and females are equal.
5.	The age-specific fecundity schedule of males of a given genotype (as measured by mating success weighted by fertilisation rate) is proportional to that of females of the same genotype. The proportionality constant is the same for all genotypes.

(The parameters have been written as dependent on time as well as age, to allow for possible effects of changes in population density or environmental factors on survival or fecundity.)

This type of model is, of course, most suitable for describing species such as iteroparous vertebrates with discrete breeding seasons, where the time interval corresponds to the interval between breeding seasons. An adequate approximation to the case when individuals reproduce continuously is obtained if the number of age classes is taken sufficiently large.

Demographic models commonly consider only the female contribution to population growth rate, and most ecological models of life-history evolution assume that r or K are determined by female life-history variables alone. In order to produce a theory of gene frequency change in which only the female population need be considered, and which therefore parallels the standard ecological models, several simplifying assumptions have to be made. These are listed in Table 6.1.

Given these assumptions, it is possible to reduce to a fairly tractable form the rather complicated difference equations describing selection in age-structured populations of infinite size (Charlesworth 1980, pp. 120–127). Even these are difficult to handle algebraically, and it has proved necessary to investigate certain informative limiting cases. The results of these investigations provide support for the use of r or K as fitness measures, as will be seen below.

Weak selection

If we are prepared to assume that selection is weak, so that second-order terms in genotypic differences in $P(x)$ and $m(x)$ are negligible, it turns out to be possible to reduce the basic equations to a form which is very similar to those for the standard discrete-generation model of population genetics (Charlesworth 1980, pp. 142–168). With density independence and frequency independence, the rate of change of gene frequency per time interval is governed by an equation in which standard discrete-generation fitnesses W_{ij} are replaced by the genotypic intrinsic rates of increase r_{ij}, where r_{ij} is defined as the root of the Lotka equation for the life-history characteristic of A_iA_j

$$\sum_x e^{-rx} l_{ij}(x) m_{ij}(x) = 1. \tag{2}$$

A very similar result holds for density dependence with frequency independence, except that r_{ij} is replaced by the carrying capacity K_{ij}, i.e. the equilibrium density that would be reached by a population with the life-history characteristics of A_iA_j. (To predict the rate of change of gene frequency per time interval, all the K_{ij} must be multiplied by a scaling factor which measures the sensitivity of $\Sigma l(x)m(x)$ to density [Charlesworth 1980, p. 167]. This does not alter the fact that the K_{ij} predict the direction of selection.)

Selection of arbitrary intensity

The above results are approximations that hold only when selection is weak. Theoretical analyses have shown that the correspondence between r or K and fitness is not exact when selection is strong (Charlesworth 1980, pp. 194–203). For example, equilibrium gene frequencies in a two-allele system with heterozygote advantage are not correctly predicted from the r values of the three genotypes with strong selection. Nevertheless, certain relationships still hold. In particular, Norton (1928) showed that (with density independence) the ultimate outcome of selection with two alleles can be predicted from the relative values of the r_{ij}'s, i.e. which allele becomes fixed, or whether there is a stable polymorphism, is determined by the r_{ij}'s in the same way as by discrete-generation fitnesses.

A possibly more important result, from the point of view of models of optimal life-histories, concerns the conditions for spread of a rare gene into a population intially fixed for an alternative allele, and their relationship with Maynard Smith's (1972) concept of an evolutionarily-stable strategy (ESS). A population is at an ESS if its mean phenotype is such that a mutant gene whose carriers deviate from the mean is always eliminated by selection. This is a necessary condition for the population mean to be at a stable point with respect to evolutionary change. In the present context, it is straightforward to show that a rare gene decreasing r or K (in density-independent and density-dependent environments respectively) is always eliminated by selection (Charlesworth 1980, pp. 168–173). Hence, the maximization of r or K enables us, in principle, to determine the form of ESS life histories (Charlesworth 1980, pp 231–232). This method has the advantage that it can easily be extended to cases in which there is frequency-dependent selection on the $P(x)$ or $m(x)$, which may be important when there is competition between individuals with different life-history phenotypes. An example of such a life-history model is discussed by Maynard Smith (1982). Extensions to the important case when sex differences in survivorship or fecundity need to be considered can also be readily made with the rare gene approach (Charlesworth 1980, p. 169; Charnov, Bull & Mitchell-Olds 1981).

POLYGENIC MODELS OF LIFE-HISTORY EVOLUTION

The models described above assume that there is variation at only a single locus with defined effects on life-history parameters. In practice, of course, such complex traits are likely to be subject to variability due to the segregation of several loci, as well as environmental factors, and it is important to take this into account in theories of life-history evolution. This is possible if we are prepared to resort to the usual assumptions of quantitative genetics (Falconer 1981), notably that life-history characters are affected by many genes with effects which are small compared with the total variability in the characters. The joint effects of the genes at the different loci affecting a given character are assumed to be approximately additive, and the genes are assumed to segregate independently of each other in the population (i.e. there is no linkage disequilibrium). It is then trivial to show that the net rate of evolutionary change in a life-history parameter is obtained by summing up the effects contributed by each locus affecting the parameter in question (Charlesworth 1980, p. 205). It follows that r and K are once again maximized by selection, assuming no frequency dependence.

Although these considerations suggest that the usual maximization criteria can be used with polygenic inheritance, they do not help us to formulate a predictive genetic theory of life-history evolution, in the sense of using information about the nature of genetic variation in life-history parameters to predict future rates of evolution, or the form of optimal life-histories. This is because we are only exceptionally able to identify the effects of the individual loci underlying life-history variation. (For an example of such an exception, see Kallman & Borkoski 1978). A model of life-history evolution which is based firmly on genetic parameters that can be estimated experimentally has recently been developed by Lande (1982), extending earlier work by Robertson (1968), Crow & Nagylaki (1976), and Hill (1977).

The assumptions of this model, and the main results, are briefly outlined below for the case of density independence. (An extension to density dependence is easily made on the lines discussed for the single-locus case above.) The assumptions are as follows:

(i) A given life-history phenotype is described by a vector $\mathbf{z}$, such that the values of the components of $\mathbf{z}$ completely determine the values of $l(x)$ and $m(x)$. For example $\mathbf{z}$ might be composed of all the $P(x)$ and $m(x)$ for each x, or of the reproductive efforts $E(x)$ for each age (if reproductive effort alone determines survival and fecundity).

(ii) $\mathbf{z}$ for a given population follows a multi-variate normal distribution whose variance–covariance matrix is the sum of genetic and environmental components. The additive genetic component has a variance-covariance matrix $\mathbf{G}$, such that g_{ii} is the additive genetic variance of z_i, and g_{ij} is the additive genetic covariance between z_i and z_j. If we write $r_A(i, j)$ for the additive genetic correlation between z_i and z_j, and $\sigma_A(i)$ for the additive standard deviation of z_i, we have

$$g_{ij} = r_A(i, j)\sigma_A(i)\sigma_A(j). \tag{3}$$

(iii) Partial derivatives $\partial r/\partial z_i$ can be obtained for each component z_i, and are evaluated at the population mean $\bar{\mathbf{z}}$. These measure the effect of a small change in each component of $\mathbf{z}$ on r, holding all other components fixed. In the case when $\mathbf{z}$ is made up of $P(x)$ and $m(x)$, we have (Hamilton 1966)

$$\frac{\partial r}{\partial P(x)} = \sum_{y=x+1} e^{-ry}l(y)m(y)/P(x)\sum_y ye^{-ry}l(y)m(y), \tag{4}$$

$$\frac{\partial r}{\partial m(x)} = e^{-rx}l(x)/\sum_y ye^{-ry}l(y)m(y). \tag{5}$$

Given these assumptions, each life-history variable obeys the following law for change in mean per time interval:

$$\Delta\bar{z}_i = \sum_j g_{ij}\left(\frac{\partial r}{\partial \bar{z}_j}\right). \tag{6}$$

The intrinsic rate of increase of the population as a whole obeys the law

$$\Delta r = \sum_{ij} g_{ij}\left(\frac{\partial r}{\partial \bar{z}_i}\right)\left(\frac{\partial r}{\partial \bar{z}_j}\right) = V_A(r) \tag{7}$$

where $V_A(r)$ is the additive genetic variance in r. This last relation shows that, as expected, the rate of change of r under selection is always non-negative, and equal to the additive genetic variance in r (Fisher's Fundamental Theorem of Natural Selection).

At equilibrium under selection, each of the $\Delta\bar{z}_i$ in eqn (6) must equal zero. It follows that either the $\partial r/\partial \bar{z}_i$ are all zero, or that the determinant of the **G** matrix is zero. The latter condition is probably the most likely; for example, if we choose the $P(x)$ and $m(x)$ as components of **z**, it is apparent from eqns (4) and (5) that the $\partial r/\partial \bar{z}_i$ are all positive. If there is additive genetic variance in at least some of the $P(x)$ and $m(x)$, it follows that there must be negative additive genetic correlations between some of the component of **z**, in order that $\Delta\bar{z} = 0$ in eqn (6). Thus, although the additive genetic variance in r or K is expected to be zero in a population at equilibrium under selection, it is perfectly possible for life-history parameters to have substantial additive genetic variance, provided that sufficient negative additive genetic correlations exist. An explicit model of how such additive genetic variance and correlations may be maintained has been provided by Rose (1982). It is interesting to note that these conclusions about equilibria do not require the assumption of frequency-independent selection needed for the dynamic equations.

This model of life-history evolution is expressed in terms of genetic parameters (additive genetic variances and correlations) that can be estimated by standard quantitative genetics methods (Falconer 1981). The derivatives $\partial r/\partial z_i$ are easily written down once the nature of **z** has been defined. Given a life history, and the values of these parameters, one can, therefore, ask whether or not its form seems to be consistent with the equations for equilibrium under selection. Furthermore, since the $\partial r/\partial z_i$ are dependent on survivorship and fecundity schedules, one can use eqn (6) to predict the nature of the response of the life history to environmental changes that alter survivorship or fecundity. In the next section, I shall consider some of the experimental evidence on genetic variation in life history parameters, in the light of these considerations.

GENETIC VARIATION IN LIFE-HISTORY PARAMETERS

There is now a sizeable number of studies in which information on genetic variation in life histories has been collected, and I do not intend to provide a comprehensive review here (cf. Law 1979; Dingle & Hegmann 1982; Rose 1983). I shall confine myself to discussing a few cases that illustrate the main conclusions that have emerged from this work. Two classes of experiment have been performed, the first involving estimation of the genetic variances and correlations in life-history traits, and the second involving inferences about genetic parameters from the response of life-history traits to artificial selection. Rather more experiments of the second class than of the first have been reported.

Analyses of genetic variance and correlation

Probably the most complete study of this kind so far reported is that of Rose & Charlesworth (1980, 1981a) on a laboratory population of *Drosophila melanogaster*. The population in question was descended from several hundred flies collected by P.T. Ives from his population at Amherst, Massachusetts, which is known to be a long-established natural population. Although the life-history parameters of the flies were estimated in highly artificial conditions, the variability detected is probably representative of that of a wild population. The design of the experiment involved mating groups of ten females to a single male, rearing two daughters from each female, and then following their life-time daily egg production from 2 days after eclosion from the pupa. The longevity and fecundity schedule of each daughter can then be determined. Since full-sib and half-sib families were studied, an analysis of variance and covariance into additive dominance and environmental components is possible (Falconer 1981). The life histories of about 1200 female progeny were followed in this way; because of mortality, estimates of genetic parameters are more reliable for early life-history components than for late ones.

Table 6.2 shows the components of genetic variability for several characters, given as ratios of additive and dominance standard deviations (σ_A and σ_D) to the mean for the purpose of expressing them on a common scale. For comparison, estimates of these parameters for egg-to-adult viability, obtained by Mukai and co-workers (Mukai 1977), are also shown. It is clear that substantial levels of additive genetic variability exist for all of the fecundity characters, and for egg-to-adult viability. Dominance variance for these characters is low or absent. Longevity shows low additive variance but high dominance variance.

TABLE 6.2. Patterns of genetic variance in life-history traits of *D. melanogaster*

Character	σ_A (% of population mean)	σ_D (% of population mean)	Heritability
Egg-to-adult viability	0.095	0.033	—
Adult female longevity	0.053	0.175	0.028
Egg laying:			
days 1–5	0.125	0.050	0.623
days 6–10	0.115	0.018	0.296
days 11–15	0.176	0.047	0.316
days 16–20	0.259	<0	0.434
days 21–25	0.123	0.222	0.040

*σ_A and σ_D are the additive and dominance genetic standard deviations respectively.

TABLE 6.3. Patterns of genetic and phenotypic correlations between life-history traits in *D. melanogaster*

Characters	Additive genetic correlation	Phenotypic correlation
Egg-laying days 1–5 *and*		
Longevity	− 1.43	− 0.22
Egg-laying days 6–10	− 0.13	+ 0.02
days 11–15	− 0.48	+ 0.50
Egg-laying days 6–10 *and*		
Longevity	+ 0.30	+ 0.17
Egg-laying days 11–15	+ 0.51	+ 0.54
Egg-laying days 11–15 *and*		
Longevity	− 0.71	+ 0.21

Table 6.3 shows estimates of additive genetic and phenotypic correlations between those characteristics for which a reasonable amount of information is available (this excludes late fecundities). Because of imbalance in the experimental design, standard errors for these correlations could not be estimated satisfactorily; only the larger values in Table 6.3 are likely to be significant. The general picture is one of mainly negative genetic correlations between life-history characters. There is little relation between genetic and phenotypic correlations, which probably reflects the fact that good environments may have a beneficial effect on several characteristics at once, thus obscuring underlying physiological trade-offs between them.

These results are consistent with the expectation that life-history traits with high levels of additive genetic variance should display negative genetic correlations, in a population at a near equilibrium under selection (see above). They imply that selection applied to any single trait would be highly effective, but at the expense of a reduction in performance among some of the other traits. The size of these reductions can be calculated from the standard equation for the correlated response in a trait z_j to direct selection for trait z_i (Falconer 1981, p. 286).

$$R_j = r_A(i, j)\frac{\sigma_A(j)}{\sigma_A(i)}R_i \tag{8}$$

where R_i is the response of z_i to selection, and R_j is the correlated response of trait j. For example, using the data of Tables 6.2 and 6.3 (taking r_A for longevity and fecundity at days 1–5 as -1), selection for longevity that succeeded in raising it by 10% of the initial value would be expected to reduce fecundity at days 1–5 by 23%.

It should be emphasized that a given pattern of genetic variation and correlation may be valid only for the population for which they have been measured, since they reflect allele frequencies at the underlying loci which are the outcome of past evolutionary processes. For example, Berven (1982) and Berven & Gill (1983) have found differences between populations of the wood frog *Rana sylvatica* in North America with respect to r_A for the traits larval body size and developmental time. Lowland populations have a value of $+0.65$, whereas mountain populations have a value of -0.86. The patterns of response to selection would be quite different in the two populations. Careful studies have revealed that developmental correlations between quantitatively varying traits may themselves display additive genetic variance, and hence can respond readily to selection (e.g. Atchley & Rutledge 1980). There is no reason to expect life-history traits to behave any differently. Hence, trade-offs between different traits may well be less rigid than is commonly imagined in models of optimal life-histories.

Artificial selection experiments

Studies of a number of species have demonstrated the ability of selection to change life-history traits (Law 1979; Dingle & Hegmann 1982), but there are few cases in which estimated components of genetic variance and correlation can be compared with the results of selection. The selection experiments of Rose & Charlesworth (1980, 1981b) constitute such a case, and will be briefly described here. Two types of experiment were performed, both involving the population used for estimating the genetic parameters discussed above.

In the first type of experiment, direct selection for increased female fecundity was practised for three generations, either on early fecundity (days 1–5) or late fecundity (days 21–25). Unselected control stocks, in which the flies reproduced at the same age, were maintained using the same number of breeding adults as in the selection lines. Female life-histories were assayed in the fourth generation, with results shown in Table 6.4. Genetic correlations, estimated using equation (8), are also shown. It will be seen that selection was effective in increasing early fecundity, but that there was little evidence for correlated responses in the other traits, so that estimated genetic correlations are near zero. In contrast, selection for late fecundity led to a highly significant drop in fecundity at days 1–5 and 6–10, together with a significant increase in longevity, despite the fact that the data apparently showed no direct effect of selection. (This oddity is the effect of small sample sizes for measurements of late fecundity, since the data for the previous three generations showed an increase in late fecundity, and there is a highly significant linear regression of response at a given age on age in generation four.

TABLE 6.4. Responses to direct selection for increased fecundity and corresponding estimates of genetic correlations (r_A)

	Character	Selected line (mean)	Control (mean)	Response (%)	r_A
(a) Selection for egg-laying at days 1–5	Egg-laying				
	Days 1–5	498	454	+ 9.7*	1
	6–10	440	445	− 1.1	− 0.14
	11–15	393	380	+ 3.2	+ 0.33
	16–20	271	267	+ 1.4	+ 0.09
	21–25	184	187	− 1.6	− 0.28
	Lifetime total	1613	1552	+ 3.9	—
	Longevity (days)	26.20	26.15	+ 0.2	+ 0.05
(b) Selection for egg-laying at days 21–25	Egg-laying				
	Days 1–5	419	519	− 19.3**	− 0.51
	6–10	406	425	− 4.4	− 0.41
	11–15	378	360	+ 5.0	− 0.16
	16–20	241	222	+ 8.3	+ 0.063
	21–25	187	193	− 2.2	1
	Lifetime total	1467	1487	− 1.4	—
	Longevity (days)	24.14	21.05	+ 14.7*	+ 5.9

*Indicates significance at $p < 0.05$

**Indicates significance at $p < 0.01$

Note that the r_A values for experiment (b) were calculated using expected selection responses in age-specific fecundity calculated from the linear regression of response at a given age against age.

The expected fecundity at age 21–25 predicted by this regression was used to compute the genetic correlations in Table 6.4.) This experiment strongly suggests the existence of negative genetic correlations between early and late fecundity, and a concomitant positive correlation between late fecundity and longevity. As a result of the negative correlations, cumulative life-time fecundity is hardly changed by the selection procedure.

The discrepancies in the patterns of genetic correlations suggested by these two experiments and the earlier one were discussed by Rose & Charlesworth (1981b), who pointed out that such inconsistencies can be explained by asymmetries in gene frequencies at the underlying loci (Bohren, Hill & Robertson 1966). In any case, the lack of replication of the experiments, and the small numbers of breeding individuals, means that the estimates of genetic correlations obtained from the selection experiments must be treated with caution (Falconer 1977), since finite population size effects may strongly influence the outcome of any single experiment.

The second type of selection experiment involved the much less laborious procedure of comparing two populations of large size ($N \approx 400$) one of

TABLE 6.5. Responses to selection for increased longevity and late fecundity

Character	*L* line (mean)	*E* line (mean)	Response (%)
Egg laying			
Days 1–5	422	551	− 23.5**
6–10	480	472	+ 1.6
11–15	393	323	+ 21.6**
16–20	287	239	+ 20.1*
21–25	183	136	+ 33.7*
Lifetime total	1700	1658	+ 2.5
Longevity	30.25	26.79	+ 12.9*

which (*E*) was maintained by the normal laboratory procedure of breeding from adults collected between 1 and 6 days from eclosion, the other of which (*L*) was maintained by breeding from flies that had lived at least 21 days from eclosion. The *L* stock is thus selected for both increased longevity and late fecundity compared with *E*. Twelve generations of *L* and twenty generations of *E* were maintained, and then female life-histories were assayed with the results shown in Table 6.5. It is clear that *L* differs from *E* in that fecundity at days 1–5 is severely depressed, whereas longevity and fecundity at days 16–25 are increased. The results agree rather well with those obtained by selecting directly for increased late fecundity.

Experiments of this sort have been reported by several other *Drosophila* workers. Lints & Hoste (1977) selected lines of *D. melanogaster* for increased abdominal bristle number in either young or old individuals for several generations. Their Fig. 7 shows a difference between the young and old strains in age-specific female fecundity schedules which resembles the difference between the *E* and *L* lines described above. Similarly, Wattiaux (1968) found an increase in male sexual activity, a decline in longevity and late female fecundity, and an increase in early fecundity in *E*-selected compared with *L*-selected *D. subobscura*. The increase in male sexual activity is interesting in view of the negative phenotypic correlation between male mating activity and longevity, observed by Partridge & Farquhar (1981) in *D. melanogaster*. Taylor & Condra (1980) found a reduced longevity of *E*-selected, compared with *L*-selected *D. pseudoobscura* females. This agreement between the results for different species is encouraging.

Similar experiments have been performed on *Tribolium* (Sokal 1970; Mertz 1975; Law, unpublished), with conflicting results. Sokal obtained a reduction in longevity in *E*-selected beetles maintained for forty generations, whereas Mertz observed an increase in early female fecundity and a change in longevity that only bordered on significance. Law (personal communication) was unable to obtain evidence for any change in the female fecundity schedule of beetles that were *E*-selected for sixteen generations.

It should be noted that this design of experiment does not provide water-tight evidence for negative genetic correlations between life-history components even if the expected correlated responses are observed. An alternative interpretation of such correlated responses is that mutations affecting life-history traits are constantly arising, with effects confined to limited age bands. In an *L* population, for example, mutations affecting early fecundity, but with no effect on late fecundity or longevity, would be unopposed by selection (cf. Edney & Gill 1968; Sokal 1970). There would thus be a gradual reduction in early fecundity, analogous to that observed for egg-to-adult viability in *Drosophila* experiments where mutations are allowed to accumulate without selection (Simmons & Crow 1977). It seems unlikely at first sight that mutation accumulation could explain an effect of the magnitude of that shown in Table 6.5, where fecundity for days 1–5 was reduced by 30% in the *L* population after twelve generations. But the rate of reduction per generation in viability due to homozygosity for new mutations on the second chromosome of *D. melanogaster* alone is about 0.004. Taking account of the fact that the heterozygous effect of such mutations is about 0.35 of their homozygous effect, and that the second chromosome constitutes about $\frac{2}{5}$ of the genome, the reduction in viability in a random mating population after twelve generations is expected to be 0.042. The rate of mutational reduction in female fecundity would thus have to be about seven times that for viability to explain the observed results. This seems somewhat unlikely, but cannot be excluded in the absence of mutational data on female fecundity. Of course, the results of the estimates of genetic correlations and direct selection for late fecundity support the alternative explanation in the case of the experiment in question, but such results are not available in other cases.

CONCLUSIONS

There is now, as described above, a fairly satisfactory basis in population genetics theory for the optimization procedures commonly employed in life-history studies, from the results of both single gene and polygenic models. The most general results assume frequency-independent selection, with no sex-differences in life-history parameters, and these conditions are tacitly assumed in ecological models that maximise *r* or *K*. Considerations of the spread of rare genes, in conjunction with the ESS concept of Maynard Smith (1972, 1982), can be used to handle the complications arising from frequency dependence or sex differences. Such cases have not been studied very much in the context of life-history evolution, but may be of considerable interest, especially in cases of intermale competition for mates (Maynard Smith 1982), or changes in sex expression with age (Leigh, Charnov & Warner 1976). The main gap in the theory at present is in dealing with life-history evolution in temporally fluctuating environments, which has attracted some interest among evolutionary ecologists (Murphy 1968; Schaffer 1974b). Although the population genetics of this has been worked out for the limiting case of a cyclical environment with a long period, no satisfactory general results are apparently available (Charlesworth 1980, p. 177).

The experimental results discussed above, and the other studies reviewed by Law (1979) and Rose (1983), show that many life-history traits in certain species show levels of additive genetic variance sufficient to permit a rapid response to selection. The existence of such high additive variance in fitness components is not inconsistent with the expectation that fitness itself should show low additive variance in an equilibrium population (Robertson 1955), provided that there are negative genetic correlations between some of the components. Evidence for this type of situation has long been known to quantitative geneticists (Dickerson 1955), and work such as that of Rose & Charlesworth (1981a, b) and Berven (1982) indicates that it may hold good for certain life-history traits. Nevertheless, there is as yet a lack of detailed documentation of the genetic properties of life-history traits. Such investigations are a promising field for future research, especially as quantitative genetics experiments can easily be carried out in any species that can be reared in sufficiently large numbers. Male life-history traits, as well as female ones, should be given attention in this context (cf. Wattiaux 1968; Partridge & Farquhar 1981; Taylor *et al.* 1981).

It may also be worth pointing out here that the trade-offs between life-history traits, commonly assumed in ecological optimization models, should strictly be formulated in terms of negative genetic correlations between the traits concerned, as purely phenotypic correlations have no evolutionary consequences (cf. eqn [8]). There is not necessarily a close correspondence between phenotypic and genetic correlations (cf. Table 6.3; Lande 1979). Much of the evidence for trade-offs has been based on phenotypic correlations (Calow 1979).

Finally, one area in which there is an almost complete lack of data concerns the contribution of newly arising mutations to life history variation. Lande's selection eqn (6) above can easily be modified by the addition of a term expressing the effect of new mutations on the mean values of the traits. The equilibrium structure of the variance–covariance matrix **G** and the corresponding life-history could be significantly modified if the mutational terms are large enough. Furthermore, one genetically plausible theory of senescence is based on the higher equilibrium frequencies of deleterious mutations which are expressed late in life compared with ones that express themselves early on (Medawar 1952; Charlesworth 1980, p. 218). It would thus be of considerable interest to examine the spectrum of age specificity of spontaneous mutations affecting life-history traits.

ACKNOWLEDGEMENTS

I wish to thank P. Calow, R. Lande, and M.R. Rose for their comments on the manuscript.

REFERENCES

Atchley W.R. & Rutledge J.J. (1980) Genetic components of size and shape. I. Dynamics of components of phenotypic variability and covariability during ontogeny in the rat. *Evolution,* **34**, 1161–1173.

Berven K.A. (1982) The heritable basis of variation in larval development patterns within populations of the wood frog *Rana sylvatica. Evolution*, **36**, 962–983.

Berven K.S. & Gill D.E. (1982) Interpreting genetic variation in life history traits. *American Zoologist*, **23**, 85–98.

Bohren B.B., Hill W.G. & Robertson A. (1966) Some observations on asymmetrical correlated responses to selection. *Genetical Research*, **7**, 44–57.

Calow P. (1979) The cost of reproduction—a physiological approach. *Biological Reviews*, **54**, 23–40.

Charlesworth B. (1980) *Evolution in Age-Structured Populations*. Cambridge University Press, London.

Charnov E.L., Bull J.J. & Mitchell-Olds S.T. (1981) A note on sex and life histories. *American Naturalist*, **117**, 814–818.

Crow J.F. & Nagylaki T. (1976) The rate of change of a character correlated with fitness. *American Naturalist*, **110**, 207–213.

Dickerson G.E. (1955) Genetic slippage in response to selection for multiple objectives. *Cold Spring Harbor Symposium on Quantitative Biology*, **20**, 213–224.

Dingle H. & Hegmann J.P. (1982) *Evolution and Genetics of Life Histories*. Springer-Verlag, New York.

Edney E.B. & Gill R.W. (1968) Evolution of senescence and specific longevity. *Nature*, **220**, 281–282.

Falconer D.S. (1977) Some results of the Edinburgh selection experiments with mice. *Proceedings of the International Conference on Quantitative Genetics* (Ed. by E. Pollak, O. Kempthorne & T.B. Bailey), pp. 101–115. Iowa State University Press, Iowa.

Falconer D.S. (1981) *An Introduction to Quantitative Genetics*, 2nd. ed. Longman, London.

Fisher R.A. (1930) *The Genetical Theory of Natural Selection*. Oxford University Press, Oxford.

Gadgil M. & Bossert W.H. (1970) Life historical consequences of natural selection. *American Naturalist*, **104**, 1–24.

Haldane J.B.S. (1927) A mathematical theory of natural and artificial selection. Part IV. *Proceedings of the Cambridge Philosophical Society*, **23**, 607–615.

Haldane J.B.S. (1962) Natural selection in a population with annual breeding but overlapping generations. *Journal of Genetics*, **58**, 122–124.

Hamilton W.D. (1966) The moulding of senescence by natural selection. *Journal of Theoretical Biology*, **12**, 12–45.

Hill W.G. (1977) Selection with overlapping generations. *Proceedings of the International*

Conference on Quantitative Genetics (Ed. by E. Pollak, O. Kempthorne & T.B. Bailey) pp. 367–378. Iowa State University Press, Iowa.

Kallman K.D. & Borkoski V. (1978) A sex-linked gene controlling the onset of sexual maturity in female and male platyfish (*Xiphophorus maculatus*), fecundity in females and adult size in males. *Genetics*, **89**, 79–119.

Lande R. (1979) Quantitative genetic analysis of multivariate evolution applied to brain: body size allometry. *Evolution*, **33**, 402–416.

Lande R. (1982) A quantitative genetic theory of life history evolution. *Ecology*, **63**, 607–615.

Law R. (1979) Ecological determinants in the evolution of life histories. *Population Dynamics* (Ed. by R.M. Anderson, B.D. Turner & L.R. Taylor) pp. 81–103, Blackwell Scientific Publications, Oxford.

Leigh E.G., Charnov E.L. & Warner R.R. (1976) Sex ratio, sex change, and natural selection. *Proceedings of the National Academy of Sciences of the United States of America*, **73**, 3656–3660.

Lints F.A. & Hoste C. (1977) The Lansing effect revisited. II. Cumulative and spontaneously reversible parental age effect on fecundity in *Drosophila melanogaster*. *Evolution*, **31**, 387–404.

MacArthur R.H. & Wilson E.O. (1967) *The Theory of Island Biogeography*. Princeton University Press, Princeton.

Maynard Smith J. (1972) *On Evolution*. Edinburgh University Press, Edinburgh.

Maynard Smith J. (1982) *Evolution and the Theory of Games*. Cambridge University Press, London.

Medawar P.B. (1952) *An Unsolved Problem of Biology*. H.K. Lewis, London.

Mertz D.B. (1975) Senescence decline in flour beetles selected for early adult fitness. *Physiological Zoology*, **48**, 1–23.

Michod R.E. (1979) Evolution of life histories in response to age-specific mortality factors. *American Naturalist*, **113**, 531–550.

Mukai T. (1977) Lack of experimental evidence supporting selection for the maintenance of isozyme polymorphisms. *Proceedings of 2nd Taniguchi International Symposium on Biophysics: Molecular Evolution and Polymorphism*. (Ed. by M. Kimura) pp. 103–126. National Institute of Genetics, Mishima.

Murphy G.I. (1968) Pattern in life history phenomena and the environment. *American Naturalist*, **102**, 52–64.

Norton H.T.J. (1928) Natural selection and Mendelian variation. *Proceedings of the London Mathematical Society*, **28**, 1–45.

Partridge L. & Farquhar M. (1981) Sexual activity reduces the lifespan of male fruitflies. *Nature*, **294**, 580–582.

Robertson A. (1955) Selection in animals: synthesis. *Cold Spring Harbor Sympsium on Quantitative Biology*, **20**, 225–229.

Robertson A. (1968) The spectrum of genetic variation. *Population Biology and Evolution*. (Ed. by R.C. Lewontin) pp. 5–16. Syracuse University Press, Syracuse.

Rose M.R. (1982) Antagonistic pleiotropy, dominance, and genetic variation. *Heredity*, **48**, 63–78.

Rose M.R. (1983) Theories of life-history evolution. *American Zoologist*, **23**, 15–23.

Rose M.R. & Charlesworth B. (1980) A test of evolutionary theories of senescence. *Nature*, **287**, 141–142.

Rose M.R. & Charlesworth B. (1981a) Genetics of life history in *Drosophila melanogaster*. I. sib analysis of adult females. *Genetics*, **97**, 173–186.

Rose M.R. & Charlesworth B. (1981b) Genetics of life history in Drosophila melanogaster. II. Exploratory selection experiments. *Genetics*, **97**, 187–196.

Schaffer W.M. (1974a) Selection of optimal life-histories: the effects of age structure. *Ecology*, **55**, 291–303.

Schaffer W.M. (1974b) Optimal reproductive effort in fluctuating environments. *American Naturalist*, **108**, 783–790.

Simmons M.J. & Crow J.F. (1977) Mutations affecting fitness in *Drosophila* populations. *Annual Review of Genetics*, **11**, 49–78.

Sokal R.R. (1970) Senescence and genetic load: evidence from *Tribolium*. *Science*, **167**, 1733–1734.

Stearns S.C. (1976) Life history tactics: a review of the ideas. *Quarterly Review of Biology*, **51**, 3–47.

Stearns S.C. (1977) The evolution of life history traits: a critique of the theory and a review of the data. *Annual Review of Ecology and Systematics*, **8**, 145–171.

Taylor C.E. & Condra C. (1980) *r*- and *K*-selection in *Drosophila pseudoobscura*. *Evolution*, **34**, 1183–1193.

Taylor C.E. Condra C., Conconi M. & Prout M. (1981) Longevity and male mating success in *Drosophila pseudoobscura*. *American Naturalist*, **117**, 1035–1039.

Wattiaux J.M. (1968) Cumulative parental age effects in *Drosophila subobscura*. *Evolution*, **22**, 406–421.

Williams G.C. (1966) *Adaptation and Natural Selection*. Princeton University Press, Princeton.

7. WHAT IS A POPULATION?

T.J. CRAWFORD
Department of Biology, University of York, Heslington, York Y01 5DD

INTRODUCTION

The term 'population' has been used in a variety of senses by both ecologists and geneticists. Ecologists often view populations as convenient aggregations of organisms, usually of the same species. The properties of populations of main interest are demographic: size and age structure, how these change with time and how they are regulated. These properties are also of interest to population geneticists, not only because they affect the number of individuals in a population, but also because they affect the structure and organization of a gene pool.

The genetic information contained in a gene pool is at any time distributed among a group of individuals as their genotypes. These individuals are merely temporary custodians of the genetic information which is redistributed with each succeeding generation. The number of individuals that share the gene pool is itself an important attribute of the gene pool because it determines the extent to which random processes affect the redistribution of genetic information among genotypes. But population number must not be assessed with respect to the distribution of the individuals themselves, but rather with respect to the limits of the gene pool.

Gene pools have both temporal continuity through parent–offspring relations and also spatial continuity through mating patterns. The limits to a gene pool and, therefore, to the population of individuals that represent it, must be defined both temporally and spatially. Temporal limits may be difficult to define if generations overlap or individuals show perennial life-cycles. Even in strictly annual plants, generations are effectively overlapping if there is a long-lived dormant seed bank. Biennials may not be clearly distinguished into even- and odd-yeared populations. In plants some individuals may reproduce in their first year or after the second year, or not all seed may germinate within a year (Wells & Wells 1980). Biennial life-cycles in animals may be superimposed upon an annual life-cycle as, for example, in *Taxomyia taxi* (Redfern 1975). It may also be difficult to define the spatial limits of gene pools. Closely adjacent individuals that interbreed with high probability carry genes belonging to the same gene pool but rare interbreeding also occurs between more distant individuals that carry genes representative of quite distinct gene pools.

The criterion commonly adopted by population geneticists is to define a population as a group of individuals genetically connected through parenthood or mating. Thus asexual organisms are connected through parenthood but not through mating. A Mendelian population (panmictic unit, local group, gamodeme, or deme) is a group of individuals, genetically isolated or semi-isolated from other groups, that mate effectively at random or at least without major internal constraints. Real populations usually differ from this ideal condition but their genetical properties can be described relative to those of Mendelian populations if the extent and nature of the differences can be determined.

Even if population limits can be defined it is difficult to assess the number of individuals within a population. Well-developed methods exist for estimating animal numbers (Blower, Cook & Bishop 1981) and it might be thought that plants are easily counted because they are stationary. But many plants, and also some animals, show clonal growth. The ramet is often the unit of interest to the ecologist whereas to the geneticist the genet is the primary individual, even though genets may be fragmented and interspersed (Harper 1977).

The aim of this chapter is to show why population sizes of relevance to population genetics are typically not equivalent to the number of individuals in a population and to emphasize that, as far as the properties of the gene pool are concerned, natural populations usually behave as if they are smaller than a simple count of individuals would indicate. The chapter is relevant to both animals and plants. The difficulties peculiar to plant studies will, however, be emphasized because a number of important and challenging problems, both theoretical and practical, remain to be solved and many of these are of interest to ecologists as well as to population geneticists.

THE WAHLUND EFFECT

The consequences of sampling across the limits of genetically differentiated populations are easily demonstrated. A sample of 600 individuals are scored for their genotypes at a locus polymorphic for a pair of alleles A_1 and A_2. Suppose that 200 of these individuals have been drawn from a population where the frequencies of A_1 and A_2 are $p = 0.3$ and $q = 0.7$ respectively, and that the other 400 have been drawn from a population where $p = 0.6$ and $q = 0.4$. For the purposes of simplicity we will further suppose that each component of the sample reflects exact agreement with Hardy–Weinberg

TABLE 7.1. Demonstration of the Wahlund effect

Sample	A_1A_1	A_1A_2	A_2A_2	Total	p	q	χ_1^2
Population 1	18	84	98	200	0.3	0.7	0
Population 2	144	192	64	400	0.6	0.4	0
Total observed	162	276	162	600	0.5	0.5	
Total expected	150	300	150	600			3.84*

*Significant at $P = 0.05$.

expectation (Table 7.1). The allele frequencies estimated from the total sample are $p = q = 0.5$ and the observed genotype frequencies differ significantly from Hardy–Weinberg expectation ($P = 0.05$) because of a deficiency of heterozygotes. This loss of heterozygotes when populations differing in allele frequency are combined is known as the Wahlund effect and will be explained later.

THE IDEALIZED POPULATION

The Hardy–Weinberg law states that in a random mating infinite population where there is no mutation, migration or selection, genotype frequencies at a diallelic autosomal locus are $p^2\,A_1A_1$, $2pq\,A_1A_2$ and $q^2\,A_2A_2$ where p and q are the frequencies of alleles A_1 and A_2. Furthermore, these frequencies are stable. The conditions are restrictive and not achieved in practice. We will continue to assume random mating, the absence of mutation, migration and selection, but relax the condition of infinite population size.

Consider a population of N *breeding* individuals. In addition to the assumptions given above, the following conditions hold:

(i) The number of breeding individuals in each generation remains constant and equal to N.

(ii) Individuals are monoecious (hermaphrodite) and self-fertilization occurs at its random frequency of $1/N$.

(iii) Each breeding individual has an equal probability of transmitting genes to future generations.

(iv) Generations are discrete and do not overlap.

A population of this type is called an *idealized* population. Natural populations do not conform to these ideal conditions and the consequences of nonconformity will be considered later. Two inter-related processes occur in an idealized population as a result of its finite size: random genetic drift and inbreeding.

Random genetic drift

An impressive mathematical theory has been developed to describe the process of random genetic drift. For present purposes it is sufficient to describe the major consequences of drift in terms of the population size. Further details may be found, for example, in Crow & Kimura (1970) and Ewens (1979).

The N zygotes in generation 1 are formed from $2N$ gametes picked in pairs at random from a gamete pool to which all individuals in the previous generation 0 have an equal chance of contributing. Let the frequency of A_1 in generation 0 be p_0. The process of zygote formation at the beginning of generation 1 is statistically equivalent to drawing a binomial sample of size $2N$ gametes where p_0 and q_0 are the probabilities that a gamete carries A_1 or A_2.

Suppose a very large number of populations, all size N, are formed independently in generation 1. The frequency of A_1 in these populations will be distributed with mean and variance

$$\bar{p}_1 = p_0$$

and

$$V(p_1) = \frac{p_0 q_0}{2N}.$$

The change in allele frequency between generation 0 and generation 1

$$\delta p = p_1 - p_0$$

will have mean and variance

$$\overline{\delta p} = 0$$

and

$$V(\delta p) = V(p_1 - p_0) = \frac{p_0 q_0}{2N} \tag{1}$$

as subtraction of the constant p_0 does not affect the variance. As a result of chance sampling effects, therefore, allele frequencies in finite populations will change from one generation to the next. The direction of the change is random and, on average, will be zero. The dispersion of allele frequencies between populations is predictable according to $V(\delta p)$ so that as N becomes smaller the dispersion of p_1 around p_0 increases. In a single population, the smaller its size the more likely is the allele frequency to change by a given amount.

Let the process of gamete formation and fertilization proceed independently in each population for a further generation. Sampling errors in the formation of generation 2 lead to further dispersal of the allele frequencies but this time with respect to each individual population's p_1. In any given population δp is a single random value from a distribution with $\Sigma(\delta p) = 0$ and $V(\delta p) = p_1 q_1/2N$. The allele frequency averaged across all populations is expected to be $\bar{p}_2 = \bar{p}_1 = p_0$. As the process continues for further generations, allele frequencies in each population fluctuate at random from generation to generation. The average of all populations remains p_0 but the variance of allele frequencies between populations increases so that in generation t

$$V(p_t) = V(q_t) = p_0 q_0 \left[1 - \left(1 - \frac{1}{2N} \right)^t \right].$$

The frequency of A_1A_1 homozygotes averaged across all populations in generation t is $\Sigma(p_t^2) = V(p_t) + (\Sigma p_t)^2 = V(p_t) + p_0^2$ and similarly for A_2A_2 homozygotes. The three genotype frequencies are, therefore

$$A_1A_1 : p_0^2 + p_0 q_0 \left[1 - \left(1 - \frac{1}{2N} \right)^t \right]$$

$$A_1A_2 : 2p_0 q_0 - 2p_0 q_0 \left[1 - \left(1 - \frac{1}{2N} \right)^t \right]$$

$$A_2A_2 : q_0^2 + p_0 q_0 \left[1 - \left(1 - \frac{1}{2N} \right)^t \right].$$

Even though each population is random mating there is a deficiency of heterozygotes when populations are considered together. This is the origin of the Wahlund effect. As $t \to \infty$, $V(p_t) \to p_0 q_0$ so that all heterozygotes are eventually lost. A proportion p_0 of populations contains only A_1A_1 homozygotes (A_1 fixed) and a proportion q_0 contains only A_2A_2 homozygotes (A_1 lost). The probability that A_1 will eventually become fixed as a result of drift alone is, therefore, equal to p_0, its frequency at the start. In particular, a single newly arisen neutral mutation will have a probability of ultimate fixation of $1/2N$.

Drift has been considered with respect to allele frequency dispersion in separate populations, all sized N and all starting from the same initial allele frequencies p_0 and q_0. It can equally be considered from the point of view of dispersion of allele frequencies at many loci, all initially p_0 and q_0, in a single population sized N.

The main consequencies of drift may be summarized as follows:

(i) In a finite population allele frequencies change at random from generation to generation.
(ii) Differentiation occurs between independent populations.
(iii) The frequency of homozygotes increases and heterozygosity decreases.
(iv) Genetic variation within populations is reduced.

All these processes occur at an increasing rate as N becomes smaller.

Inbreeding in a finite idealized population

There are two types of homozygote. The two genes may simply be functionally equivalent (*identical in state*) or may be copies by recent replication of a single ancestral gene (*identical by descent*). Inbreeding leads to an increase in the frequency of the latter and the inbreeding coefficient F is defined as the probability that the two genes at any locus in a random individual are identical by descent.

Within a line subject to regular inbreeding, homozygotes increase in frequency at the expense of heterozygotes. Allele frequencies remain constant in the absence of selection, migration and mutation so that each homozygote at a particular locus is expected to increase in frequency to the same extent. The genotype frequencies may be expressed as

$$A_1A_1: p^2 + pqF$$
$$A_1A_2: 2pq - 2pqF$$
$$A_2A_2: q^2 + pqF.$$

In any generation an empirical estimate of F may be obtained as

$$F_t = 1 - \frac{H_t}{2pq}$$

where H is the observed frequency of heterozygotes.

The inbreeding effect in an idealized population as a result of its finite size is different. Even though mating is at random, the frequency of heterozygotes is expected to decline, but for a different reason. Allele frequencies change as a result of drift and there is an eventual increase in the frequency of one or other homozygote depending upon which allele achieves frequencies nearer to fixation. The frequency of heterozygotes decreases because, under random mating, it is at a maximum of $\frac{1}{2}$ when $p = q = \frac{1}{2}$. Nevertheless, inbreeding occurs because there is an increase in the frequency of homozygotes that are identical by descent.

Let the inbreeding coefficient of an idealized population, size N, in generation $t - 1$ be F_{t-1}. When pairs of gametes are picked at random to form zygotes at the beginning of generation t they may carry at a particular locus alleles that are identical by descent for two reasons. They may both be copies by replication of the same gene in generation $t-1$ with probability $1/2N$. Alternatively, they will be copies of different genes in generation $t-1$ with probability $1-(1/2N)$ and identical by descent with probability F_{t-1}. Therefore

$$F_t = \frac{1}{2N} + \left(1 - \frac{1}{2N}\right)F_{t-1}.$$

This expression shows that the inbreeding in any generation is a function of that in the previous generation plus an increment $\Delta F = 1/2N$. The inbreeding effect arising from random mating in a finite population may be related to the change of allele frequencies under random genetic drift (eqn 1) as

$$V(\delta p) = \frac{pq}{2N} = pq\Delta F \qquad (2)$$

in any generation.

DRIFT IN RELATION TO ISOLATION AND SELECTION

Random genetic drift leads to divergence of allele frequencies between isolated populations at a rate inversely proportional to their size. If populations are incompletely isolated, so that some gene flow occurs between them, the rate of differentiation will be slower. Large amounts of migration can completely swamp the effects of drift. The extent to which divergence occurs will depend on a balance between population sizes and migration rates. Allele frequencies may also be influenced by selection. In large populations the dispersive effects of drift may be so small that even slight selective pressures may primarily determine allele frequencies. This may lead to genetic similarity greater than expected under drift alone for populations experiencing similar selective conditions or to genetic differentiation between populations subject to opposite selective effects. Again, the critical feature is the balance between population sizes and the magnitudes of selective pressures.

How small a population size is necessary before drift predominates? The solution depends to some extent upon the nature of the model assumed but a 'rule-of-thumb' answer is that if $4Nm \ll 1$ or $4Ns \ll 1$ allele frequencies will be determined largely by drift whereas if $4Nm \gg 1$ or $4Ns \gg 1$ migration or selection will have the major influence. A rather small level of migration, particularly between large populations will, therefore, be effective at preventing random differentiation. It should be remembered, however, that such migrations as occur will be largely between adjacent populations which are more likely to have similar allele frequencies than are more distant populations. Selective differentials of the order of 1% will be effective in populations of only a few hundred individuals. Greater selective differentials will be effective in even smaller populations.

EFFECTIVE POPULATION NUMBER

The theory outlined so far has been in relation to the number of breeding individuals in an idealized population as defined earlier. Natural populations will, however, differ from idealized populations in a number of important respects. Wright (1931, 1938) showed that differences from the ideal structure could be allowed for by assessing the *effective* size of a real population. The effective number of a population, N_e is defined as the number of breeding individuals in an idealized population that would show the same amount of dispersion of allele frequencies under random genetic drift or the same amount of inbreeding as the population under consideration. N_e can, therefore, be defined in two ways from eqn (2)

$$N_e = \frac{pq}{2V(\delta p)} \quad \text{or} \quad N_e = \frac{1}{2\Delta F},$$

where $V(\delta q)$ = the variance of allele frequency change from drift and ΔF = the inbreeding increment between parents and offspring in successive generations in the population under consideration. The two effective numbers are known respectively as the variance effective number and the inbreeding effective number. They are usually considered as being nearly equal and we shall treat them as such. But they can differ, particularly when population size is changing (Kimura & Crow 1963). Inbreeding effective number is related to the size of the parental generation whereas variance effective number is related to the progeny. Therefore, if population size is increasing, inbreeding effective number will be less than variance effective number. If the population size is decreasing, the opposite will be true. If, for example, a single heterozygote produces a very large number of offspring by self-fertilization the inbreeding effective number is one but the variance effective number is very large. Inbreeding effective number is retrospective whereas variance effective number is prospective and sometimes one may be of more interest than the other. Effective numbers that allow for some of the simpler deviations from the idealized structure may be found as follows.

Unequal numbers in different generations

If the number of breeding individuals changes from one generation to the next the effective number is approximately the harmonic mean of the numbers in each generation:

$$\frac{1}{N_e} \approx \frac{1}{t}\sum_{i=1}^{t}\frac{1}{N_i}.$$

The effective number is dominated by the size of the population when it is small. Following an example of Wright, let population size increase tenfold per generation from N_1 to $10^5 N_1$ five generations later. $1/N_e \approx (1.11111 N_1)/6$ and, therefore, $N_e \approx 5.4 N_1$. This contrasts sharply with the arithmetic mean of 18 518N_1.

Separate sexes

Random self-fertilization is excluded with separate sexes and this has a trivial (unless N is very small) influence upon effective number:

$$N_e = N + \tfrac{1}{2}.$$

If the sex-ratio differs from 1 : 1

$$N_e = \frac{4N_m N_f}{N_m + N_f} + \tfrac{1}{2},$$

where N_m and N_f are the numbers of breeding males and females, respectively. If $N_m = N_f = N/2$ then $N_e = N + \frac{1}{2}$. But if $N_m \neq N_f$ the effective number is nearer to that of the minority sex. For example, if $N_m = 1$ and $N_f = 100$, $N_e \approx 4.5$.

For an X-linked locus with equal numbers of males and females ($N_m = N_f = N/2$)

$$N_e = 0.75N.$$

If $N_m \neq N_f$

$$N_e = \frac{9N_{XY}N_{XX}}{4N_{XY} + 2N_{XX}},$$

where N_{XY} and N_{XX} are the numbers of the heterogametic and homogametic sexes. N_e tends to 4.5 N_{XY} when N_{XX} is greatly in excess, and to 2.25 N_{XX} when N_{XY} is greatly in excess.

Differential contributions to the next generation

In an idealized population every breeding individual has an equal chance of producing progeny that become parents in the next generation. This does not mean that each parent contributes to the same number of eventually successful progeny but rather, that progeny number is a random variable following a binomial distribution. If the population is constant in size each parent will on average contribute to two progeny (i.e. contribute two gametes to the next generation) and the distribution will approximate closely to a Poisson distribution with both mean $\bar{k}$ and variance $V(k) = 2$. Natural variation in progeny number is often considerably greater than that expected from sampling variation alone and

$$N_e = \frac{4N - 2}{2 + V(k)}.$$

If $V(k) > 2$, $N_e < N$ and this is probably the most important reason for reduction in effective number. Kimura & Crow (1963) and Crow & Kimura (1970) consider complications arising from changing population size and different contributions from males and females in dioecious populations. Nei & Murata (1966) have shown that effective size is decreased further when fertility differences are heritable.

It should be remembered that in a population of constant size $\bar{k} = 2$ when the breeding progeny reach sexual maturity. If progeny are counted at an earlier stage, as is usually the case, $\bar{k}$ may be considerably greater than 2 and, as $V(k)/\bar{k}$ is correlated with $\bar{k}$, the observed $V(k)/\bar{k}$ must be adjusted to a ratio corresponding to $\bar{k} = 2$. The problem has been discussed by Crow & Morton (1955) in relation to two different models for survival from young stages to sexual maturity. If the survival of each individual is random and is independent of the fate of other individuals from the same family

$$\frac{V(k)}{\bar{k}} = 1 + \frac{\bar{k}}{\bar{k}'}\left[\frac{V(k')}{\bar{k}'} - 1\right], \tag{3}$$

where $\bar{k}'$ and $V(k')$ are the mean and variance of progeny number enumerated at an early stage of the life-cycle. For example, if $\bar{k}' = 10$ and $V(k') = 100$, $V(k)/\bar{k} = 2.8$ and $N_e/N = 0.53$ assuming constant population size ($\bar{k} = 2$). At the other extreme entire families survive or perish at random and the fate of an individual is solely determined by the fate of its family. Then

$$\frac{V(k)}{\bar{k}} = \frac{V(k')}{\bar{k}'} + \bar{k}' - \bar{k}. \tag{4}$$

Again if $\bar{k}' = 10$ and $V(k') = 100$, $V(k)/\bar{k} = 18$ and $N_e/N = 0.11$ assuming constant population size.

The two models lead to quite different results and the true situation in nature must be intermediate depending upon the length of time over which progeny remain in their family units. If the progeny average is decreased from $\bar{k}'$ to $\bar{k}^*$ by the death of entire families and subsequently to $\bar{k}$ by individual death irrespective of family

$$\frac{V(k)}{\bar{k}} = 1 + \frac{\bar{k}}{\bar{k}^*}\left[\frac{V(k')}{\bar{k}'} + \bar{k}' - \bar{k}^* - 1\right]$$

by combining eqns (3) and (4). If $\bar{k}' = 10$ and $V(k') = 100$, $V(k)/\bar{k} = 3.75$ and $N_e/N = 0.42$ when $\bar{k}^* = 8$. When $\bar{k}^* = 4$, $V(k)/\bar{k} = 8.5$ and $N_e/N = 0.21$.

Inbreeding

The effective size of an inbred population is

$$N_e = \frac{N}{1 + F}, \tag{5}$$

where $F = 1$—(heterozygote frequency/$2pq$). When F is at its maximum value of 1, $N_e/N = 0.5$.

Overlapping generations

Many populations do not show discrete generations; rather individuals of different ages are present at the same time. Hill (1972, 1979) has shown that for a population of constant size and stable age distribution

$$N_e = \frac{(4N_T - 2)L_T}{2 + V(k)},$$

where N_T = the number of individuals born per convenient time period T, L_T = generation interval (mean age of parents when their progeny are produced) measured in units of T, and $V(k)$ = lifetime variance of progeny number. A discussion of effective size in age-structured populations where life-table information is available is given by Charlesworth (1980).

Estimates of effective population number

When a population differs for several reasons from idealized structure effective number can be computed by applying in turn the appropriate formulae. Even if it is not possible to determine the number of breeding individuals in a population and, therefore, an estimate of effective number it is still of interest to calculate N_e/N. This ratio gives an indication of the extent to which effective population size will be less than the actual size. A selection of estimates of both N_e/N and N_e, where possible, is given in Table 7.2. Some studies have allowed for more deviations from idealized structure than have others.

TABLE 7.2. Estimated ratios of effective population number to actual number (N_e/N) and/or estimated effective population numbers (N_e)

	N_e/N	N_e	Reference
Insects			
Dacus oleae	0.1	400	Nei & Tajima 1981
Drosophila subobscura	0.04	400	Begon 1977
Drosophila subobscura		> 4000	Begon *et al.* 1980
**Drosophila* ♀	0.71–0.90		Crow & Morton 1955
**Drosophila* ♂	0.48		Crow & Morton 1955
**Tribolium castaneum*	0.76–1.0		Wade 1980
Molluscs			
**Lymnaea columella*	0.75		Crow & Morton 1955
Cepaea nemoralis	0.5	95–6000	Greenwood 1974
Amphibians			
Notophthalmus viridescens		25	Gill 1978
Rana pipiens	0.01–0.67	46–112	Merrell 1968
Reptiles			
Uta stansburiana stejnegeri	0.7	16	Tinkle 1965
Sceloporus olivaceus	0.2	225–270	Kerster 1964
Birds			
Zonotrichia leucophrys	0.36	36	Baker 1981
Mammals			
Mus musculus		6–80	Petras 1967
Sciurus caroliniensis ♀	0.59		Charlesworth 1980
†*Alces alces*	0.2–0.4		Ryman *et al.* 1981
†*Odocoileus virginianus*	0.3–0.4		Ryman *et al.* 1981
Homo sapiens ♀	0.69–0.95		Crow & Morton 1955
Homo sapiens ♀	0.52		Nei & Murata 1966
Homo sapiens ♀	0.34		Felsenstein 1971
Homo sapiens ♀	0.60		Charlesworth 1980

*Laboratory populations.
†Computer simulation.

Two features emerge from this table. First, estimates of N_e may appear small but, for purely practical reasons, there may be a tendency to study small isolated populations. Furthermore, isolation may be less complete than is sometimes assumed. The surprisingly high levels of migration found for *Drosophila pseudoobscura* (Jones & Parkin 1977) and *D. subobscura* (Inglesfield & Begon 1981), for example, would suggest very large effective population sizes for these species. It is significant that values of N_e/N indicate that effective population numbers are usually within an order of magnitude of the numbers of breeding individuals. Secondly, information on plant populations is sadly absent.

Some progress towards an estimate of N_e/N has, however, been made by Mackay (1980) in his study of an ephemeral population of *Papaver dubium*.

The population contained 2316 plants with a mean seed output per plant of 245 and a variance of seed output of 1 488 630. Fifty per cent of all seed were produced by only 2% of the population and 4.6% were produced by the most fecund plant. The ratio $V(k')\bar{k}' = 6076$, and Mackay adjusted this to the ratio, $V(k)/\bar{k} = 25.8$, consistent with a mean seed output of 1 seed plant^{-1} assuming random survival. This represents variation between plants in their contribution to the next generation via female gametes only. That via male gametes may be quite different. Mackay preferred not to proceed to an estimate of N_e/N, but if we assume that male gamete variation is similar a value of $N_e/N \approx 0.07$ is indicated. This figure illustrates how variation in progeny number, which is a common feature of natural plant populations, can cause a significant reduction in effective number. Other deviations from the idealized population structure lead to further reductions and it seems likely that effective numbers for many plant species may be rather small.

Alternative information on the size of plant populations has been provided by estimates of neighbourhood size. A neighbourhood defines a group equivalent to a panmictic unit within a continuous array of individuals. Neighbourhood number may be regarded as an upper limit below which the effective population number will be decreased depending upon the deviation from idealized structure.

NEIGHBOURHOOD SIZE

Wright (1946) introduced the idea of a neighbourhood and defined it as an area from which the parents of central individuals may be treated as if drawn at random. For an explanation of the theory see Wright (1969).

Individuals are considered to be distributed at a uniform density either along a linear range, for example a river bank, or throughout an area. The length or area of a neighbourhood depends upon σ^2, the variance of the parent–offspring dispersal distribution, where dispersal distances are measured between corresponding stages of the life-cycle in parents and offspring. The most simple situation is a monoecious population mating at random, including self-fertilization at its appropriate random frequency ($1/N$). If dispersal distances are normally distributed a linear neighbourhood has a length $L = 2\pi^{\frac{1}{2}}\sigma$ and is expected to contain about 92.4% of the parents of central individuals, provided σ^2 is equivalent for male and female parents. If dispersal distributions in an areal neighbourhood follow a bivariate normal distribution with equal variances σ^2 along two orthogonal axes, the neighbourhood is a circle of radius 2σ and has an area of $A = 4\pi\sigma^2$. As the proportion of random observations from a circular normal distribution that lie within a radius $r\sigma$ is given by $1 - \exp(-r^2/2)$ about 86.5% of the parents of central individuals will lie within the neighbourhood.

The number of individuals within a neighbourhood is $N = 2\pi^{\frac{1}{2}}\sigma d$ (linear) or $N = 4\pi\sigma^2 d$ (areal) where d is the density of breeding individuals. The genetically effective number of a neighbourhood will usually be smaller and is obtained by using the effective density. It is, however, subject to the same problems of estimation as is effective population number.

Estimation of neighbourhood area

Confusion has sometimes arisen over the correct determination of neighbourhood area, particularly with respect to the variance of the dispersal distribution. We shall concentrate on the areal model for a monoecious population practising random mating, including self-fertilization, and we shall assume, initially, that the female parent–offspring and male parent–offspring dispersal distributions are identical with variance σ^2.

Wright (1946) did not emphasize that the dispersion variance should be measured along a single axis, although he made the point in a later paper (Dobzhansky & Wright 1947). The correct form of the variance has been discussed by Kerster (1964) and by Greenwood (1976). Wright (1978) refers to the required variance as the 'one-way variance'. It is also relevant that Wright's model assumes that the dispersal distribution has a mean of zero, i.e. there is no net displacement of populations from one generation to the next. There are three situations to consider.

The distances between offspring and their parents may be measured relative to an arbitrary pair of rectangular axes, x and y. The one-way dispersal variance is estimated as the average of the variance along each axis:

$$\sigma^2 = (s_x^2 + s_y^2)/2.$$

Remembering that $\bar{x}$ and $\bar{y}$ are assumed to be zero,

$$s_x^2 = \Sigma(x - \bar{x})^2/n = \Sigma x^2/n$$

and similarly for s_y^2. Therefore

$$\sigma^2 = (\Sigma x^2 + \Sigma y^2)/2n.$$

In practice, dispersal is often measured as absolute radial distance h along lines at angles θ to one of the axes. The zero mean assumption requires that θ takes random values between 0° and 360°. As

$$h^2 = x^2 + y^2,$$
$$\sigma^2 = \Sigma h^2/2n$$

i.e. the one-way variance is estimated as one half of the variance of absolute dispersal distances.

Finally, it is sometimes convenient to measure dispersion along a single axis, say x, in a positive direction only. This is often the case when measuring seed dispersal under artificial conditions (e.g. Levin & Kerster 1969a). The zero mean assumption requires that each observation is balanced by one of negative sign, but equal absolute value, along the same axis. Then

$$\sigma^2 = 2\Sigma x^2/2n = \Sigma x^2/n.$$

Parent–offspring dispersal is easier to measure in plants than in animals because there are usually only two components, through pollen and seed. Let any observed pollen or seed dispersals be of length p or s from samples of n_p or n_s observations respectively. Then assuming zero means, the one-way variances are estimated as $\sigma_p^2 = \Sigma p^2/n_p$ and $\sigma_s^2 = \Sigma s^2/n_s$. If either pollen or seed dispersal has been measured absolutely the appropriate one-way variance is one-half of this. A number of approaches to the combination of pollen and seed dispersal have been proposed.

Levin & Kerster (1968) simply summed the two components and halved the pollen dispersal variance to allow for absolute measurement so that neighbourhood area

$$A = 4\pi\left(\frac{\sigma_p^2}{2} + \sigma_s^2\right).$$

In later papers (Kerster & Levin 1968; Levin & Kester 1969a, b, 1971, 1975) they averaged the pollen and seed variances so that

$$A = 4\pi\left(\frac{\sigma_p^2}{4} + \frac{\sigma_s^2}{2}\right).$$

Again, the pollen variance has been divided by two to correct for absolute measurement. In other papers (Levin & Kerster 1974; Levin 1978, 1979) the same formula is proposed for apparently different reasons. Once again pollen and seed variances are averaged. No correction is made for absolute measurement of pollen dispersal, but the pollen variance is subsequently halved to allow for pollen being haploid and seed diploid:

$$A = 4\pi\left(\frac{\sigma_p^2}{4} + \frac{\sigma_s^2}{2}\right). \tag{6}$$

This distinction is, however, irrelevant because as far as any progeny locus is concerned both the paternal and maternal contributions are haploid.

The important asymmetry is that paternal dispersal is through pollen and seed whereas maternal dispersal is through seed only. Indeed, seed dispersal could be more properly regarded as progeny dispersal as it is a post-fertilization event. Male gamete dispersal variance, $\sigma_{\male}^2 = \sigma_p^2$ and female gemete dispersal variance, $\sigma_{♀}^2 = 0$. The average gamete variance is $\frac{1}{2}(\sigma_p^2 + 0)$ to which is added the progeny dispersal variance, σ_s^2 so that

$$A = 4\pi\left(\frac{\sigma_p^2}{2} + \sigma_s^2\right). \tag{7}$$

As in eqn (6) no corrections are included for absolute measurements and the variances should be adjusted as appropriate.

Equation (7) has the advantage of avoiding a logical difficulty that is inherent in eqn (6). It is possible for the pollen dispersal variance to be greater than the seed dispersal variance to a sufficient extent to make the total dispersal variance, $\sigma^2 < \sigma_p^2$. It is not, however, possible for σ^2 to be less than σ_s^2. From eqn (6) we find $\sigma^2 < \sigma_s^2$ when $\sigma_s^2 > \sigma_p^2/2$. Equation (7) never leads to $\sigma^2 < \sigma_s^2$ because $\sigma^2 - \sigma_s^2 = \sigma_p^2/2$.

Wright (1978, p. 75) reanalysed the data of Levin & Kerster (1968) using eqn (7) for similar reasons. He corrected the seed dispersal variance to allow for the zero mean assumption but, unfortunately, divided by two to correct for absolute measurement where seed dispersal was in fact measured axially.

Mixed selfing and outcrossing

The theory outlined above has assumed a population of hermaphrodites mating at random including self-fertilization at a frequency $1/N$. If self-fertilization is excluded the effect on neighbourhood area and effective size is negligible unless N is very small. This need not be the case of selfing occurs at a rate greater than its random frequency.

Levin & Kerster (1971) considered the effects of a variety of plant mating systems, including mixed selfing and outcrossing, upon neighbourhood structure. They proposed that the neighbourhood area is reduced by $(1 - t)$, the proportion of progeny produced by self-fertilization in excess of that expected under random mating (effectively equivalent to the actual selfing rate):

$$A = 4\pi t\left(\frac{\sigma_p^2}{4} + \frac{\sigma_s^2}{2}\right).$$

No reduction of effective neighbourhood number except that resulting from smaller neighbourhood area was suggested. This formulation seems to have been accepted (e.g. Schmitt 1980) even though it leads to zero neighbourhood area and number in the absence of outcrossing ($t = 0$).

Even with total self-fertilization, however, parent–offspring dispersal still occurs to the extent that seed are dispersed so that adjacent individuals may have different parents (Wright 1946). Only gamete (pollen) dispersal is affected by self-fertilization. The observed pollen dispersal variance $\Sigma p^2/n_p$ represents the proportion t of progeny that result from outcrossing. The total number of pollen dispersals, including the proportion $1 - t$ of zero dispersal self-fertilizations is n_p/t and, therefore, provided the distribution of outcrossed pollen dispersal remains the same as under random mating, the total pollen dispersal variance is $t\Sigma p^2/n_p = t\sigma_p^2$. The average gamete variance is $\frac{1}{2}(\sigma_♂^2 + \sigma_♀^2) = \frac{1}{2}t\sigma_p^2$ and

$$A = 4\pi\left(\frac{t\sigma_p^2}{2} + \sigma_s^2\right).$$

According to this expression neighbourhood area becomes $4\pi\sigma_s^2$, rather

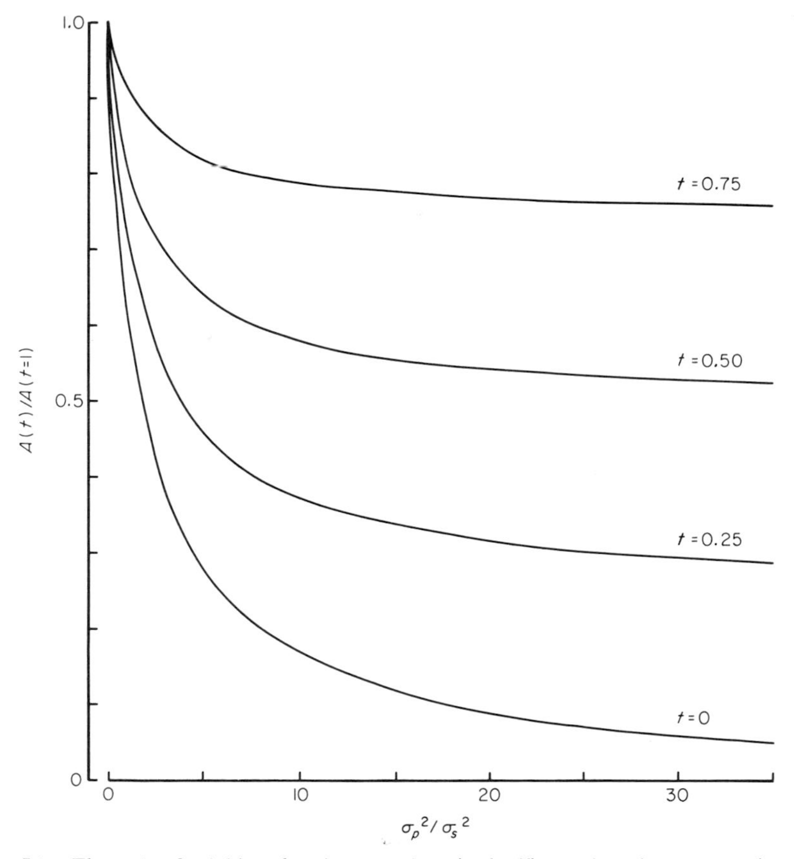

FIG. 7.1. The ratio of neighbourhood area under mixed selfing and random outcrossing (proportion t) to that under zero selfing ($t = 1$), $A(t)/A(t = 1)$, plotted against σ_p^2/σ_s^2, the ratio of pollen to seed dispersal variances. Four levels of outcrossing are shown: $t = 0.75$, 0.50, 0.25 and 0 (complete selfing).

than zero, under complete self-fertilization. The reduction in neighbourhood area is a function, not only of the frequency of self-fertilization, but also of the ratio of pollen to seed dispersal variance. The greater the seed dispersal variance compared to that of the pollen the smaller is the reduction in neighbourhood area. The ratio of neighbourhood area under a given level of outcrossing $A(t)$ to the area under zero selfing $A(t=1)$ is

$$\frac{A(t)}{A(t=1)} = \frac{t(\sigma_p^2/\sigma_s^2)+2}{(\sigma_p^2/\sigma_s^2)+2}$$

and is shown as a function of σ_p^2/σ_s^2 for different levels of outcrossing in Fig. 7.1.

Self-fertilization leads to a reduction in neighbourhood number, not only because neighbourhood area is reduced, but also because of the inbreeding effects upon effective density of a neighbourhood. If progeny are produced by a mixture of random outcrossing with frequency t and of self-fertilization with frequency $1-t$ an equilibrium level of inbreeding is achieved where $F=(1-t)/(1+t)$ (Haldane 1924). The effective size of an inbred population is $N_e = N/(1+F)$ (eqn 5) so it follows that the effective density under mixed random outcrossing and selfing is

$$d_e = \frac{d}{1+F} = \frac{d(1+t)}{2},$$

where d is the density of reproducing individuals and no other factors influencing effective density are considered. The effective neighbourhood number is

$$N_e = 4\pi\left(\frac{t\sigma_p^2}{2}+\sigma_s^2\right)\frac{d}{2}(1+t).$$

The ratio of effective neighbourhood number for a given level of outcrossing $N_e(t)$ to that for zero selfing $N_e(t=1)$ is

$$\frac{N_e(t)}{N_e(t=1)} = \frac{[t(\sigma_p^2/\sigma_s^2)+2](1+t)}{2[(\sigma_p^2/\sigma_s^2)+2]} \tag{8}$$

and is shown as a function of σ_p^2/σ_s^2 for different levels of selfing in Fig. 7.2. At high levels of selfing (small t) effective neighbourhood number is greatly reduced, particularly if the pollen dispersal variance substantially exceeds the seed dispersal variance. The equivalent expression from Levin & Kerster (1971) is, however, independent of σ_p^2/σ_s^2:

$$\frac{N_e(t)}{N_e(t=1)} = t. \tag{9}$$

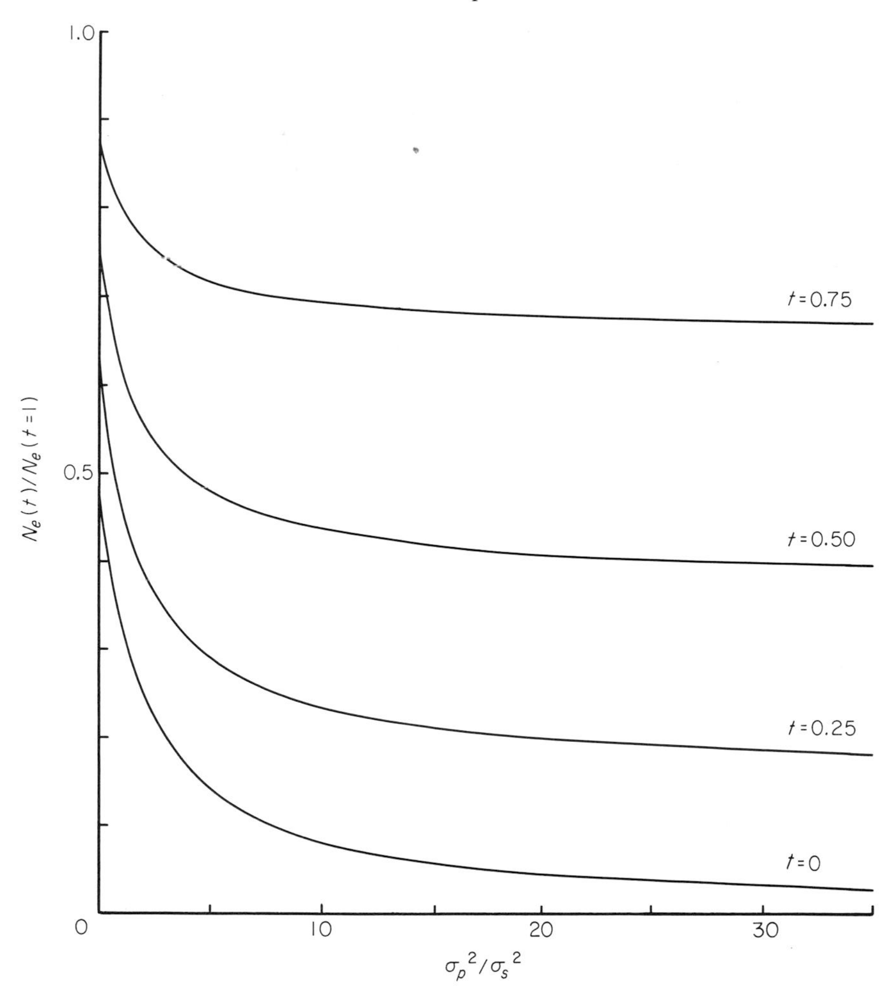

FIG. 7.2. The ratio of neighbourhood effective number under mixed selfing and random outcrossing (proportion t) to that under zero selfing, $N_e(t)/N_e(t = 1)$, plotted against σ_p^2/σ_s^2, the ratio of pollen to seed dispersal variances. Four levels of outcrossing are shown: $t = 0.75$, 0.50, 0.25 and 0 (complete selfing).

If $\sigma_p^2/\sigma_s^2 > 2/t$, eqn (8) leads to a greater reduction in effective neighbourhood number than does eqn (9) except for the limiting case when $t = 0$. It also follows from eqn (8) that if $\sigma_p^2/\sigma_s^2 > 2/t$ then $N_e(t)/N_e(t = 1) < t$. Thus, if 75% of progeny are produced by self-fertilization ($t = 0.25$) effective neighbourhood number is 0.27 of that without selfing when $\sigma_p^2/\sigma_s^2 = 6(< 2/t)$ but 0.23 when $\sigma_p^2/\sigma_s^2 = 10\ (> 2/t)$.

TABLE 7.3. Estimates of neighbourhood area (A) and number (N) for a variety of animals, trees and herbaceous plants

	$A(m^2)$	N	Reference
Insects			
Drosophila pseudoobscura	308×10^3	25 700*	Dobzhansky & Wright 1943; Wright 1978
Drosophila pseudoobscura	598×10^3	2212*	Dobzhansky & Wright 1947; Wright 1978
Drosophila pseudoobscura	$(184–1410) \times 10^3$	3239–6479*	Crumpacker & Williams 1973
Drosophila pseudoobscura	658×10^3	$(16–79) \times 10^3$*	Powell *et al.* 1976
Drosophila subobscura ♂	133×10^3	5333*	Begon 1976
Drosophila subobscura ♀	84×10^3	7166*	Begon 1976
Molluscs			
Cepaea nemoralis	1400	2800*	Lamotte 1951 (cited in Wright 1978)
Cepaea nemoralis	1200–1900	95–6000*	Greenwood 1974, 1976
Reptiles			
Sceloporus olivaceus	10 000	225–270*	Kerster 1964
Birds			
Zonotrichia leucophrys	506m	100*	Baker 1981
Troglodytes aedon	21.8×10^6	7679	Barrowclough 1980
Thryomanes bewickii	35.6×10^6	564	Barrowclough 1980
Phoenicurus phoenicurus	4.2×10^6	151	Barrowclough 1980
Parus major	11.5×10^6	1806	Barrowclough 1980
Parus major	8.3×10^6	770	Barrowclough 1980
Emberiza schoeniclus	30.4×10^6	1193	Barrowclough 1980
Melospiza melodia	1.5×10^6	891	Barrowclough 1980
Coereba flaveola	6.9×10^6	3290	Barrowclough 1980

TABLE 7.3 *continued*

Trees			
Fraxinus pennsylvanica	42m	16	Wright 1953
Fraxinus americana	1766	4.4	Wright 1953
Ulmus americana	1681m	253	Wright 1953
Populus deltoides	1528m	230	Wright 1953
Pseudotsuga taxifolia	2101	26	Wright 1953
Cedrus atlantica	33 623	208	Wright 1953
Pinus cembrioides	1766	11	Wright 1953
Pinus radiata	29 550	1–3200	Bannister 1965
Herbs			
Linanthus parryae	2.6–260	14–27*	Wright 1943b
Linanthus parryae	30	10–100*	Wright 1978
Phlox pilosa	11–21	75–282	Levin & Kerster 1968
Lithospermum caroliniense	4–26	2–7	Kerster & Levin 1968
Liatris aspersa	17–30	30–191	Levin & Kerster 1969a
Liatris cylindracea	33	165	Schaal & Levin 1978
Viola spp.	19–57	167–547	Beattie & Culver 1979
Primula veris	20–30	5–200	Richards & Ibrahim 1978
Lupinus texensis	2.8–6.3	42–95	Schaal 1980
Senecio spp. (bees)	0.7–7.3	8–24	Schmitt 1980
Senecio spp. (butterflies)	57–572	993–6154	Schmitt 1980
Lupinus amplus		36†	Zimmerman 1982
Thermopsis montana		132†	Zimmerman 1982

*Some allowance made for effective density.
†Spike counts of clonal species.
m = length of neighbourhood on linear model.

Observed neighbourhood sizes

Examples of published neighbourhood areas and numbers are listed in Table 7.3 for a range of animals, trees and herbaceous plants. Except where indicated, no attempts have been made to estimate effective densities and N refers to census neighbourhood number rather than to effective neighbourhood number. Neighbourhood areas for herbaceous plants have been obtained by various methods of combining pollen and seed dispersal. Application of eqn (7) would, in most cases, lead to some increase in area.

Significance of neighbourhood structure in plant populations

The subdivision of a population into smaller neighbourhoods leads to random genetic drift if the effective neighbourhood number is small. If the entire population is not much larger than the neighbourhood, drift will affect the whole population. If, however, there are many neighbourhoods within the population, genetic drift may lead to local random differentiation between neighbourhoods. Wright (1946) concluded that considerable differentiation would occur if neighbourhood effective number was twenty, or less, and a moderate amount with about 200 individuals. With effective numbers in excesses of 1000 there would be virtually no differentiation. The results of computer simulation studies are in broad agreement. Rohlf & Schnell (1971) simulated a population of 10 000 individuals representing a single neighbourhood or divided into neighbourhoods of twenty-five or nine individuals. Variation in allele frequencies throughout the area was very small (range 0.4–0.6) when the neighbourhood size was 10 000 and differences did not persist for many generations. When neighbourhood size was twenty-five, and more so with nine individuals, great variation in allele frequencies was soon established and ranged from 0 to 1 in different areas. Furthermore, particular patterns of allele frequencies persisted for many generations. These conclusions refer to areal neighbourhoods. Greater local differentiation occurs between neighbourhoods in a linear habitat because gene flow is restricted to two directions only.

The neighbourhood numbers for the majority of herbaceous plants listed in Table 7.3 are small and indicate considerable scope for random differentiation between neighbourhoods. Densities have been estimated from the standing crop of reproducing individuals and effective neighbourhood numbers will be even smaller. Neighbourhood area does not, however, suffer from the same problems of estimation and is itself informative.

Gene flow between neighbourhoods occurs at a rate that is independent of neighbourhood area. The genetic isolation between two parts of a population depends not on the physical distance involved but rather on the number of intervening neighbourhoods. If neighbourhood areas are small compared with the scale of selectively significant environmental heterogeneity, gene frequencies may vary between neighbourhoods as a result of locally different selection. If effective neighbourhood numbers are sufficiently small, however, random local differentiation will swamp the effects of weak selection.

Plant density affects neighbourhood area because flight parameters of pollinating insects are a function of density. Levin & Kerster (1969b) found that the means and variances of bee flight lengths decreased with increasing density for all nine plant species investigated. Similar results were obtained by Beattie (1976) for bees visiting *Viola* spp. although lepidopteran behaviour failed to show clear relationships with plant density. Provided the pollen dispersal variance is substantially greater than the seed dispersal variance, which is density independent, two points in high density arrays will be separated by more neighbourhoods and there will be greater opportunities for selective differentiation. Levin & Kerster (1969b) drew attention to an important consequence of this. In an unfavourable period when a population passes through a bottleneck in numbers, or in the proportion of flowering individuals in perennials, the reduction in density leads to an expansion in neighbourhood areas. Gene flow between previously more isolated areas of the population will assemble genes in new combinations. If flowering individuals produce a smaller than usual crop of flowers pollen carry-over will be promoted leading to a further expansion in neighbourhood area (Levin, Kerster & Niedzlek 1971).

Assumptions in the neighbourhood model

The basic model involves a number of assumptions that are unlikely to be true in nature. The most important are that dispersal distributions are normal, that these distributions have zero means and that they adequately reflect the form of gene dispersal between parents and offspring.

If each parent–offspring dispersal distance is the sum of a number of independent random movements the total dispersal distribution will, on account of the central limit theorem, tend to a normal distribution irrespective of how each of the component dispersal events is distributed. Thus Dobzhansky & Wright (1947) found that dispersal distances of *Drosophila pseudoobscura* were highly leptokurtic after 1 day but that the leptokurtosis declined daily until, after about 5 days, the distribution was approximately normal. However, as individual flies would make several flights per day, Bateman (1950) supported an alternative view that the observed decrease in leptokurtosis reflected a daily increasing number of flies dispersing beyond the sample area. Nevertheless, it is clear that leptokurtosis of dispersal distributions is the most frequently encountered deviation from normality. This is particularly the case in plants where pollinator flight lengths, wind-dispersed pollen, and seed (whether dispersed aerially, ballistically or passively) all show distributions that are usually leptokurtic to some degree (Levin & Kerster 1974).

Kurtosis occurs when a compound distribution is formed in unequal proportions by two separate normal distributions with similar means but different variances (Richardson 1970). The compound distribution becomes more leptokurtic as the distribution with the smaller variance predominates. Bateman (1947) suggested that the leptokurtosis of pollinator flight lengths reflected the relative frequencies of two quite distinct types of flight: a majority of short foraging flights interspersed by occasional long exploratory or escape flights. Levin & Kerster (1968) found, for example, that the majority of butterfly flights on *Phlox pilosa* were less than 2 m and usually between neighbouring plants but about 2% of flights were between 7 m and 29 m. The leptokurtosis of wind-dispersed pollen and seeds has been explained in relation to eddy diffusion and atmospheric turbulence (Bateman 1947; Levin & Kerster 1974). Ballistic seed dispersal shows lower levels of leptokurtosis (Levin & Kerster 1968; Beattie & Culver 1979; Schaal 1980) more consistent with the exponential distribution expected if the probability of dispersal declines at a constant rate per unit distance from the parent. In contrast to the ballistic dispersal of *Viola* seeds, Beattie & Culver (1979) found that the subsequent ant dispersal was sometimes slightly platykurtic and might reflect territoriality between different ant colonies.

Few attempts have been made to allow for the effects of leptokurtic dispersal distributions upon neighbourhood area estimates. Usually it has been ignored: sometimes distributions have been truncated at a distance more consistent with a normal distribution. Although the extended tail of the distribution is the most striking feature of leptokurtosis, it must be remembered that there is also an excess of observations near the zero mean.

Wright (1969, corrected 1977) provides a method by which neighbourhood area may be estimated from the observed leptokurtic distribution. The normal distribution

$$y = \frac{1}{\sigma(2\pi)^{\frac{1}{2}}} \exp\left(-\frac{1}{2\sigma^2}\cdot x^2\right)$$

may be regarded as a special case of the more general distribution

$$y = y_0 \exp(-b \cdot x^{1/a}).$$

If $a = 0.5$ the distribution is normal, if $a < 0.5$ platykurtic and if $a > 0.5$

leptokurtic. The observed leptokurtosis is estimated as

$$\gamma_2 = \frac{n\Sigma(x-\bar{x})^4}{[\Sigma(x-\bar{x})^2]^2} - 3 = \frac{n\Sigma x^4}{(\Sigma x^2)^2} - 3,$$

remembering the zero-mean assumption. Leptokurtosis is underestimated if the true mean is not zero. For a normal distribution $\gamma_2 = 0$ and is greater than zero for a leptokurtic distribution.

The parameter a may be estimated from the observed leptokurtosis by trial-and-error application of the relationship

$$\gamma_2 = [\Gamma(a)\Gamma(5a)\Gamma^{-2}(3a)] - 3.$$

Now log a is a sigmoid function of log γ_2 but is almost linear over the range of values of leptokurtosis commonly encountered for dispersal distributions. Therefore an alternative and easier method of estimating a employs the following empirical relationships:

$$\log_{10}a = 0.341 \log_{10}\gamma_2 - 0.156 \quad \text{for} \quad 1 < \gamma_2 < 15,$$
$$\log_{10}a = 0.295 \log_{10}\gamma_2 - 0.097 \quad \text{for} \quad 15 < \gamma_2 < 50.$$

The errors involved in this approximate method are of the order of 1%, or less.

The estimated value of a is substituted into the following expressions to obtain lengths or areas of neighbourhoods:

$$L_l = 2^{a+1}\Gamma(a+1)\Gamma^{\frac{1}{2}}(a)\Gamma^{-\frac{1}{2}}(3a)\sigma$$
$$A_l = 2^{2a}\Gamma(2a+1)\Gamma(a)\Gamma^{-1}(3a)\pi\sigma^2.$$

As $\Gamma(n) = (n-1)\Gamma(n-1)$ and $\Gamma(\frac{1}{2}) = \pi^{\frac{1}{2}}$ these expressions reduce to $L = 2\pi^{\frac{1}{2}}\sigma$ and $A = 4\pi\sigma^2$ when $a = \frac{1}{2}$, the values appropriate for a normal distribution.

The effect of leptokurtosis upon neighbourhood length or area is shown in Fig. 7.3 where L_l/L and A_l/A are plotted against γ_2. Even moderate amounts of leptokurtosis significantly reduce the length of a linear neighbourhood. The reduction in an areal neighbourhood is less marked and, in fact, when $0 < \gamma_2 < 3$ there is a slight expansion in area.

The neighbourhood area may be expressed as

$$A_l = k\pi\sigma^2$$

where k is a constant that corrects for leptokurtosis. $k = 4$ if the dispersal distribution is normal; if $\gamma_2 > 3$, $k < 4$ and decreases as γ_2 increases (see Fig. 7.3 where $A_l/A = k/4$). Pollen and seed dispersal distributions will usually differ in leptokurtosis and require their own constants k_p and k_s,

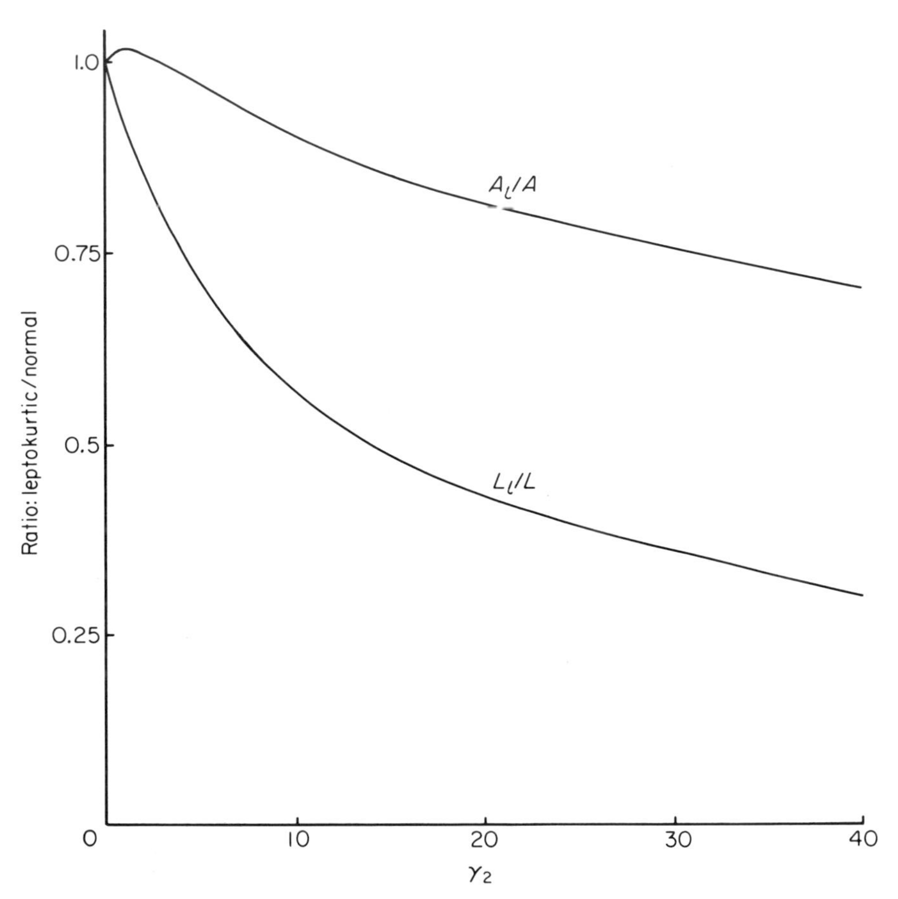

FIG. 7.3. Ratios of neighbourhood lengths or areas when dispersal distributions are leptokurtic to lengths or areas when distributions are normal (L_l/L and A_l/A, respectively) plotted against leptokurtosis, γ_2.

respectively:

$$A_l = \pi\left(\frac{k_p\sigma_p^2}{2} + k_s\sigma_s^2\right).$$

It is clear that the total reduction in neighbourhood will depend on the relative contributions of pollen and seed to the dispersal variance as well as on their deviations from normality.

Beattie & Culver (1979) have allowed for leptokurtosis when estimating neighbourhood areas for *Viola* spp. Variances and leptokurtosis are shown in Table 7.4 for their *V. pedata* population A. This species is not cleistogamous and probably does not self-fertilize. Assuming dispersal is normally

TABLE 7.4. Dispersal variances and leptokurtosis for *Viola pedata* population A (Beattie & Culver 1979)

Distribution	Method of dispersal	$\sigma^2(\mathrm{m}^2)$	γ_2	k	$k\sigma^2(\mathrm{m}^2)$
Pollen	Insects	11.07	42.9	2.82	31.22
Seed	Ballistic	0.46	0.8	4.06	1.87
Seed	Ants	0.18	1.3	4.07	0.73

γ_2 = leptokurtosis. k corrects for leptokurtosis in $A = k\pi\sigma^2$.

distributed

$$A = 4\pi\left(\frac{11.07}{2} + 0.46 + 0.18\right) = 77.6\ \mathrm{m}^2.$$

Allowing for leptokurtosis

$$A - \pi\left(\frac{31.22}{2} + 1.87 \mid 0.73\right) = 57.2\ \mathrm{m}^2.$$

It is assumed that dispersal distributions are symmetrical about a mean of zero, i.e. there is no net displacement of populations from one generation to the next. If the real mean is not zero this assumption will lead to an over-estimate of the variance, even if the distribution is normal, because by definition the variance takes a minimum value when estimated around the true mean. If there is an element of directionality in dispersion the distribution may be skewed around a mode, rather than a mean, of zero.

Wind dispersal of pollen and seeds will be primarily downwind and skewed dispersal distributions will result if the wind prevails mainly from one direction. At the upwind end of Drws y Coed mine, for example, McNeilly (1968) found a sharp decline in copper tolerant *Agrostis tenuis* in a distance of 1 m across the mine boundary. Downwind of the mine, however, copper tolerant individuals were still frequent 150 m from the mine boundary.

Pollinator flight patterns may also show sequential directionality, if one flight tends to be in the same direction as the previous flight, and overall directionality when the majority of flights tend to be in similar directions. Bateman (1947) observed a significant excess of bumble-bee flights on radish to be in the same direction as the immediately preceding flight although there was no overall directionality. He concluded that under normal wind conditions pollen would be dispersed equally in all directions if plants were evenly distributed. Similarly, Levin *et al.* (1971) found that both bees and butterflies foraging on *Lythrum salicaria* showed high correlations in direction between consecutive flights, but directionality was largely restricted to successive pairs of flights. Similar conclusions were drawn by Pyke (1978) observing bumble-bees visiting *Aconitum columbianum* and *Delphinium nelsoni*. In contrast, Woodell (1978) found that bumble-bees foraging on *Armeria maritima* and *Limonium vulgare* showed sequential directionality and also strong overall directionality into the wind, whatever its direction. If winds tend to blow primarily in one or two directions, as is often the case for example in coastal regions, overall directionality would lead to non-zero mean dispersal distributions.

The flight distances of pollinators may not adequately reflect the actual distances travelled by pollen to the ovule parents on which fertilization occurs. Only a proportion of the available pollen is deposited on the stigmas of any single flower leading to 'pollen carry-over' between flowers separated by more than one flight. The level of carry-over between plants will depend on a number of factors: the proportion of pollen grains carried on an insect that are deposited or acquired with each flower visit, the number of flowers visited in sequence on a single plant, and the mating system. Heteromorphic incompatibility systems and dioecy will promote carry-over as many pollinator flights will be to plants of the same morph or sex (Levin & Kerster 1971). Furthermore, homomorphic incompatibility together with the formation of clones leads to effective carry-over because pollen grains that are carried to a member of the same clone fail to fertilize ovules. In the absence of an outbreeding mechanism or of apomixis Levin & Kerster (1971) suggest that carry-over will be negligible, but there is little evidence for or against this view.

Several models of pollen carry-over have been proposed (e.g. Bateman 1947; Mulcahy & Caporello 1970; Levin & Kerster 1971; Primack & Silander 1975). Although the models differ in detail the general conclusion is clear. Pollen deposition on a sequence of plants declines in a geometric progression so that long range dispersal involving several pollinator flights occurs, but at relatively low frequency. This may be illustrated by considering a simple model where the number of available pollen grains on a pollinator is in dynamic equilibrium so that each time it visits a flower a proportion p of its pollen grains is exchanged. It is assumed that in a given flower pollen grains are deposited on the stigma before fresh pollen grains are collected from that flower. There is no self-incompatibility and n flowers are visited in sequence on each plant. Consider the total complement of pollen collected from a given plant. A proportion

$$1 - \frac{1}{np}[1 - (1 - p)^n]$$

of this pollen will be deposited as self pollen on the same plant.

TABLE 7.5. Proportion (%) of pollen grains collected from a single plant that are deposited as self grains on that plant or as outcross grains on the *i*th successive plant for different values of *p*, the proportion of the pollen load that is deposited on, and collected from each flower and *n*, the number of flowers visited per plant

			Proportion of grains on the *i*th next plant													
p	*n*	Percentage of self grains	*i* = 1	2	3	4	5	6	7	8	9	10	11	12	13	...
0.05	5	10	20	16	12	9	7	6	4	3	3	2	2	1	1	...
0.05	10	20	32	19	12	7	4	2	1	1	1					
0.1	5	18	34	20	12	7	4	2	1	1						
0.1	10	35	42	15	5	2	1									
0.2	5	33	45	15	5	2	1									
0.2	10	55	40	4	1											

A proportion

$$\frac{1}{np}[1-(1-p)^n]^2(1-p)^{n(i-1)}$$

will be deposited on the *i*th subsequent plant visited. The deposition schedule is shown for various values of p and n in Table 7.5. Substantial levels of carry-over are indicated when p and n are small.

A considerable amount of information is available on gene flow between groups of plants, particularly in relation to the isolation distances required for the maintenance of genetic purity in crops (Levin & Kerster 1974). Observations have also been made on the rate of decline of a full pollinator load of marked pollen when an insect visits a series of flowers. Gerwitz & Faulkner (1972), for example, allowed bees to collect pollen from plants labelled with ^{32}P. The level of radioactivity measured in a sequence of flowers visited on unlabelled plants fell on average by 30% per flower. The presence of radioactive pollen was still detectable at the tenth flower. The pattern of deposition of full pollen loads is not, however, equivalent to the fate of the pollen collected from a single plant when a pollinator is already in a pollen gain/loss equilibrium. Further work is required along the lines of that reported by Levin & Berube (1972) and Primack & Silander (1975) where actual counts of pollen loads on insects and deposition schedules of pollen from known sources have been determined. Thomson & Plowright (1980) used pollen grain colour dimorphism, red or yellow, to follow the deposition pattern of *Erythronium americanum* pollen on a sequence of isolated flowers. Bumble-bees with loads of yellow pollen were allowed to visit a single flower producing red grains followed by a sequence of protogynous flowers that were pre-anthesis. The majority of red grains were deposited on the next six or seven flowers but this represents an under-estimate of carry-over because the pollen load was not replenished as it would be in nature. Handel (1976) labelled staminate spikelets of individuals of *Carex plantaginea* and *C. platyphylla* with the normally absent element samarium. After anthesis the level of samarium on pistillate spikelets of surrounding plants was estimated by neutron-activation analysis. Rather restricted wind dispersal of pollen in these woodland herbs was indicated. This technique could be used to examine insect dispersal of pollen from individual plants.

The increase in pollen dispersal variance over the observed flight distance variance will be even greater when carry-over is associated with sequential and especially overall directionality. Levin *et al.* (1971) have provided a correction to the pollen dispersal variance that allows for carry-over with sequential directionality. The distribution is, however, assumed to be normal and it is unclear how carry-over and directionality, sequential or overall, will affect leptokurtosis. Schaal (1980) measured pollinator flight distances in an artificial plot of *Lupinus texensis*. Actual gene flow through pollen was also estimated by the spread of an electrophoretic marker. Mean gene flow was almost twice as far as mean pollinator flight distance resulting in neighbourhood area estimates of 6.3 and 2.8 m^2 respectively, when seed dispersal was included. She attributed the difference to pollen carry-over and, although she does not comment on directionality, it is interesting to note that the leptokurtosis of the pollinator flight distribution was 11.7 whereas that for gene flow was only 3.0.

Even if adequate allowance can be made for pollen carry-over and directionality other factors may well lead to realized parent–offspring distances being different from those estimated. Short-range pollen dispersal may be relatively under-represented in the successful progeny if neighbouring plants are close relatives or members of the same clone. Even in the absence of a self-incompatibility system, cryptic self-incompatibility may occur when outcross pollen is competitively superior in the style (Bateman 1956). Furthermore, inbred progeny may be competitively excluded as a result of inbreeding depression. Short range seed dispersal in perennials may lead to germination close to a well-established parent and subsequent competitive exclusion of the seedling. Ovule parent–offspring dispersal may be greater than indicated by seed-dispersal estimates, particularly as these reflect immediate dispersal and do not allow for movement of seeds between dispersal and germination (Harper 1977, Chapter 2).

VARIATION IN FLOWER NUMBER BETWEEN PLANTS

Wright's neighbourhood model of population subdivision is appropriate where individuals are continuously distributed. There are other models that deal with discrete subdivision. The island model (Wright 1943a) assumes that a population is divided into a number of discontinuous colonies, each random mating within itself, with occasional migration between colonies. Migrants are assumed to be drawn at random from the whole population range and there is no tendency for migration to occur more frequently between adjacent colonies. More appropriate for many cases is the stepping-stone model of Kimura & Weiss (1964) where migration is mainly between adjacent colonies. Kimura & Ohta (1971) discuss all three models.

It is important to remember that neighbourhoods are overlapping and not discontinuous. Any estimate for neighbourhood area is only typical for a population. In principle there is a neighbourhood around any point in a population, its area depending upon local biotic and abiotic circumstances. The dependence of neighbourhood area and effective number on a variety of factors has already been discussed. The consequences of variation between plants in flower number seems, however, not to have been explored in this context.

Consider, for example, a typical self-compatible, insect-pollinated herbaceous perennial. Individuals will differ in age, size and the number of flowers that they bear. The distribution is likely to be L-shaped (Levin & Wilson 1978) with a majority of small individuals and fewer large ones. The number of flowers visited per plant by pollinating insects will tend to increase with increasing number of flowers (Bowers 1975; Lloyd 1980). It has already been suggested that pollen carry-over and, therefore, pollen dispersal is reduced when insects visit several flowers in sequence on a single plant. Furthermore, large plants may well be more attractive to pollinators and, therefore, become local foci for gene flow. All of these effects will lead to local reductions in neighbourhood area.

In self-compatible species the level of self-fertilization is expected to increase with the number of flowers because of greater opportunities for geitonogamy (selfing arising from pollen being transferred between different flowers on the same individual). My own observations on *Malva moschata* L., a protandrous, self-compatible perennial, support this hypothesis. An allelic pair of α-acid phosphatase electrophoretic markers was used to estimate natural selfing rates in two artificial plots where a plant homozygous for the faster migrating band (*FF*) was surrounded by six plants homozygous for the slower band (*SS*). For each flower on the central plants the sepals, which persist and form a seed capsule, were marked with spots of coloured paint to indicate the date on which that flower was pollinated. The number of flowers in anthesis and the total number of flowers on each plant were scored for each day. Most pollinators were bees, both hive bees and bumble-bees. Capsules on the central plants were harvested separately, progeny grown and their genotypes at the marker locus scored. Self progeny were *FF* and outcross progeny were *FS*. The daily proportion of selfed progeny, weighted by the inverse of its binomial variance, was regressed against the number of flowers in anthesis and the total number of flowers on the central plant that day. Both regressions were highly significantly indicating that selfing increased with the number of flowers. Figure 7.4 shows

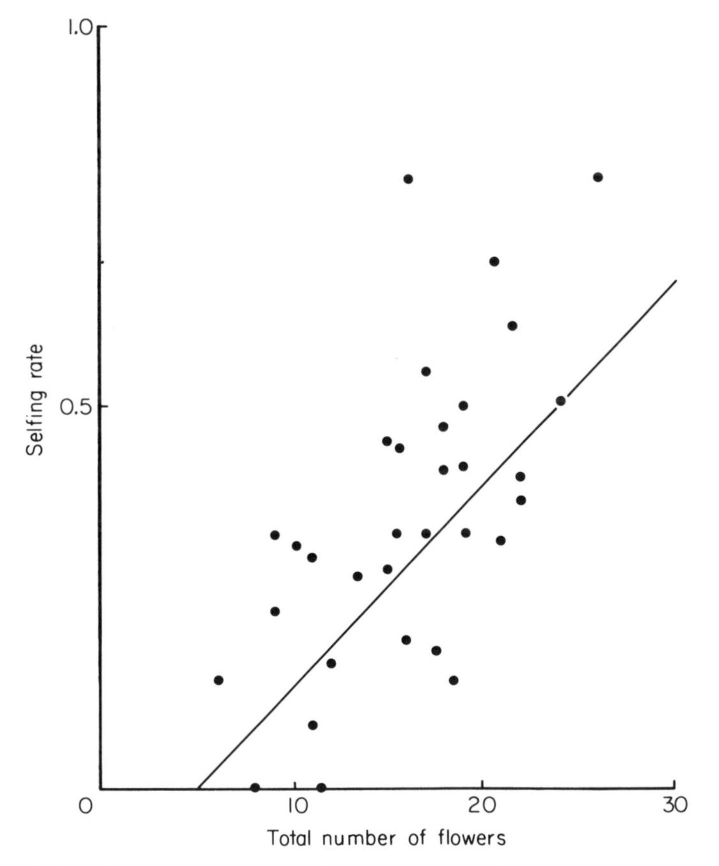

FIG. 7.4. Daily selfing rate, s, plotted against total number of flowers, n, on a single *M. moschata* at the centre of an experimental open-pollinated plot. Weighted regression line is $s = 0.026n - 0.135$ (significant at $P < 0.001$).

results for one plot. The regression equations were surprisingly similar to those from another experiment where all plants were very much larger throughout most of their flowering period. It was decided, therefore, to include a second independent variable in the regressions, the number of flowers on all six surrounding plants. Selfing rate was inversely related to the number of flowers in anthesis, or total flowers, on outer plants. These experiments suggest that the level of selfing depends on the balance between the number of flowers on a plant and the number on adjacent plants.

Methods are available for estimating selfing rates in natural populations (Vasek 1968; Brown 1975; Jain 1979; Ritland & Jain 1981). They all yield a single average estimate for a population based upon genotype frequencies at one or more loci in progeny from parents of known genotype. There is no method for estimating selfing rates for individual plants because some outcross pollen will produce progeny of the same genotype as self pollen. All that can be estimated with certainty is a minimum outcrossing (or maximum selfing) rate. Any heterozygous progeny from a homozygous parent have arisen from outcrossing. The greater the number of loci for which the parent is homozygous the more likely is the estimated minimum outcrossing rate to be near the true outcrossing rate.

A population of *M. moschata* at Rosedale Abbey, North Yorkshire (National Grid Reference 45/724960) was polymorphic for the α-acid phosphatase locus and also for pink or white petal colour, determined by a pair of alleles at a single locus. No other electrophoretic variations could be found. The size of each plant was estimated by counting the number of seed capsules at the end of the flowering season. Seed was collected from parents homozygous at both loci. Any progeny heterozygous for at least one locus indicates a known outcross. Progeny with the maternal genotype may be outcrossed or selfed progeny. Minimum outcrossing estimates are plotted against maternal size in Fig. 7.5 for progeny scored to date. Some small parents, but no large parents, yielded high estimates of minimum outcrossing. It is likely that those small plants for which the minimum outcrossing rate is low were surrounded by a majority of plants of the same genotype at these two loci and, therefore, many outcross progeny cannot be distinguished from selfed progeny. It is reasonable to conclude that selfing increases with number of flowers in this natural population. It should be noted that reliable population estimates of the proportion of progeny produced by selfing would be obtained by sampling a constant *proportion* of the total seed output of each parent, and not a constant *number* of seed per parent.

Increased selfing on larger plants will lead to a further reduction in neighbourhood area. The effective number of individuals in a neighbour-

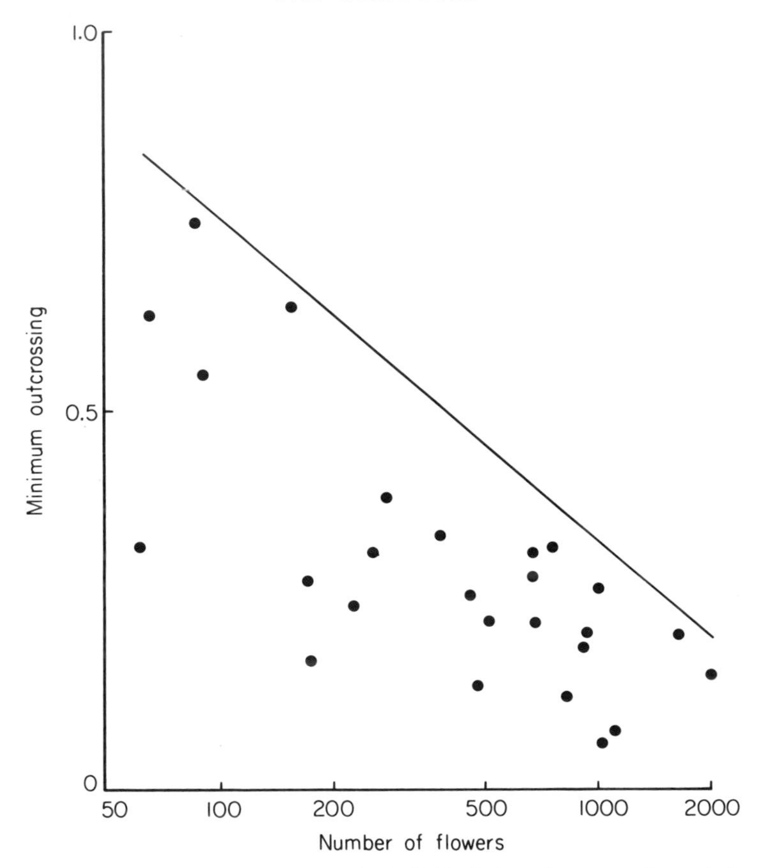

FIG. 7.5. Minimum outcrossing estimates for individual plants in a natural population of *M. moschata* as a function of the total number of flowers produced by each plant (on a logarithmic scale). The line does not imply a fitted relationship.

hood depends on both neighbourhood area and also on effective density within the neighbourhood. Effective density will be decreased when individuals differ in their fecundity. The number of capsules per plant at Rosedale Abbey varied from 0 to 2110 with a mean of 460 and a variance of 207 220. On the basis of ten capsules from each of ten plants the mean seed output was 7245 and the variance 51 403 552 so that the ratio $V(k')/\bar{k}' = 7095.4$. Adjusted to a female mean of one seed per plant, assuming random survival (eqn 3), $V(k)/\bar{k} = 1.98$ leading to $N_e/N = 0.67$. Although $V(k')/\bar{k}'$ is similar to that obtained by Mackay (1980) it is relative to a much larger average seed output and the reduction in effective number is not so dramatic.

These estimates refer to female fecundity only. Plants will also differ in their male fecundity, although this is more difficult to estimate. Anthers are easier to count than are pollen grains, but individuals may differ in the number of pollen grains per anther or in the proportion of viable pollen grains. There is increasing interest in the possibility of gender specialization in hermaphrodites (Lloyd 1979; Beach 1981) but male fecundity will, in general, be highly correlated with female fecundity.

A further problem concerns the fate, self v. outcross, of pollen grains. Estimates obtained from seed refer to female outcrossing and give no information on the outcrossed proportion of male gametes. Male outcrossing estimates are technically difficult to obtain but Horovitz & Harding (1972) showed that they were independent of female outcrossing in *Lupinus nanus*. Nevertheless, the proportion of pollen grains that produce self-fertilized progeny may increase with plant size.

There is high probability, therefore, that not only do a few plants contribute the majority of progeny to the next generation but also that these progeny are more likely to be the products of self-fertilization than are those of smaller plants. Species differing in details of life-cycle or breeding system from the example discussed above will show similar effects to greater or lesser degree. There are too few relevant observations available for a full assessment of the effect of variation in plant size upon effective neighbourhood number, let alone the other factors that will cause further reductions. It seems clear, however, that effective sizes of plant populations may well be very small. How small is still a matter for conjecture.

ACKNOWLEDGMENTS

I would like to thank the Revd H.J. Le Page for permission to work on the mallows at Rosedale Abbey and also Mary Crawford, Maureen Laycock, Simon Laycock, Sally Lewis, Keith Partridge, Carol Stansfield and David Stansfield for fieldwork and technical assistance.

REFERENCES

Baker M.C. (1981) Effective population size in a songbird: some possible implications. *Heredity*, **46**, 209–218.

Bannister M.H. (1965) Variation in the breeding system of *Pinus radiata*. *The Genetics of Colonizing Species* (Ed. by H.G. Baker & G.L. Stebbins), pp. 353–372. Academic Press, New York.

Barrowclough G.F. (1980) Gene flow, effective population sizes, and genetic variance components in birds. *Evolution*, **34**, 789–798.

Bateman A.J. (1947) Contamination in seed crops. III. Relation with isolation distance. *Heredity*, **1**, 303–336.

Bateman A.J. (1950) Is gene dispersal normal? *Heredity*, **4**, 353–364.

Bateman A.J. (1956) Cryptic self-incompatibility in the Wallflower: *Cheiranthus cheiri* L. *Heredity*, **10**, 257–261.

Beach J.H. (1981) Pollinator foraging and the evolution of dioecy. *American Naturalist*, **118**, 572–577.

Beattie A.J. (1976) Plant dispersion, pollination and gene flow in *Viola. Oecologia*, **25**, 291–300.

Beattie A.J. & Culver D.C. (1979) Neighbourhood size in *Viola. Evolution*, **33**, 1226–1229.

Begon M. (1976) Dispersal, density and microdistribution in *Drosophila subobscura* Collin. *Journal of Animal Ecology*, **45**, 441–456

Begon M. (1977) The effective size of a natural *Drosophila subobscura* population. *Heredity*, **38**, 13–18.

Begon M., Krimbas C.B. & Loukas M. (1980) The genetics of *Drosophila subobscura* populations. XV. Effective size of a natural population estimated by three independent methods. *Heredity*, **45**, 355–350.

Blower J.G., Cook L.M. & Bishop J.A. (1981) *Estimating the size of Animal Populations*. George Allen & Unwin, London.

Bowers K.A.W. (1975) The pollination ecology of *Solanum rostratum* (Solanaceae). *American Journal of Botany*, **62**, 633–638.

Brown A.H.D. (1975) Efficient experimental designs for the estimation of genetic parameters in plant populations. *Biometrics*, **31**, 145–160.

Charlesworth B. (1980) *Evolution in Age-structured Populations*. Cambridge University Press, Cambridge.

Crow J.F. & Kimura M. (1970) *An Introduction to Population Genetics Theory*. Harper & Row, New York.

Crow J.F. & Morton N.E. (1955) Measurement of gene frequency drift in small populations. *Evolution*, **9**, 202–214.

Crumpacker D.W. & Williams J.S. (1973) Density, dispersion, and population structure in *Drosophila pseudoobscura*. *Ecological Monographs*, **43**, 499–538.

Dobzhansky Th. & Wright S. (1943) Genetics of natural populations. X. Dispersion rates in *Drosophila pseudoobscura*. *Genetics*, **28**, 304–340.

Dobzhansky Th. & Wright S. (1947) Genetics of natural populations. XV. Rate of diffusion of a mutant gene through a population of *Drosophila pseudoobscura*. *Genetics*, **32**, 303–324.

Ewens W.J. (1979) *Mathematical Population Genetics*. Springer, Berlin.

Felsenstein J. (1971) Inbreeding and variance effective numbers in populations with overlapping generations. *Genetics*, **68**, 581–597.

Gerwitz A. & Faulkner G.J. (1972) *National Vegetable Research Station 22nd Annual Report, 1971*, p. 32. Wellesbourne, Warwick.

Gill D.E. (1978) Effective population size and interdemic migration rates in a metapopulation of the red-spotted newt, *Notophthalmus viridescens* (Rafinesque). *Evolution*, **32**, 839–849.

Greenwood J.J.D. (1974) Effective population numbers in the snail *Cepaea nemoralis*. *Evolution*, **28**, 513–526.

Greenwood J.J.D. (1976) Effective population number in *Cepaea:* a modification. *Evolution*, **30**, 186.

Handel S.N. (1976) Restricted pollen flow of two woodland herbs determined by neutron-activation analysis. *Nature*, **260**, 422–423.

Haldane J.B.S. (1924) A mathematical theory of natural and artificial selection. II. The influence of partial self-fertilization, inbreeding, assortative mating, and selective fertilization on the composition of Mendelian populations, and on natural selection. *Proceedings of the Cambridge Philosophical Society*, **1**, 158–163.

Harper J.L. (1977) *Population Biology of Plants*. Academic Press, London.

Hill W.G. (1972) Effective size of populations with overlapping generations. *Theoretical Population Biology*, **3**, 278–289.

Hill W.G. (1979) A note on effective population size with overlapping generations. *Genetics*, **92**, 317–322.

Horovitz A. & Harding J. (1972) The concept of male outcrossing in hermaphrodite higher plants. *Heredity*, **29**, 223–236.

Inglesfield C. & Begon M. (1981) Open ground individuals and population structure in *Drosophila subobscura* Collin. *Biological Journal of the Linnean Society*, **15**, 259–278.

Jain S.K. (1979) Estimation of outcrossing rates: some alternative procedures. *Crop Science*, **19**, 23–26.

Jones J.S. & Parkin D.T. (1977) Attempts to measure natural selection by altering gene frequencies in natural populations. *Measuring Selection in Natural Populations* (Ed. by F.B. Christiansen & T.M. Fenchel), pp. 83–96. Springer, Berlin.

Kerster H.W. (1964) Neighborhood size in the rusty lizard, *Sceloporus olivaceus*. *Evolution*, **18**, 445–457.

Kerster H.W. & Levin D.A. (1968) Neighborhood size in *Lithospermum caroliniense*. *Genetics*, **60**, 577–587.

Kimura M. & Crow J.F. (1963) The measurement of effective population number. *Evolution*, **17**, 279–288.

Kimura M. & Ohta T. (1971) *Theoretical Aspects of Population Genetics*. Princeton University Press, New Jersey.

Kimura M. & Weiss G.H. (1964) The stepping stone model of population structure and the decrease of genetic correlation with distance. *Genetics*, **49**, 561–576.

Levin D.A. (1978) Pollinator behaviour and the breeding structure of plant populations. *The Pollination of Flowers by Insects* (Ed. by A.J. Richards), pp. 133–150. Symposia of the Linnean Society of London, vol. 6. Academic Press, London.

Levin D.A. (1979) Pollinator foraging behaviour: genetic implications for plants. *Topics in Plant Population Biology* (Ed. by O.T. Solbrig, S. Jain, G.B. Johnson & P.H. Raven), pp. 131–153. Macmillan, London.

Levin D.A. & Berube D.E. (1972) *Phlox* and *Colias*: the efficiency of a pollination system. *Evolution*, **26**, 242–250.

Levin D.A. & Kerster H.W. (1968) Local gene dispersal in *Phlox*. *Evolution*, **22**, 130–139.

Levin D.A. & Kerster H.W. (1969a) Density-dependent gene dispersal in *Liatris*. *American Naturalist*, **103**, 61–74.

Levin D.A. & Kerster H.W. (1969b) The dependence of bee-mediated pollen and gene dispersal upon plant density. *Evolution*, **23**, 560–571.

Levin D.A. & Kerster H.W. (1971) Neighborhood structure in plants under diverse reproductive methods. *American Naturalist*, **105**, 345–354.

Levin D.A. & Kerster H.W. (1974) Gene flow in seed plants. *Evolutionary Biology*, **7**, 139–220.

Levin D.A. & Kerster H.W. (1975) The effect of gene dispersal on the dynamics and statics of gene substitution in plants. *Heredity*, **35**, 317–336.

Levin D.A., Kerster H.W. & Niedzlek M. (1971) Pollinator flight directionality and its effect on pollen flow. *Evolution*, **25**, 113–118.

Levin D.A. & Wilson J.B. (1978) The genetic implications of ecological adaptations in plants. *Structure and Functioning of Plant Populations* (Ed. by A.H.J. Freysen & J.W. Woldendorp), pp. 75–100. North-Holland, Amsterdam.

Lloyd D.G. (1979) Parental strategies of angiosperms. *New Zealand Journal of Botany*, **17**, 595–606.

Lloyd D.G. (1980) Demographic factors and mating patterns in Angiosperms. *Demography and Evolution in Plant Populations* (Ed. by O.T. Solbrig), pp. 67–88. Blackwell Scientific Publications, Oxford.

Mackay I.J. (1980) *Population Genetics of Papaver dubium*. Unpublished PhD. thesis, University of Birmingham.

McNeilly T. (1968) Evolution in closely adjacent plant populations. III. *Agrostis tenuis* on a small copper mine. *Heredity*, **23**, 99–108.

Merrell D.J. (1968) A comparison of the estimated size and the 'effective size' of breeding populations of the leopard frog *Rana pipiens*. *Evolution*, **22**, 274–283.

Mulcahy D.L. & Caporello D. (1970) Pollen flow within a tristylous species: *Lythrum salicaria*. *American Journal of Botany*, **57**, 1027–1030.

Nei M. & Murata M. (1966) Effective population size when fertility is inherited. *Genetical Research*, **8**, 257–260.

Nei M. & Tajima F. (1981) Genetic drift and estimation of effective population size. *Genetics*, **98**, 625–640.

Petras M.L. (1967) Studies on natural populations of *Mus*. I. Biochemical polymorphisms and their bearing on breeding structure. *Evolution*, **21**, 259–274.

Powell J.R., Dobzhansky Th., Hook J.E. & Wistrand H.E. (1976) Genetics of natural populations. XLIII. Further studies on rates of dispersal of *Drosophila pseudoobscura* and its relatives. *Genetics*, **82**, 493–506.

Primack R.B. & Silander J.A. Jr (1975) Measuring the relative importance of different pollinators to plants. *Nature*, **255**, 143–144.

Pyke G.H. (1978) Optimal foraging: movement patterns of bumblebees between inflorescences. *Theoretical Population Biology*, **13**, 72–98.

Redfern M. (1975) The life history and morphology of the early stages of the yew gall midge *Taxomyia taxi* (Inchbald) (Diptera: Cecidomyiidae). *Journal of Natural History*, **9**, 513–533.

Richards A.J. & Ibrahim H. (1978) Estimation of neighbourhood size in two populations of *Primula veris*. *The Pollination of Flowers by Insects* (Ed. by A.J. Richards), pp. 165–174. Symposia of the Linnean Society of London, vol. 6. Academic Press, London.

Richardson R.H. (1970) Models and analyses of dispersal patterns. *Mathematical Topics in Population Genetics* (Ed. by K. Kojima), pp. 79–103. Springer, Berlin.

Ritland K. & Jain S. (1981) A model for the estimation of outcrossing rate and gene frequencies using *n* independent loci. *Heredity*, **47**, 35–52.

Rohlf F.J. & Schnell G.D. (1971) An investigation of the isolation-by-distance model. *American Naturalist*, **105**, 295–324.

Ryman N., Baccus R., Reuterwall C. & Smith M.H. (1981) Effective size, generation interval, and potential loss of genetic variability in game species under different hunting regimes. *Oikos*, **36**, 257–266.

Schaal B.A. (1980) Measurement of gene flow in *Lupinus texensis*. *Nature*, **284**, 450–451.

Schaal B.A. & Levin D.A. (1978) Morphological differentiation and neighborhood size in *Liatris cylindracea*. *American Journal of Botany*, **65**, 923–928.

Schmitt J. (1980) Pollinator foraging behavior and gene dispersal in *Senecio* (Compositae). *Evolution*, **34**, 934–943.

Thomson J.D. & Plowright R.C. (1980) Pollen carry over, nectar rewards, and pollinator behaviour with special reference to *Diervilla lonicera*. *Oecologia*, **46**, 68–74.

Tinkle D.W. (1965) Population structure and effective size of a lizard population. *Evolution*, **19**, 569–573.

Vasek F.C. (1968) Outcrossing in natural populations: a comparison of outcrossing estimation methods. *Evolution and Environments* (Ed. by E.T. Drake), pp. 369–385. Yale University Press, New Haven.

Wade M.J. (1980) Effective population size: the effects of sex, genotype, and density on the mean and variance of offspring numbers in the flour beetle, *Tribolium castaneum*. *Genetical Research*, **36**, 1–10.

Wells H. & Wells P.H. (1980) Are geographic populations equivalent to genetic populations in biennial species? A study using *Verbascum virgatum* (Scrophulariaceae). *Genetical Research*, **36**, 17–28.

Woodell S.R.J. (1978) Directionality in bumblebees in relation to environmental factors.

The Pollination of Flowers by Insects (Ed. by A.J. Richards), pp. 31–39. Symposia of the Linnean Society of London, vol. 6. Academic Press, London.

Wright J.W. (1953) Pollen-dispersion studies: some practical applications. *Journal of Forestry*, **51**, 114–118.

Wright S. (1931) Evolution in Mendelian populations. *Genetics*, **16**, 97–159.

Wright S. (1938) Size of population and breeding structure in relation to evolution. *Science*, **87**, 430–431.

Wright S. (1943a) Isolation by distance. *Genetics*, **28**, 114–138.

Wright S. (1943b) An analysis of local variability of flower color in *Linanthus parryae*. *Genetics*, **28**, 139 156.

Wright S. (1946) Isolation by distance under diverse systems of mating. *Genetics*, **31**, 39–59.

Wright S. (1969) *Evolution and the Genetics of Populations. Vol. 2. The Theory of Gene Frequencies*. University of Chicago Press, Chicago.

Wright S. (1977) *Evolution and the Genetics of Populations. Vol.* 3. *Experimental Results and Evolutionary Deductions*. University of Chicago Press, Chicago.

Wright S. (1978) *Evolution and the Genetics of Populations. Vol.* 4. *Variability within and among Natural Populations*. University of Chicago Press, Chicago.

Zimmerman M. (1982) The effect of nectar production on neighborhood size. *Oecologia* **52**, 104–108.

8. DENSITY AND INDIVIDUAL FITNESS: ASYMMETRIC COMPETITION

MICHAEL BEGON
Department of Zoology, University of Liverpool, Liverpool L69 3BX

INTRODUCTION

Evolutionary ecology has emerged as a maturing discipline as a result, at least in part, of the increasingly overt recognition by ecologists of genetics. Ecologists have therefore had to recognize (as geneticists do by definition) that individuals, even within a population, differ from one another in their genotypes, and in the phenotypic manifestations of these genotypes. It would seem to follow from this that evolutionary ecologists should devote attention not only to effects of different environments on the average phenotype in a population, but also to their effects on the range and distribution of phenotypes. In this chapter I shall examine the effects of a single environmental factor on the ranges and distributions of various fitness components within populations. This 'environmental factor' is an extremely broad one: the population density of the species being studied and the intraspecific competition thereby induced. I take this to be the single most important aspect of the environment, because it will modify an individual's response to most other aspects, and because it is one of the primary driving forces of evolution (individual fitness being most sensibly defined relative to other, potentially competing, individuals in the population) (Wallace 1970; and Christiansen in this volume).

The chapter is set out in the following way. First, the historical and prevailing attitudes of ecologists to changes in phenotypic distributions with density will be discussed. Then, a frequently found, shared pattern of change with density will be established. This will be followed by an examination of the likely mechanism generating the pattern, which will then be incorporated into a more general description of the underlying phenomenon. Finally, some of the consequences and implications of the phenomenon will be explored.

HISTORICAL BACKGROUND

Plant ecologists have undoubtedly paid more attention than their zoological counterparts to the effects of density (or crowding) on individual differences within populations. The subject is, and probably always has been of particular interest to horticulturalists, because they are especially concerned with the proportion of their population which grows large enough to exceed a harvestable threshold; and since the pioneering studies of Japanese workers in the 1950s (cf. Koyama & Kira 1956), there has been a continuing interest in

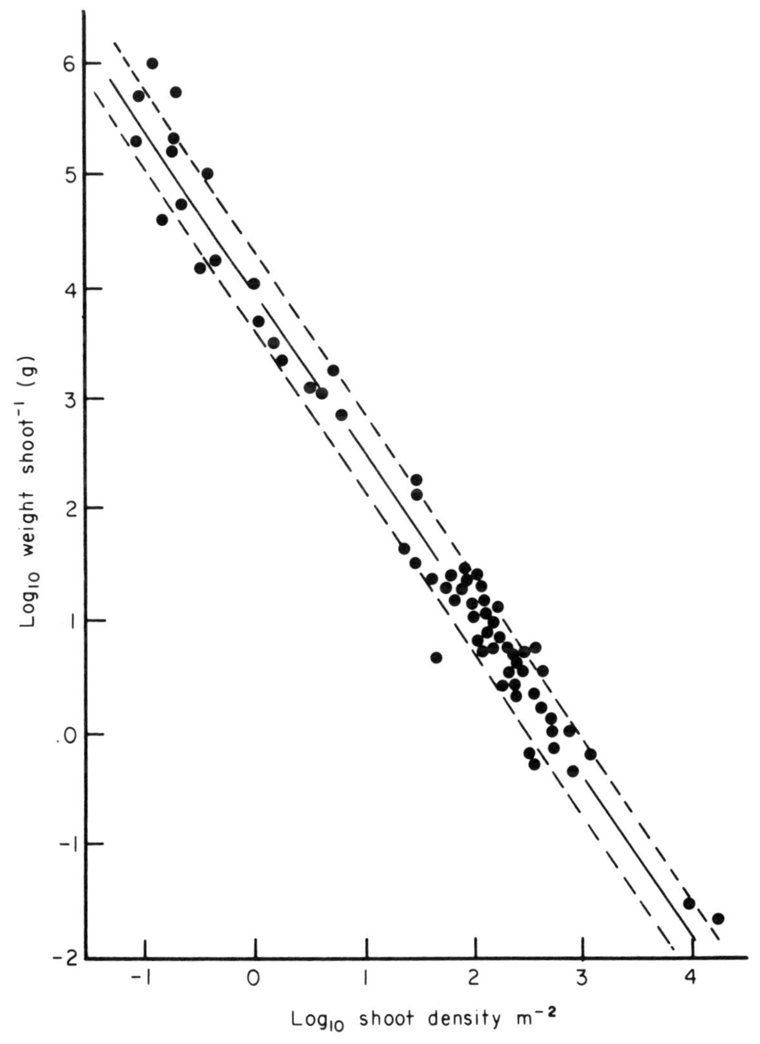

FIG. 8.1. The relationship between mean shoot dry weight and shoot density in sixty-five single-species populations (twenty-nine different species), ranging from non-woody annuals and perennials to saplings and full-sized trees' Dashed lines represent plants two and half times as large as those described by the equation $W = 9670d^{-1 \cdot 49}$. (From Harper 1981, after Gorham 1979.)

the subject amongst population ecologists working on plants (e.g. Ford & Newbould 1970, 1971; Ford 1975; Harper 1967, 1977; Rabinowitz 1979; and the theoretical work of Aikman & Watkinson 1980).

In comparison, animal ecologists have examined the effects of density on individual differences only sporadically. The work has concentrated on fish (see especially Rubenstein 1981) where attention has focused on social dominance hierarchies (Magnuson 1962; Yamagishi, Maruyama & Mashiko 1974; Li & Brockson 1977), and on amphibian larvae (Rose 1960; Wilbur & Collins 1973; Wilbur 1976, 1977; Steinwascher 1978). Łomnicki (1978, 1980) has examined the topic from a theoretical standpoint.

However, in spite of these studies, it appears in general that ecologists—even plant ecologists—have tended not to study individual variation within populations. It is, of course, difficult to provide evidence of what ecologists have not done. Yet a representative text on theoretical ecology (May 1981a) contains no models investigating the importance of individual differences; a typical study of density effects (Fig. 8.1) limits itself to average, populational traits (like mean weight); while a typical study of life-history characteristics (Table 8.1) compares two species in terms of averages. It is true, no doubt, that much recent work has recognized other sorts of heterogeneity within populations, especially spatial heterogeneity. Overall, however, the typical attitude of ecologists to individual variation appears to have been indiffer-

TABLE 8.1. A study of life-history characteristics in two *Typha* species, which are compared in terms of 'average' traits and matched against the ecosystems in which they grow: they fit the standard r/K dichotomy. (From Begon & Mortimer 1981, after McNaughton 1975).

	T. angustifolia	*T. domingensis*
Plant trait:		
Days before flowering	44	70
Mean foliage height (cm)	162	186
Mean genet weight (g)	12.64	14.34
Mean number of rhizomes per genet	3.14	1.17
Mean weight of rhizomes (g)	4.02	12.41
Mean number of fruits per genet	41	8
Mean weight of fruit (g)	11.8	21.4
Mean total weight of fruits (g)	483	171
	Growing season	
	Short	Long
Ecosystem property:		
Climatic variability ($s^2/\bar{x}$ frost-free days)	3.05	1.56
Competition (biomass above ground [g m^{-2}])	404	1336
Annual recolonization (winter rhizome mortality [%])	74	5
Annual density variation ($s^2/\bar{x}$ shoot numbers m^{-2})	2.75	1.51

ence. One must presume either that many ecologists have not even considered such variation, or, more probably, that they have implicitly assumed it to be simple and normally distributed, and therefore validly characterized by a mean. It is the thesis of this chapter that such an attitude is untenable in view of the evidence, and that it may stand in the way of our increased understanding of population processes.

SKEWED DISTRIBUTIONS

A typical and classic, though admittedly exemplary, study of the effects of crowding on the distribution of plant sizes within a population is shown in Fig. 8.2 (Obeid, Machin & Harper 1967). Crowding and intraspecific competition increase as the sowing density of flax increases (from left to right in Fig. 8.2) and as time progresses and the flax plants grow larger (from top to bottom in Fig. 8.2). It is clear that an initially normal distribution of plant weights in the top-left of the figure becomes an increasingly right-skewed distribution towards the bottom-right. There is, therefore, a tendency, which increases with intraspecific competition, for such populations to contain a few very large (and presumably very fit) individuals, and many small unfit individuals. In Harper's (1977) words, these populations become dominated, numerically, by 'suppressed weaklings'. Such changing distributions are, of course, superimposed on an underlying tendency for average plant weight to decline with density but increase over time (Fig. 8.2).

A zoological example expressing a similar pattern is shown in Fig. 8.3 (Wilbur & Collins 1973). As density increases, the average weight of frog larvae declines (asymptotically); but the high density populations are again skewed, with many small and a few relatively large tadpoles.

If this phenomenon is at all common, it is likely that many workers have observed it without recognizing it as such, and a probable example is shown in Fig. 8.4 and Table 8.2. Figure 8.4 (Watson & Miller 1971) contains maps of red grouse territories in an area of Scottish moorland for four successive years: 1958–61. Territories are established in August and September, and held until re-establishment of territories the following year. There is a high rate of mortality among non-territorial birds, especially in late winter, and hatching of chicks occurs in May and June. Watson & Miller (1971) provide evidence for believing that territory size is a component of red grouse fitness (much as body size was for the flax plants and frog larvae), and, with this in mind, Table 8.2 (Jenkins, Watson & Miller 1963; Watson & Miller 1971) explores the effect of intraspecific competition on the distribution of territory size within the population. The level of intraspecific competition has been assessed in two ways: as survivorship (the numbers alive in and around the

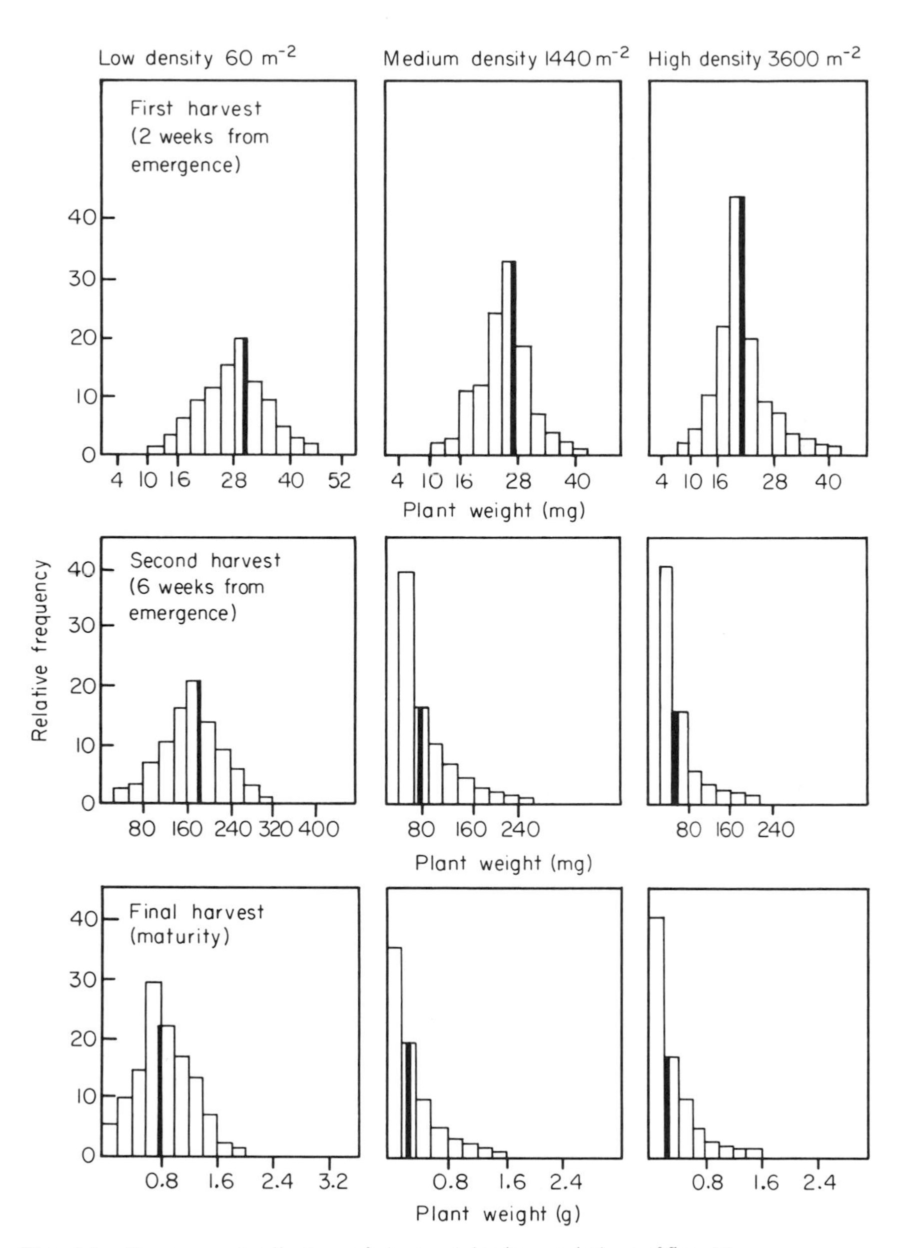

FIG. 8.2. Frequency distributions of plant weights in populations of flax, *Linum usitatissimum* sown at three densities. (Data from Obeid, Machin & Harper 1967, after Begon & Mortimer 1981.) Position of the means is indicated by the black columns.

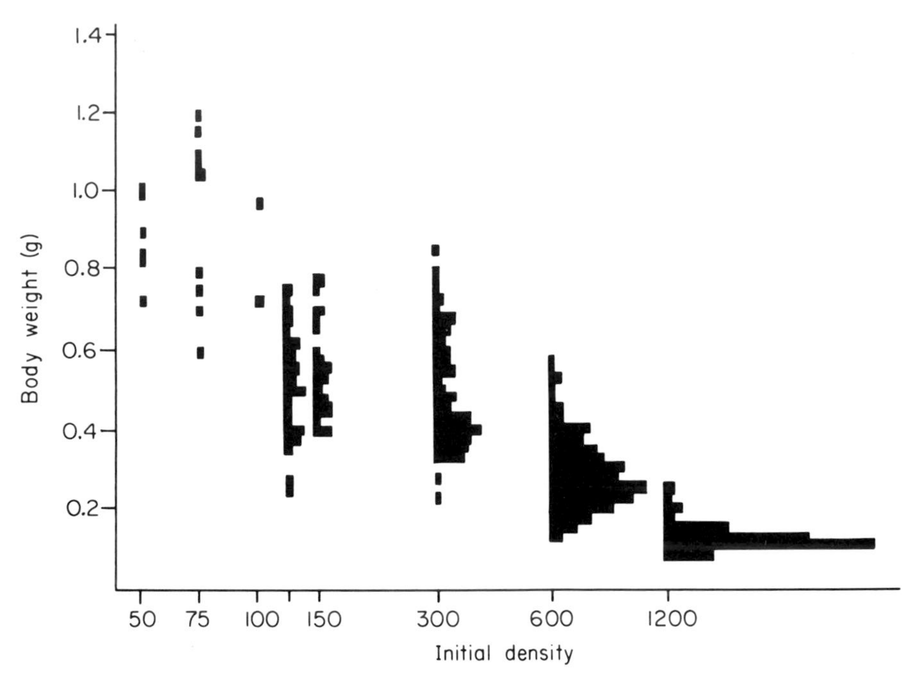

FIG. 8.3. Frequency histograms of body weights of *Rana sylvatica* larvae after 50 days growth at various initial densities. (From Wilbur & Collins 1973.)

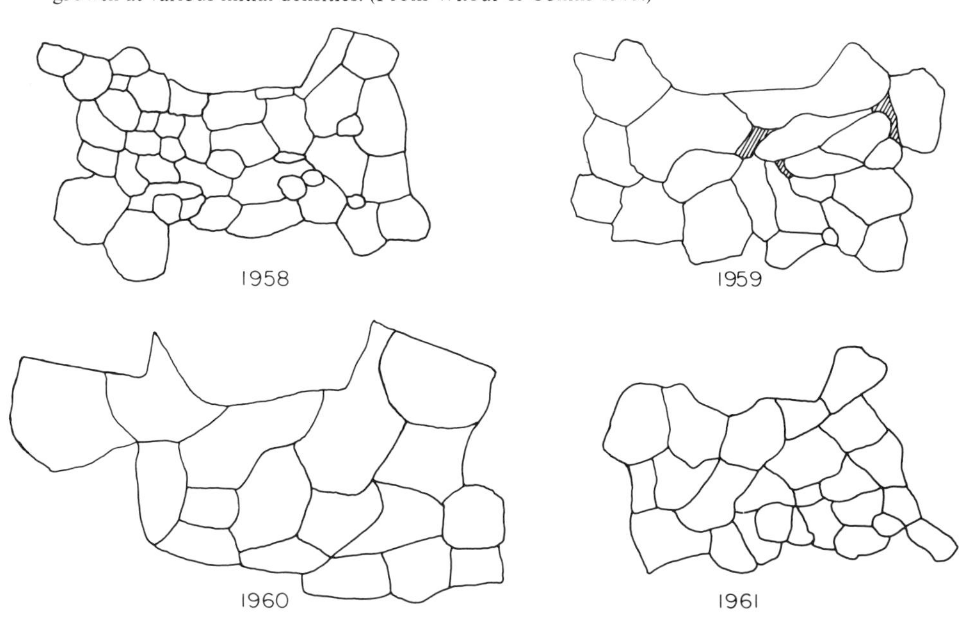

FIG. 8.4. Grouse territory maps for an area of moorland at Glen Esk, Scotland for the years 1958–1961. (From Watson & Miller 1971.) ▨, Unoccupied ground.

TABLE 8.2. Influence on grouse territory size at Glen Esk, Scotland (after Jenkins *et al.* 1963 Watson & Miller 1971)

Season	Mean territory size (ha)	Skewness in territory size-distribution	Survivorship (April nos./ September nos.)	Fecundity in August (nos. young/ nos. old)
1958–59	2.48	1.07	0.55*	0.46
1957–58	1.18	0.98	0.72	0.93
1960–61	1.42	0.62	0.85	1.93
1959–60	3.92	0.57	1.04	2.46

*Overestimate (based on October population size)

study area in April as a proportion of those alive the previous September) and as fecundity (the number of recently hatched young birds per old bird in August). It seems reasonable to expect both survivorship and fecundity to decline with increasing intraspecific competition. It is, therefore, interesting to note from Table 8.2 that, whereas neither survivorship nor fecundity show any relationship with mean territory size, both are perfectly negatively correlated with the (right-handed) skewness of the territory-size distributions. Once again, the skewness in the distribution of a fitness component within the population is seen to increase with intraspecific competition. (Mean territory size is probably more closely related to the quality and quantity of the food resource within the study area.)

It seems then, from these and other examples (e.g. Koyama & Kira 1956; Rose 1960), that there is a common pattern of distributions. Despite a general tendency to ignore individual differences therefore, they do exist; and distributions of certain fitness components, rather than being normal, become increasingly skewed as the intensity of intraspecific competition increases.

MECHANISM

Workers who have uncovered such skewed distributions appear to be in good agreement, in general terms, as regards an underlying mechanism (e.g. Harper 1967; Wilbur & Collins 1973). Taking the most common case of size in a cohort of growing individuals as the best example, this mechanism can be divided into two main components, plus a possible third.

The first component is the assumption that there is some, probably small amount of inherent variation present in all populations at birth or germination. This could be variation in size itself. Alternatively, it could be variation in time of emergence, which, by the time all individuals in a cohort have emerged, will translate itself into size variation, since some individuals will have been growing for longer than others. In either case, such variation may be genetic; but it may equally well be environmental and fortuitous: unpredictable variation in a seed's micro-site, for instance. The initial variation might also be in relative growth rate (RGR), that is, the rate of growth per unit size (g day^{-1} g^{-1}, for instance). This, too, would soon translate itself into size-variation.

In any case, the second, most important component of the suggested mechanism (Fig. 8.5a) is that competition induces or exaggerates variation in RGR in the following manner. Large individuals, by virtue of their size, are little affected by competition even at relatively high densities. Not only,

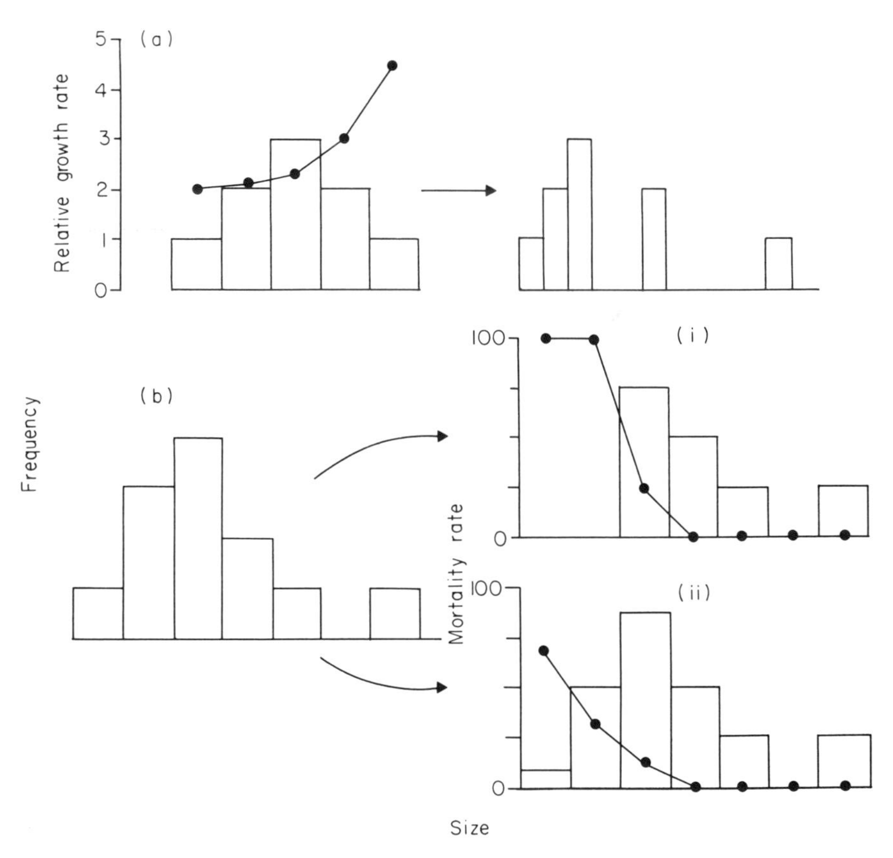

FIG. 8.5. Stylized representation of a mechanism by which skewed distributions of weights can be generated in a population. (a) Increases in relative growth rate with size convert a symmetrical frequency-distribution into a skewed one. (b) Increases in survivorship with size can either exaggerate this skewness (i), or counteract it (ii).

therefore, do they have high absolute growth rates (because they are large): they also have high relative growth rates. Small individuals, on the other hand, are much affected by competition, and have growth rates which are not only low in absolute terms, but even lower in relative terms. There is asymmetric competition or, in Harper's (1967) words, 'a hierarchy of exploitation'. The discrepancies in RGR between large and small individuals, therefore, lead to even greater discrepancies in size, which lead to greater discrepancies still in RGR, and so on. An increase in RGR with size thus generates the sort of skewed distribution seen in the previous section (Fig. 8.5a). The degree of skewness is determined by the extent of the size-dependent discrepancy in RGR, which itself increases with the level of competition.

A possible third component of the mechanism (Fig. 8.5b) is that the smallest individuals also have the lowest rate of survivorship. Depending on the exact nature of the survivorship–probability function, this could exaggerate the skewness (Fig. 8.5bi) or counteract it (Fig. 8.5bii).

The idea that skewed distributions can be generated by size-dependent variations in RGR mediated by competition is supported (Fig. 8.6) by the data of Ford (1975). Further support for this mechanism is provided by the fact that a simulation model incorporating a version of it (Aikman & Watkinson 1980) shows a pleasing correspondence with some more of Ford's (1975) field data (Fig. 8.7).

In addition, the mechanism elicits four important comments. The first is that the present account does not seek to provide a blanket explanation for

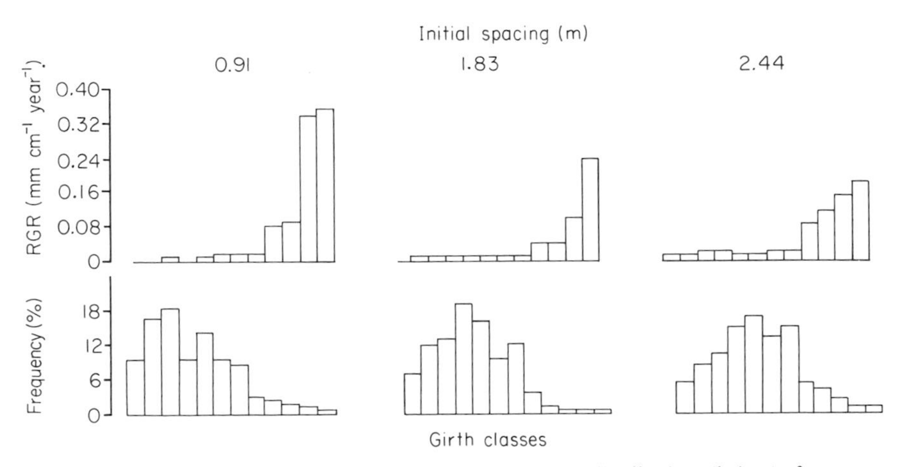

FIG. 8.6. Mean relative growth rates (above) and frequency distributions (below) of *Picea sitchensis* classified into twelve equal-interval girth classes, having been planted at three densities 29 years previously at Gwydyr, Wales. (From Ford 1975.)

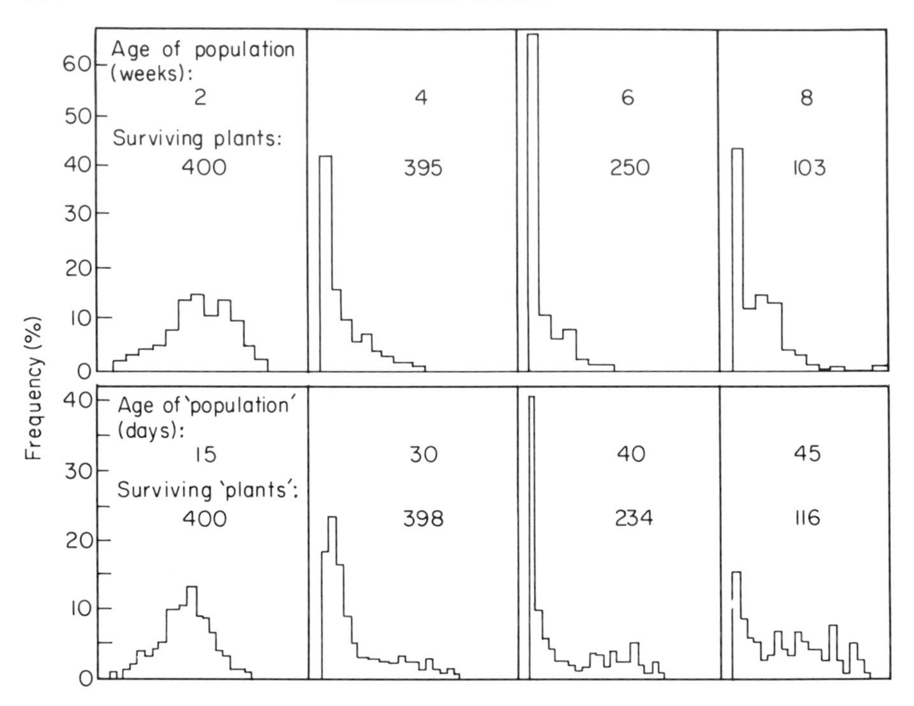

FIG. 8.7. Plant weight distributions in a population after growth and mortality for various lengths of time. Above: data from a population of *Tagetes patula* (Ford 1975). Below: results of a simulation with model parameters chosen appropriately (Aikman & Watkinson 1980).

all skewed distributions. It seeks rather to draw attention to the potential role of competition in generating increases in skewness (but see the following section). It also seeks to examine the important consequences of such density-dependent increases.

The second point is that the large, fit individuals consistently behave (grow) in a manner which would be appropriate to a much lower density if the interaction was considered in terms of averages. To this extent, such individuals perceive density as being much lower than it is; and, in a comparable way, the smallest individuals perceive density as being much higher than it is. Thus, while ecologists, especially animal ecologists, usually choose to characterize a population by a single density, the individual organisms may perceive density differentially; and in populations where the distribution of fitness is strongly skewed, a large proportion of the population may behave in a way which is quite inappropriate to the abstracted average density.

The third point is that the phenomenon, as described, will be most significant in populations which are even-aged, or approximately so—as in the case of any population with non-overlapping generations. Where there is continuous breeding or considerable overlap of generations, there will also be important asymmetries in competition; but most of these are likely to be age-dependent, as is the case with 'space-capture' in perennial plants (cf. Begon & Mortimer 1981). Nevertheless, the consequence will still often be a skewed distribution, with a few older, fitter, established individuals, and may younger individuals doomed never to become established.

ASYMMETRIC COMPETITION

The fourth point is that the skewed distributions may be seen, in the light of the suggested mechanism, as a specific manifestation of the following more general phenomenon. Those individuals that are inherently or fortuitously fittest prior to the action of intraspecific competition are least affected by it; while those that are least fit are most affected. Such a description can, and in the previous section has been framed in terms of size (as a measure of fitness) and RGR (as a measure of preceived competition and a means of achieving greater fitness); but the description can also be applied to other situations where skewed distributions are not necessarily generated. Table 8.3, for instance, (Wall, unpublished) summarizes data on egg-to-adult developmental times in the grasshopper *Chorthippus brunneus*, reared in the laboratory at three densities. At the lowest density, the difference between the fittest (fastest) individuals and the least fit was not great (Table 8.3, column 2). At higher densities, however, this difference progressively increased. (Fitness and developmental rate can be equated, because, if anything, the faster individuals are also larger.) The increasing difference is made apparent by the '(more dense)/(less dense)' developmental time ratios in columns 5, 6 and 7 of Table 8.3: with only one exception, the ratios increase down the columns from the fastest to the slowest members of the populations.

TABLE 8.3. Rate of development of grasshoppers (*Chorthippus brunneus*) at different densities (Wall, unpublished)

	Day on which $x\%$ are adult					
Percentage (x)	High density	Medium density	Low density	High/ low	High/ medium	Medium/ low
20	24.5	25.0	20.0	1.23	0.98	1.25
40	26.75	26.75	21.5	1.24	1.00	1.24
60	29.0	28.25	22.5	1.29	1.03	1.26
80	35.0	30.5	23.5	1.49	1.15	1.30
90	39.0	32.0	24.0	1.63	1.22	1.33

Thus, while there were no skewed distributions, the underlying process was clearly the same: asymmetric intraspecific competition, in which the fitnesses of the fittest (fastest) individuals were affected far less than those of the least fit. Moreover, the pattern generated by this process in the grasshoppers can be described in a way which is equally appropriate to the skewed distributions discussed previously: at low levels of competition, individual fitnesses are distributed symmetrically over a narrow range; at high levels of competition, the fittest individuals still lie close to or within this same range, but other individuals have fitnesses which are very much lower.

The important underlying phenomenon, therefore, appears to be asymmetric intraspecific competition and the increased and asymmetric variation in individual fitness that it induces. This has already been demonstrated for size and territory size (where skewed distributions were generated), and for developmental rate; and it has also been demonstrated by, amongst others, Rubenstein (1981) for growth rate, egg size and number, and male reproductive activity in the Everglades pygmy sunfish, and by Sohn (1977) for maturation in the platyfish. Asymmetric competition and the patterns it generates are apparently widespread and not uncommon.

CONSEQUENCES AND IMPLICATIONS

However, even if there are important patterns of individual variation within populations which ecologists have tended to neglect, it remains to the established whether or not these patterns have consequences which are sufficiently important for the neglect to be regrettable. A number of consequences in a variety of fields are therefore presented here. There are undoubtedly others (cf. Rubenstein 1981). The present selection, however, should be sufficient to establish the importance of the patterns described in previous sections.

Predator–prey dynamics

If a population of prey exhibited a skewed distribution of size, it might, as a consequence, consist of a large number of weak individuals susceptible to predation, plus a very small number of very fit individuals almost certain to escape. Many mobile prey species living in full view of their predators are likely to correspond to this discription. In such a case, even heavy predation would have little influence on the total productivity of the prey population, since productivity would be largely determined by the few, fittest individuals. Skewed distributions would then provide an important reason (there are others: cf. Hassell 1978; Begon & Mortimer 1981) why a predator and prey species, even when they are closely linked trophically, might nevertheless have only a limited influence on one another's population dynamics.

Life-history strategies

In the study of life-history strategies, attention is usually focused on average differences between species or (more rarely) populations (Stearns 1976, 1977). The previous sections might suggest, however, that here, as elsewhere, a proper understanding of the subject can only be achieved by accepting and studying differences between individuals. The following example appears to support this suggestion.

Figure 8.8 (Moeur & Istock 1980) shows the responses of pitcher plant

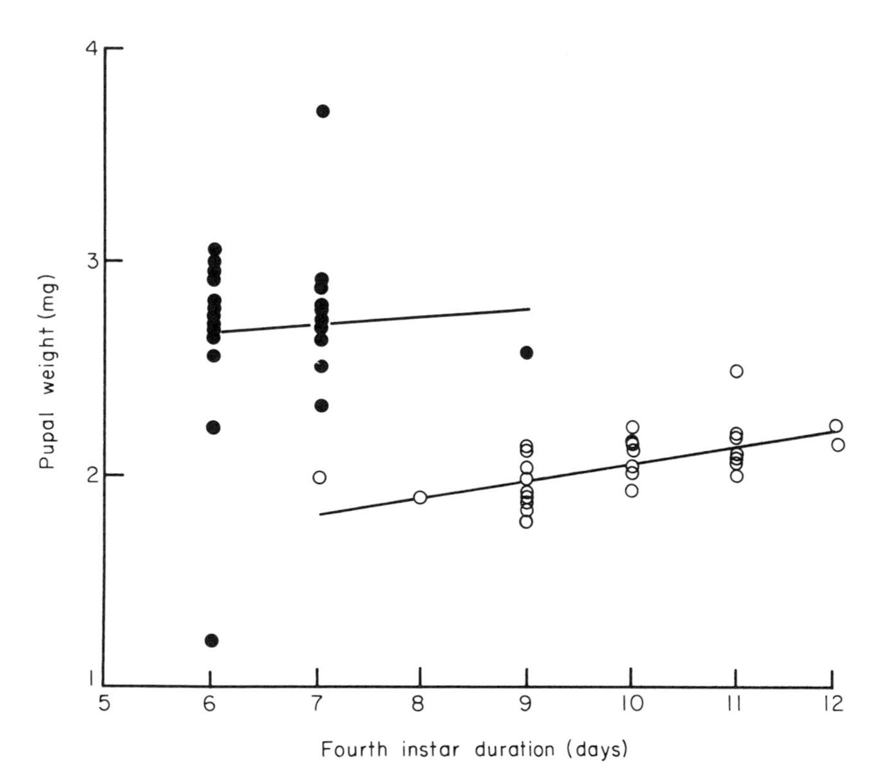

FIG. 8.8. Relationship between duration of the fourth larval instar and weight at pupation in female *Wyeomyia smithii*. Individuals fed a high ration during the third and fourth instars are shown as filled circles ($y = 0.042x + 2.42$; $r^2 = 0.005$; $P > 0.5$). Individuals fed a low ration are shown as open circles ($y = 0.083x + 1.23$; $r^2 = 0.39$; $P < 0.001$). (From Moeur & Istock 1980.)

mosquitoes to two regimes: high and low food ration (only partly analogous to low and high levels of competition since individuals were reared separately). These responses were measured in terms of two important components of fitness (Moeur & Istock 1980): fourth instar duration (the major determinant of age at first reproduction) and pupal weight (an important determinant of lifetime fertility). A comparison of averages would indicate that the high ration individuals reproduce earlier (traditionally an 'r'-character) and are larger (traditionally a 'K'-character). But such a comparison, and such dogmatic conclusions, are singularly uniformative: they tell us only that one 'population' had more resources at its disposal than the other. Far more informative is a comparison of individuals within the low ration population. There the regression of pupal weight on fourth instar duration is significant, and suggests a life-history trade-off at the individual level. Individuals that reproduce early are small: large individuals reproduce late. Food shortage has, therefore, uncovered important phenotypic differences; and these are only apparent when attention is diverted away from averages and towards individuals.

Effective population size

The concept and importance of effective population size has been discussed in detail by Crawford (in this volume). In the present context, however, it is interesting to consider the consequences for effective population size of asymmetric competition. A skewed distribution of fitness leads to a non-random distribution of family sizes. A few individuals may, therefore, make a highly disproportionate contribution to the next generation; and this is well known as a potent force for the reduction of effective population size (cf. Crawford in this volume). But if the distribution becomes increasingly skewed with density, then the reducing effect will rise as density itself rises. It is quite likely, therefore, that effective population size will frequently be insensitive to apparent changes in actual density. Indeed, contrary to expectations, and as a consequence of the skewed patterns of previous sections, effective population size may actually decline as apparent density increases.

The stability and regulation of populations

The stability of single-species populations, and the regulation of their size, are subjects to which a great deal of attention has been devoted, especially by theoreticians (cf. May 1981b for a review). In the present context, however, the effect of the patterns of previous sections on population stability may be deduced without recourse to mathematical models, since these would simply formalize the deductive process. The patterns, and the asymmetric competition from which they stem, are essentially restatements of the fact that there are commonly, in relative terms, 'winners' and 'losers' within a competing population. Indeed, the patterns suggest that as density increases the proportion of quasi-losers rises while the proportion of quasi-winners falls. Thus, these populations increasingly approach a situation of 'pure contest' (Nicholson 1954; Begon & Mortimer 1981). It is well known that such situations (in which changes in reproductive-rate compensate perfectly for changes in density) represent the apogee of population stability: populations approach a stable equilibrium in the most direct and most rapid way. It is, therefore, reasonable to conclude that asymmetries in intraspecific competition will be a force, and perhaps a potent force, in the stabilization and regulation of populations.

In a different context, De Jong (1979; cf. Hassel 1980) has derived a mathematical model which may be used to reinforce this conclusion. She imagined a population 'laying eggs' in a patchy environment, such that different patches had different numbers of eggs and subsequently different intensities of intraspecific competition (depending simply on intrapatch density). If 'different-density patches' are replaced by 'classes within the population within which density is perceived differently', then this model becomes applicable in the present context. De Jong considered (amongst others) the special case in which the distribution of eggs followed the negative binomial, a distribution defined by two parameters: the mean, and the degree of contagion (aggregation). Increased contagion leads to a greater proportion of the population being confined to fewer patches and thus suffering more intense competition. In the present case, therefore, increased contagion can be equated with increased skewness; and De Jong's pertinent conclusion is that increases in the level of aggregation lead to increases in the resemblance to 'pure contest' and thus to increases in stability.

Finally, in support of the same theme, Łomnicki (1978, 1980) has suggested that unequal partitioning of resources is a necessary condition for the stability of single-species populations. Asymmetric competition inevitably leads to unequal partitioning of resources.

Cannibalism

For obvious reasons, cannibalism is one particular process with considerable potential for regulating populations. Its frequency throughout the animal kingdom, however, is commonly underrated (Polis 1981). Cannibalism is very often density-dependent (references in Polis 1981), and it very often requires a minimum size-ratio to be exceeded before the larger animal eats the smaller (references in Polis 1981). In addition, the previous sections suggests that we may often be able to link these two elements by a density-dependent increase in the disparity between the largest and smallest members of a a population. This would then add to the regulatory powers of cannibalism, and make its neglect all the more regrettable.

Intrapopulational arms races

Finally, in considering consequences, we turn to one of the 'arms races' studied by sociobiologists: that which is presumed to occur within a population, over evolutionary time. Here the costs of investing in 'personal armaments' are balanced against the benefits acruing to (resources gained by) heavily armed individuals. (For example, at its simplest, the high costs to an individual of investing energy in growing large may be offset against the large benefits gained by winning much of a resource; the low costs of small

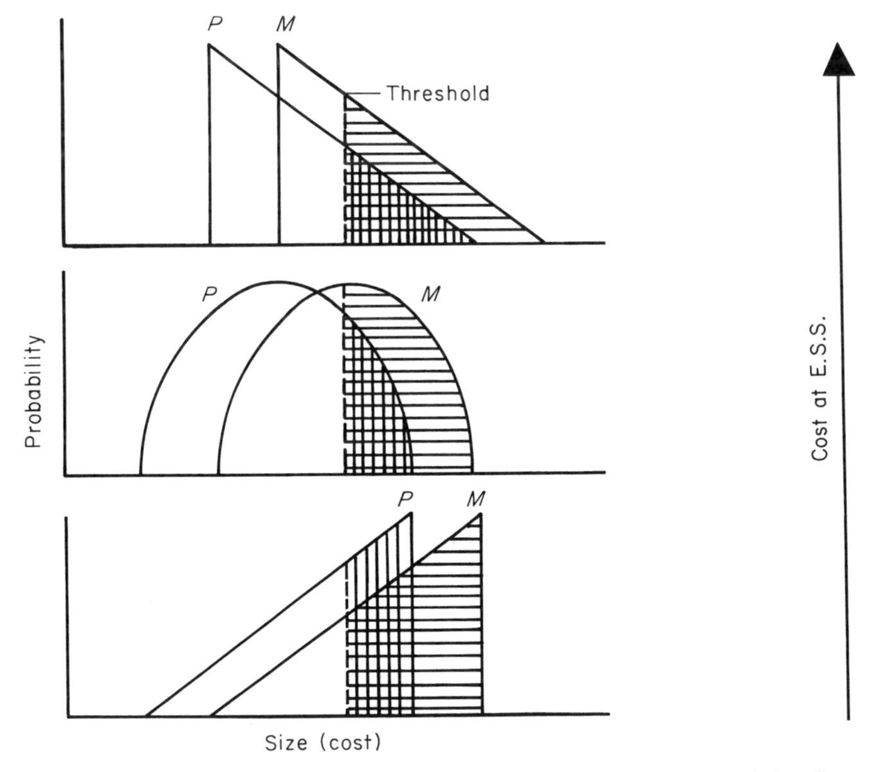

FIG. 8.9. Figurative outline of a model for predicting the outcome of intrapopulational, evolutionary arms races (after Parker 1982). For further explanation, see text. ⊟, ▥, Mutaut's increase in benefit.

size, on the other hand, are matched by the small benefits of winning only rarely.) Early models of such arms races (Parker 1979) were unable to predict any evolutionarily stable strategy, that is, they were unable to predict a size or level of armament which could never be bettered adaptively—whether this be by the decreased costs of small size or the increased benefits of large. Parker (1982), however, has recently produced a new model directed at this problem, and the essentials of the model are shown in Fig. 8.9.

Taking the upper graph in Fig. 8.9, the '*P*' curve is a probability distribution appropriate to all typical, genetically identical individuals in a population: it is the probability, given that genotype, that an individual will have the phenotype (size, level of armament) indicated. Moreover, by its very nature, the same curve is a frequency-distribution appropriate to the population of typicals as a whole. In the case of the upper graph in Fig. 8.9 each individual has a high probability of being small and a small probability of being large; and there will, therefore, be many small individuals and a few large ones. The typical cost to individuals is given by the location of the curve on the 'size' axis. The expected benefit is based on the assumption that some threshold size (Fig. 8.9) needs to be exceeded before resources are acquired. The benefit is given by the probability of exceeding that threshold, that is, the area under the curve to the right of the threshold.

The '*M*' curve is the equivalent probability distribution for a mutant individual. It is the same shape as the '*P*' curve, but is displaced to the right (greater size). Cost is therefore increased by the extent of the displacement, and benefit increased by the extent of the increase in area beyond the threshold. Parker's (1982) model then seeks to determine the evolutionarily stable location of the '*P*' curve on the axis, which no mutant can better (whether by decreased costs or increased benefits). There are two aspects of his results that are particularly pertinent in the present context.

The first is that such an evolutionarily stable location can indeed be found, whereas previously it could not: the difference, of course, is that the new model acknowledges and incorporates individual phenotypic variation within the population. The second point (Fig. 8.9) is that the exact nature of the stable strategy is crucially dependent on the shape of the distribution of individuals. The existence of differences, and the form they take, are both important.

CONCLUSIONS

It seems clear, in short, that in a wide range of fields there are significant consequences stemming from the existence of asymmetric intraspecific competition, and from the dynamic patterns of phenotypic variation to which it leads. Conclusions can therefore be drawn which echo the sentiments of Harper (1981); (and the strength of these conclusions is reinforced by the fact that Harper was concerned with a different, though not independent subject: the construction of a population biology appropriate to modular organisms). Developments are needed in evolutionary population biology which take account of the behaviour and characteristics of individuals. This is particularly true because the ubiquitous concept of density disguises the real level at which organisms interact. And if we spend our time, in essence, weighing populations and dividing by the number present, then what we are doing is hiding what must be one of the most interesting and important aspects of the natural world: individual variation.

Finally, the approach of this paper may be contrasted with a more 'adaptationist' approach, in which an explanation for population variation is sought in terms of adaptation to a broad or heterogeneous or unpredictable niche, or in terms of a balance between stabilizing selection and destablizing forces like immigration and mutation (cf. Grant & Price 1981 for a review). The two approaches—'variation as a constructive, adaptive evolutionary response' and 'variation as consequence of a differential response to the environment' are not, of course, contradictory: both processes can apply simultaneously. From a methodological point of view, however, it seems that workers taking the adaptationist approach should first appreciate the variation generated in their populations by more immediate responses. This, then, may be the reason why, in the application of the adaptationist approach to population variation, '... empirical studies lag far behind theory' (Grant & Price 1981).

ACKNOWLEDGEMENTS

Amongst many valuable comments on this paper, and the symposium presentation on which it is based, I should like to express my particular gratitude for those of Professor M.P. Hassell and Dr J. White.

REFERENCES

Aikman D.P. & Watkinson A.R. (1980) A model for growth and self-thinning in even-aged monocultures of plants. *Annals of Botany*, **45**, 419–427.

Begon M. & Mortimer A.M. (1981) *Population Ecology: A Unified Study of Animals and Plants.* Blackwell Scientific Publications, Oxford.

De Jong G. (1979) The influence of the distribution of juveniles over patches of food on the dynamics of a population. *Netherlands Journal of Zoology*, **29**, 33–51.

Ford E.D. (1975) Competition and stand structure in some even-aged plant monocultures. *Journal of Ecology*, **63**, 311–333.

Ford E.D. & Newbould P.J. (1970) Stand structure and dry weight production through the sweet chestnut (*Castanea sativa* Mill.) coppice cycle. *Journal of Ecology*, **58**, 275–296.

Ford E.D. & Newbould P.J. (1971) The leaf canopy of a coppice deciduous woodland. I. Development and structure. *Journal of Ecology*, **59**, 843–862.

Gorham E. (1979) Shoot height, weight and standing crop in relation to density of monospecific plant stands. *Nature*, **279**, 148–150.

Grant P.R. & Price T.D. (1981) Population variation in continuously varying traits as an ecological genetics problem. *American Zoologist*, **21**, 795–811.

Harper J.L. (1967) A Darwinian approach to plant ecology. *Journal of Ecology*, **55**, 247–270.

Harper J.L. (1977) *Population Biology of Plants*. Academic Press, London.

Harper J.L. (1981) The concept of population in modular organisms. *Theoretical Ecology*, 2nd ed. (Ed. by R.M. May). Blackwell Scientific Publications, Oxford.

Hassell M.P. (1978) *The Dynamics of Arthropod Predator–Prey Systems*. Princeton University Press, Princeton.

Hassell M.P. (1980) Some consequences of habitat heterogeneity for population dynamics. *Oikos*, **35**, 150–160.

Jenkins D., Watson A. & Miller G.R. (1963) Population studies on red grouse, *Lagopus lagopus scoticus* (Lath.) in north-west Scotland. *Journal of Animal Ecology*, **32**, 317–376.

Koyama H. & Kira T. (1956) Intraspecific competition among higher plants. VIII. Frequency distribution of individual plant weight as affected by the interaction between plants. *Journal of the Polytechnic Institute, Osaka City University*, Series D, **7**, 73–94.

Li H.W. & Brockson R.W. (1977) Approaches to the analysis of energetic costs of intraspecific competition for space by rainbow trout (*Salmo gairdneri*). *Journal of Fish Biology*, **11**, 329–341.

Łomnicki A. (1978) Individual differences between animals and the natural regulation of their numbers. *Journal of Animal Ecology*, **47**, 461–475.

Łomnicki A. (1980) Regulation of population density due to individual differences and patchy environment. *Oikos*, **35**, 185–193.

McNaughton S.J. (1975) *r*- and *K*-selection in *Typha*. *American Naturalist*, **109**, 251–261.

Magnuson J.J. (1962) An analysis of aggressive behaviour, growth, and competition for food and space in medaka (*Oryzias latipes*) (Pisces, Cyprinodontidae). *Canadian Journal of Zoology*, **40**, 313–363.

May R.M. (1981a) *Theoretical Ecology*, 2nd Ed. Blackwell Scientific Publications, Oxford.

May R.M. (1981b) Models for single populations. *Theoretical Ecology*, 2nd ed. (Ed. by R.M. May). Blackwell Scientific Publications, Oxford.

Moeur J.E. & Istock C.A. (1980) Ecology and evolution of the pitcher-plant mosquito. IV. Larval influence over adult reproductive performance and longevity. *Journal of Animal Ecology*, **49**, 775–792.

Nicholson A.J. (1954) An outline of the dynamics of animal populations. *Australian Journal of Zoology*, **2**, 9–65.

Obeid M., Machin D. & Harper J.L. (1967) Influence of density on plant to plant variations in Fiber Flax, *Linum Usitatissimum*. *Crop Science*, **7**, 471–473.

Parker G.A. (1979) Sexual selection and sexual conflict. *Sexual Selection and Reproductive Competition in Insects*. (Ed. by M.S. Blum & N.A. Blum). Academic Press, New York.

Parker G.A. (1982) Arms races in evolution—an ESS to opponent-independent costs game. *Journal of Theoretical Biology* (in press)

Polis G.A. (1981) The evolution and dynamics of intraspecific predation. *Annual Review of Ecology and Systematics*, **12**, 225–251.

Rabinowitz D. (1979) Bimodal distributions of seedling weight in relation to density of *Festuca paradoxa* Desv. *Nature*, **277**, 297–298.

Rose S.M. (1960) A feedback mechanism of growth control in tadpoles. *Ecology*, **41**, 188–199.

Rubenstein D.I. (1981) Individual variation and competition in the everglades pygmy sunfish. *Journal of Animal Ecology*, **50**, 337–350.

Sohn J.J. (1977) Socially induced inhibition of genetically determined maturation in the platyfish, *Xiphophorus maculatus*. *Science*, **195**, 199–201.

Stearns S.C. (1976) Life history tactics: a review of the ideas. *Quarterly Review of Biology*, **51**, 3–47.

Stearns S.C. (1977) The evolution of life history tactics: a critique of the theory and a review of the data. *Annual Review of Ecology and Systematics*, **8**, 145–171.

Steinwascher K. (1978) Interference and exploitation competition among tadpoles of *Rana utricularia*. *Ecology*, **59**, 1039–1076.

Wallace B. (1970) *Genetic Load*. Prentice-Hall, Englewood Cliffs.

Watson A. & Miller G.R. (1971) Territory size and aggression in a fluctuating red grouse population. *Journal of Animal Ecology*, **40**, 367–383.

Wilbur H.M. (1976) Density-dependent aspects of metamorphosis in *Ambystoma* and *Rana sylvatica*. *Ecology*, **57**, 1289–1296.

Wilbur H.M. (1977) Density-dependent aspects of growth and metamorphosis in *Bufo americanus*. *Ecology*, **58**, 196–200.

Wilbur H.M. & Collins J.P. (1973) Ecological aspects of amphibian metamorphosis. *Science*, **182**, 1305–1314.

Yamagishi H., Maruyama T. & Mashiko K. (1974) Social relation in a small experimental population of *Odontobutis obscurus* (Temminck et Schlegel) as related to individual growth and food intake. *Oecologia*, **17**, 187–202.

9. THE POPULATION AS A UNIT OF SELECTION

JOHN MAYNARD SMITH
School of Biological Sciences, University of Sussex, Brighton BN1 9QG

The question of whether the individual or the population is the unit of selection was raised most clearly for ecologists by Wynne-Edwards (1962). Ten years ago, most ecologists had accepted that few, if any, species meet the extremely stringent conditions (i.e. small and almost isolated demes) required if group selection as understood by Wynne-Edwards was to operate. Why, then, should the organizers of this symposium have asked me to speak on this topic? I do not know, but I suspect that three things may have contributed. The first is the discussion of 'trait-group selection' by Wilson (1975, 1980). The second is the proposal by Gould & Eldredge (1977) and by Stanley (1979) that 'species selection' is a major determinant of macroevolutionary trends. The third, probably stimulated by the first two, is the renewed interest in Wright's (1931, 1970) 'shifting balance' theory of evolution. I will discuss these in turn, but in the historical sequence, taking Wright's ideas first.

Wright was much influenced by the ubiquitous epistatic interactions which he found when analysing the genetics of coat colour in the guinea-pig. Suppose that the present genotype of a species is *ab* (I shall use a haploid sexual model, because nothing of importance is lost), and that *AB* is superior, but both *Ab* and *aB* are inferior in fitness. In a large random-mating sexual population, the evolutionary change, $ab \rightarrow AB$, is impossible. Wright suggests that it could take place if the species were subdivided into a number of small demes, with little gene flow between them. Thus one deme might cross the 'adaptive valley' by chance. Then, by virtue of the greater fitness of its members, it might by degrees transform the rest of the species, by sending out migrants which infected other demes, and by founding new demes.

We are still uncertain about whether such a process has often been important. I will make one general comment, and mention the strongest argument for, and the strongest against Wright's proposal. The comment is as follows. The undoubted fact that hybrids between species, and often between subspecies are often less fit than their parents is no proof that an adaptive valley has been crossed. It is perfectly possible for genotype 1 to be replaced by genotype 2 in a series of steps, each step being favoured by natural selection in a large random-mating population, and yet for hybrids between 1 and 2 to be of low fitness; an example is shown in Fig. 9.1.

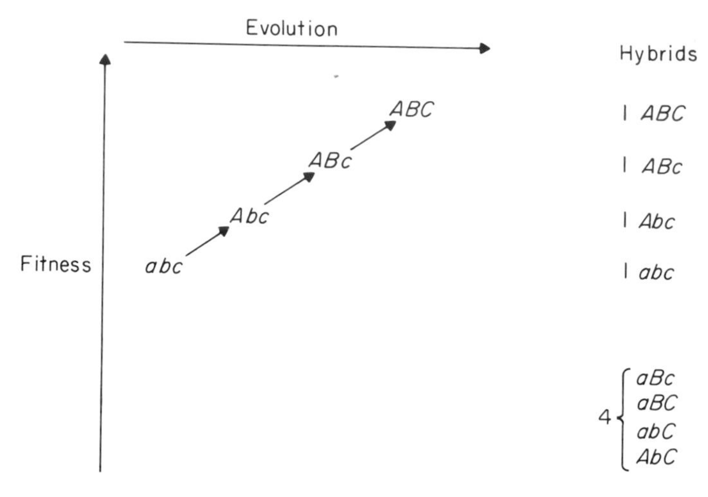

FIG. 9.1. A case in which evolution could proceed by natural selection in a large population, and yet hybrids between the initial and final forms would show lower mean fitness than their parents. The organisms are haploid, with three relevant loci. It is assumed that fitness increases if *A* replaces *a*, if *B* replaces *b* only if A is present, and if *C* replaces *c* only if *A* and *B* are already present. The relative numbers of the different types of *F*1 hybrid are shown on the right. Their fitnesses are indicated by their vertical position; note that half the *F*1 hybrids are less fit than either parent.

The strongest argument for the importance of Wright's model is that something very like it seems to be required to explain some kinds of chromosomal evolution. Here we are concerned with the low fitness of heterozygotes, rather than low fitness caused by epistasis, but if one kind of valley-crossing can occur, why not the other? The strongest argument against is that few species seem to be divided into demes in quite the way needed if Wright's mechanism is to work. For example, although many mammals live in social groups, these groups are usually exogamous. However, the requirements for Wright's model are less stringent than for Wynne-Edwards'. Thus in Wright's model, once a deme has made the transition $ab \rightarrow AB$, it would not be converted back to *ab* by a single *ab* immigrant, whereas in Wynne-Edwards' model a deme of altruists can be converted by a single selfish immigrant.

One extension of Wright's model is to suggest that small peripheral isolates may, by genetic drift, cross adaptive valleys, and thus become the starting point of new species. This is an entirely plausible idea. However, not every peripheral isolate which happens to make such a transition will found a new species; it is far more likely to be swamped by renewed contact with the parental species. Also, of course, peripheral isolates can give rise to new species without crossing adaptive valleys.

The trait-group model of Wilson (1975, 1980) and Matessi & Jayakar

TABLE 9.1. Additive and synergistic fitness interactions in trait groups

Trait group	C,C	C, D	D, D
Fitnesses (additive)	$(a+b-c), (a+b-c)$	$(a-c), (a+b)$	a, a
Example ($a=2$, $b=2$, $c=1$)	3, 3	1, 4	2, 2
Fitnesses (synergistic)	5, 5	1, 4	2, 2

Payoff Matrices

additive

	C	D
C	3	1
D	4	2

synergistic

	C	D
C	5	1
D	4	2

(1976) has little in common with Wright's. Its essential feature is that the fitness of an individual is determined, not in isolation, but in interaction with other conspecifics. Wilson has usually imagined that individuals interact in groups—'trait groups'—but this is not essential. What is essential is that they should interact with a finite set of neighbours. These trait groups are *not* breeding groups; in most of his models he supposes that mating is panmictic.

The logic of trait groups can best be illustrated by considering groups of two (Table 9.1). There are two kinds of individual, which I will call 'co-operators', C, and 'defectors', D. A co-operator, regardless of the nature of its partner, performs acts which lower its fitness by $-c$ ('cost'), and increases the fitness of its partner(s) by $+b$ ('benefit'). If we assume: (a) association into pairs is random; and (b) fitnesses are additive, as defined in Table 9.1, then it will be apparent that co-operation is never favoured by selection. This conclusion extends to groups of more than two (Wilson, 1975). It also follows from Hamilton's (1964) argument, that a gene which causes an individual to lower its own fitness by $-c$ and to increase by $+b$ the fitness of other, randomly chosen, members of the population will not increase in frequency.

There are two situations in which selection might favour co-operation:
(i) Association into groups non-random. Much the most likely reason for this is that members of groups are relatives. This is the mechanism (kin selection) proposed by Hamilton (1964).
(ii) Fitnesses non-additive, i.e. the fitness of two (or more) co-operating individuals is greater than would be predicted from interactions between one co-operator and one defector. I have labelled such interactions 'synergistic' in Table 9.1. It would be nice to call them 'mutualistic', although this term is sometimes reserved for interspecific interactions.

Even when interactions are synergistic, co-operation will not increase when rare. As Charlesworth (1979) pointed out, it may need kin selection to get co-operation started.

It is not hard to think of situations in which fitnesses will be non-additive. For example, Bygott, Bertram & Hanby (1979) and Packer & Pusey (1982) show that male lions have higher individual fitnesses if they co-operate with others in holding a group of females. Interestingly, groups of co-operating males are not always relatives, although they often are.

I cannot see that Wilson's trait-group model has added anything conceptually new. It does provide an elegant way of revealing kin selection and synergistic effects as the two causes of the evolution of co-operation. Wilson, however, does not see his model in that light. He prefers to say that co-operation spreads because, although there is individual selection against co-operation in mixed groups, there is selection in favour of groups of co-operators. It seems to me confusing to call this 'group selection', since the groups referred to are not breeding groups. Whether selection will or will not favour a particular type depends on the fitness of that type relative to others in the breeding population, and not relative to others with which it happens to interact.

Wilson (1980) has extended his argument to cover interactions between different species. It is clear that mutualism will evolve if individuals of species A which co-operate with individuals of species B are fitter than A individuals which do not, and if the same is true of B individuals which co-operate with A. Wilson develops a model in which the beneficiary of an A individual which co-operates with a B is not the A individual itself, but its relatives, for example, its offspring. Logically, this can lead to mutualism, but it requires extremely viscous populations. Thus an individual A which, by helping members of species B, helps future generations of species A which are only distantly related to itself will not enhance its inclusive fitness. Hence mutualistic interactions between species involving more than one generation can evolve, but it will require low costs, high benefits, and little mobility of either species. The idea is interesting, but provides no justification for resurrecting the concept of the community as a super-organization.

The concept of 'species selection' has been proposed by Gould & Eldredge (1977) and by Stanley (1979) as something which follows from a supposed feature of the palaeontological record. This feature is that most morphological change is concentrated into relatively brief periods, associated with the splitting of lineages—i.e. with speciation. Between such 'punctuational' events are long periods of morphological 'stasis'. There is controversy about how far this is true. I shall not comment on this empirical issue. Instead, I shall discuss the further claim that it follows from this pattern of change that large-scale macro-evolutionary events can be 'decoupled' from changes occurring within populations, and are in fact determined by a different process, namely species selection.

The first point to establish is that, even if stasis and punctuation are accepted as the typical pattern, the role of species selection would not follow. Thus suppose that speciation occurs when a new ecological opportunity arises, or when, as a result of geographical isolation or of the crossing of some selective threshold, a population is enabled to occupy some previously empty niche. Rapid change would then be associated with speciation events, but would be driven by natural selection within the population. I think this corresponds fairly closely with the views of both Wright and Simpson; in no sense would it represent a new evolutionary paradigm.

What, then, is meant by 'species selection'? To grasp this, we must first distinguish between properties of individuals and emergent properties of species. 'Ability to run fast' is a property of an individual. 'Ability to evolve fast' is a property of a species, although it will depend on properties of individuals, such as chiasma frequency and mutation rate. Gould (1982) now accepts that the term species selection should be confined to the evolution of emergent properties of species. For example, ability to evolve fast may make a species less likely to go extinct. Some characteristics, such as low mobility or narrow food specialization, may make speciation more likely.

I see no logical difficulty with species selection of this kind, but it is not independent of selection at the individual level. The point can best be illustrated with an example. Vrba (1982) discusses two groups of antelopes, the Alcephalini (blesbuck–hartebeest–wildebeest group) and the Aepycertatini (impalas). Since the Miocene there has been only one species of impala, which has changed only slightly. There are seven extant species of alcelaphines; both speciation and extinction have been relatively frequent, and there is considerable diversity among existing species. Individual impala take a wide range of plant food, and do not migrate, whereas alcelaphine species are food specialists, and usually migratory.

The explanation of these facts which Vrba favours is that the wide food range of the impala enabled that species to remain relatively unchanged despite wide environmental fluctuations. In the alcelaphines, food specialization favoured speciation. Interestingly, the low gene flow between impala populations, compared to the high gene flow in the migratory alcelaphines, did not have the expected effect in causing speciation in the former group and inhibiting it in the latter. Instead, ecological relationships were apparently more important in determining speciation rates.

Since stenotopy (narrow food range) has favoured speciation, so that there are now more stenotopic than eurytopic species, we might wish to say that species selection has favoured stenotopy. But, as Vrba points out, the single impala species may leave a comparable number of genes to future generations as all the diverse alcelaphine species together. When we think about selection at the individual level, we usually assume that the relative *frequencies* of two types is what matters, since we assume (tacitly) that the total *number* of the two types together remains constant. But in Africa it is the number of individual antelopes, not the number of species, that is resource-limited. More species means fewer individuals. Ignoring human interference, the ultimate outcome would depend on competition between individuals of the various species for limiting resources.

Hence some traits, such as stenotopy, may facilitate speciation, but this will not lead to the replacement of eurytopes by stenotopes unless the latter can, as individuals, outcompete the former for the same limiting resources.

This brings me to the major difficulty with species selection as originally formulated. Consider, for example, the set of traits which distinguish the mammals from the mammal-like reptiles. These include changes in the teeth, skull roof, palate, jaw musculature and jaw articulation, associated with chewing; in the limbs, girdles and vertebral column, associated with the evolution of the characteristic mammalian gaits; and, presumed rather than observed in fossils, the evolution of hair and homeothermy, of a double circulation, and of viviparity and lactation. Now species do not chew, gallop or lactate: individual animals do these things. The usual explanation is that these various changes occurred by selection within populations. The alternative that seems to be preferred by the punctuationists is that, when new species arose, they differed from their ancestral species in ways randomly related to the major evolutionary trends. That is, new species were as likely as not to have smaller secondary palates, more sticking-out elbows and less hair (indeed, on this theory I see no reason why, usually, new species should not have been identical to their parental species in these respects, and have differed from them in ways unrelated to the major trends). If there is an overall trend, it is because, once they had arisen, new species with, for example, less hair went extinct in competition with species with more hair.

I find it hard to imagine why anyone should hold such an odd opinion. It may have something to do with the fact that, unavoidably, the best-known fossil lineages are of the shells of marine invertebrates. Often, we have little idea of the adaptive significance, if any, of the changes observed. Yet the first task of an adequate theory of evolution is to explain complex adaptations.

It does not follow that palaeontology has nothing to contribute to evolution theory. Indeed, often it provides the only source of evidence we can look to to decide between alternative theories. Let me give one example. For some years, Nils Stenseth and I have been trying to formalize Van

Valen's (1973) 'Red Queen hypothesis'. This is the idea that each species in an ecosystem is forced to evolve continuously to meet the changes in its biotic environment caused by the evolution of other species in the system. It seems to us that one can in fact draw two quite different conclusions from a model of the evolution of a multi-species ecosystem, each consistent with what we know of ecology and population genetics. According to one picture, evolutionary change continues indefinitely in a uniform physical environment. The system reaches a steady state of flux, with constant rates of extinction and speciation, and continuous evolutionary change of the constituent species. This corresponds to the Red Queen hypothesis.

According to the alternative picture, in a uniform physical environment evolution would gradually slow down and stop, resulting in a system with high species diversity, and no extinction or speciation. Continued evolution would depend on changes in the physical environment.

Which of these models is correct depends on assumptions about the relationship between species in an ecosystem which, it seems to me, it would be difficult if not impossible to settle directly. However, it may be possible for palaeontologists to tell us which pattern of behaviour corresponds to what they observe. There is, of course, no guarantee that the answer will be the same for different systems; for example, marine and terrestrial systems may differ.

CONCLUSIONS

To conclude, the theory of evolution by natural selection asserts that any population of entities having the properties of multiplication, heredity, and variation will evolve. Usually, we think of the relevant entities as being individual organisms. Recently there has been much discussion of the idea that the appropriate entities are genes. This chapter, however, is concerned with whether it is ever appropriate to take entities larger than the individual as the units in models of evolution. It is certainly relevant for evolution that the fitness of individuals is often determined in interactions with others, either in 'trait groups' or with neighbours in a continuously distributed population, but in such cases the units in terms of which evolution is modelled are still the individual organisms. Local populations are probably seldom sufficiently isolated for interdemic selection to be an important force, but this is still a matter of controversy. Species are, by definition, reproductively isolated, and the processes of speciation and extinction correspond to the birth and death of individuals. Species selection is, therefore, a logical possibility. However, it is best to reserve the term for the evolution of characteristics, such as the ability to evolve fast or a tendency to speciate, which are emergent properties of species rather than properties of the component individual.

REFERENCES

Bygott J.D., Bertram B.C.R. & Hanby J. (1979) Male lions in large coalitions gain reproductive advantages. *Nature*, **282**, 839–841.

Charlesworth B. (1979) A note on the evolution of altruism in structured demes. *American Naturalist*, **113**, 601–605.

Gould S.J. (1983) The meaning of punctuated equilibrium and its role in validating a hierarchical approach to macroevolution (in press).

Gould S.J. & Eldredge N. (1977) Punctuated equilibria: the tempo and mode of evolution reconsidered. *Palaeobiology*, **3**, 115–151.

Hamilton W.D. (1964) The genetical evolution of social behaviour. I and II. *Journal of Theoretical Biology*, **7**, 1–16; 17–23.

Matessi C. & Jayakar S.D. (1976) Conditions for the evolution of altruism under Darwinian selection. *Theoretical Population Biology*, **9**, 360–387.

Packer C. & Pusey A.E. (1982) Cooperation and competition within coalitions of male lions: kin selection or game theory? *Nature*, **296**, 740–742.

Stanley S.M. (1979) *Macroevolution*. W.H. Freeman, San Francisco.

Van Valen L. (1973) A new evolutionary law. *Evolution Theory*, **1**, 1–30.

Vrba E.S. (1983) Macroevolutionary trends: new perspectives. (in press).

Wilson D.S. (1975) A theory of group selection. *Proceedings of the National Academy of Sciences*, **72**, 143–146.

Wilson D.S. (1980) *The Natural Selection of Populations and Communities*. Benjamin Cummings, Menlo Park, California.

Wright S. (1931) Evolution in Mendelian populations. *Genetics*, **16**, 97–159.

Wright S. (1970) Random drift and the shifting balance theory of evolution. In *Mathematical Topics in Population Genetics* (Ed. by K. Kojima). Springer, Berlin.

Wynne-Edwards V.C. (1962) *Animal Dispersion in relation to Social Behaviour*. Oliver & Boyd, Edinburgh.

10. STRONG PRESENT-DAY COMPETITION BETWEEN THE *ANOLIS* LIZARD POPULATIONS OF ST MAARTEN (NETH. ANTILLES)

JONATHAN ROUGHGARDEN, STEPHEN PACALA* AND JOHN RUMMEL

Department of Biological Sciences, Stanford University, Stanford, CA 94305, USA

INTRODUCTION

There has been controversy about whether interspecific competition is an important cause of community structure in terrestrial vertebrates, especially for birds and lizards (Wiens 1977; Birch 1979; Connor & Simberloff 1979; Strong, Szyska & Simberloff 1979; Connell 1980). The issues concern both existence and generality. Are there *any*, and, are there *many* such systems in which competition is important?

During the last 5 years we have demonstrated that strong present-day competition occurs under natural conditions between populations of *Anolis* lizards on one island in the eastern Caribbean. This competition is one of the principal causes of present-day community structure on that island, and has been very important in the evolution and biogeography of all the *Anolis* populations of the eastern Caribbean.

Mathematical competition theory provides a picture of the population dynamics, evolution and biogeography of *Anolis* populations in the eastern Caribbean. The predictions of competition theory seem to explain the biogeography of *Anolis* in the eastern Caribbean (Roughgarden, Heckel & Fuentes 1983); but this is not the primary reason for using competition theory. We use competition theory because we have shown that interspecific competition really occurs in this system. It is not the only force; particularly predation from birds (Adolph & Roughgarden 1983) and long-term changes in the habitat (Pregill & Olson 1981) seem important too, and surely influence some aspects of the community structure and biogeography of *Anolis* populations.

*Present address: Biological Sciences Group, Ecology Section -U-42, University of Connecticut, Storrs, CT 06268, USA

We offer here a review of our evidence for interspecific competition on an eastern Caribbean island. These results put to rest doubts of the *existence* of interspecific competition as an important process in this group of terrestrial vertebrates. Results from only one system cannot speak to doubts of the generality of competition as an important process. However, Schoener (1983) has reviewed 164 replicated competition experiments performed under field conditions, and notes that 148 of these detected interspecific competition.

Yet even if competition were important only in the community structure of Caribbean anoles, this finding would, itself, be biologically significant. The Anoles of the Caribbean are an important biological system. The genus, including the continental populations, has about 300 species and about half of these are on Caribbean islands (E. Williams, personal communication). The genus contains between 5 to 10% of the lizard species in the world today. In the Caribbean, anoles are very abundant; 50–100 lizards 100 m^{-2} is a common value (see Tables later). Moreover, the Caribbean islands, including the Greater Antilles and the Bahamas, are about 235 000 km^2 in area. This is about $\frac{1}{4}$ of Oceania and about one and a half times the size of England. In the terrestrial communities of the Caribbean, anoles substitute for the ground-feeding insectivorous bird guild (Terborgh & Faaborg 1980; Adolph & Roughgarden 1983), and are major components of the animal biomass in these communities ($\frac{1}{4}$ to $\frac{1}{2}$ kg 100 m^{-2}). As principal top carnivores, they may regulate the abundance of insects, including herbivorous insects, and thus control the terrestrial ecosystem transport processes of the region.

DISTRIBUTIONAL EVIDENCE OF COMPETITION

Two *Anolis* populations are native to St Maarten. *Anolis gingivinus* is a light brown form with adult males about 60 mm SV and with females about 45 mm SV. *Anolis wattsi pogus* is a darker brown form with males about 45 mm SV and females about 40 mm SV. The body sizes of these species are more similar than on any other island with two species in the eastern Caribbean. Both species are insectivorous, territorial as adults, and somewhat arboreal in habit. They eat the same kinds of insects and perch on the same trees. *Anolis gingivinus* typically perches higher in the vegetation than *A. wattsi* but there is substantial overlap. Eggs are laid singly on the ground and there is some reproductive activity throughout the year but mostly during the rainly season (September to January). Maturity is attained in about 1 year and the longevity is typically 3 years or more. Our initial evidence of competition is distributional.

Anolis wattsi pogus occurs only in the central hills of St Maarten as illustrated in Fig. 10.1. *A. gingivinus* occurs throughout St Maarten including

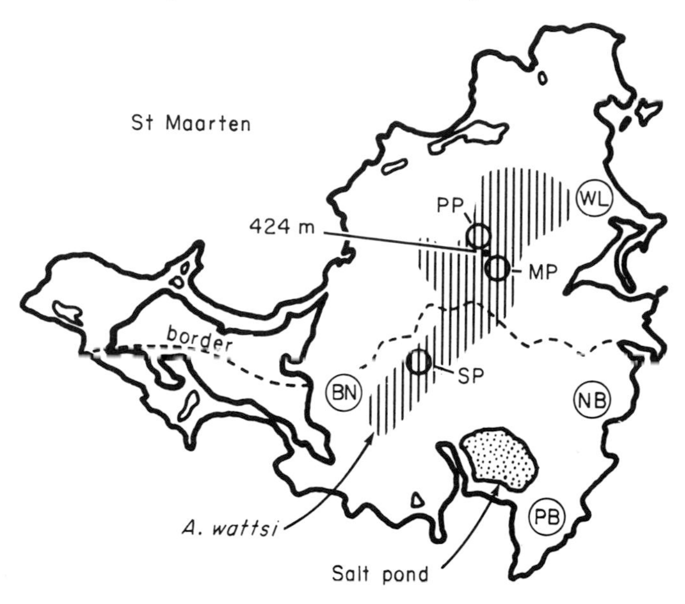

FIG. 10.1 Map of St Maarten. The distribution of *A. wattsi* is restricted to hills in the centre of the island. *Anolis gingivinus* occurs throughout the island, including the central hills. The sites from Table 10.1 are indicated: Boundary (BN), Naked Boy Mt (NB), Well (WL), Point Blanc (PB), Pic du Paradis (PP), Medium Pic (MP), and St Peter Mt (SP).

the central hills. This observation is interesting because all the relatives of *A. wattsi* on nearby islands do occur throughout all elevations and habitats, including sea-level habitat.

Anolis gingivinus suffers a generally lower abundance where it co-occurs with *A. wattsi* relative to its abundance where it is alone. Table 10.1 presents

TABLE 10.1. Number of lizards per 100 m^2 on St Maarten in July 1977

Sites	*A. gingivinus*	*A. wattsi*
With *A. gingivinus* alone		
Boundary (scrub)	50.1 (2.2)*	
Naked Boy Mt (scrub)	43.2 (3.4)	
Well (woods)	42.4 (1.9)	
Point Blanc (scrub)	6.4 (0.5)	
With both species		
Pic du Paradis (woods)	8.4 (2.6)	10.5 (0.7)
Medium Pic (woods)	10.2 (2)	17.7 (0.7)
St Peter Mt (woods)	19.7 (1.4)	16.9 (0.8)

*Standard error of estimate given in parentheses.

TABLE 10.2. Number of lizards per 100 m² on St Maarten over 4 years

Date	Naked Boy Mt		Boundary				Pic du Paradis			
	A. gingivinus		*A. gingivinus*		*A. wattsi*		*A. gingivinus*		*A. wattsi*	
July 1977	43.2	(3.4)*	50.1	(2.2)	0		8.4	(2.6)	10.5	(0.7)
July 1978	50.1	(5.4)	65.0	(3.0)	0		14.2	(5.5)	24.9	(3.3)
July 1979	32.3	(4.2)	64.8	(2.8)	0		7.6	(6.2)	21.2	(2.2)
March 1980	89.8	(6.6)	129.8	(5.5)	0		14.6	(3.1)	43.9	(5.2)
November 1980	75.5	(3.1)	95.4	(3.0)	4.4	(1.8)	39.8	(4.7)	56.8	(2.7)
March 1981	45.8	(3.5)	72.1	(3.3)	2.0	(1.0)	13.3	(2.9)	39.4	(2.5)
October 1981			122.3	(6.5)	1.3	(0.7)	15.7	(4.8)	54.3	(6.1)

*Standard error of estimate given in parentheses.

census data from 1977. Some of these sites have been censused for 5 years since 1977 and for those sites Tables 10.2 shows that the same rank order of abundances were observed. The censusing here and in the experiments presented later was done using three visits per site according to the method of Heckel & Roughgarden (1979).

Anolis gingivinus shifts its perch position where it co-occurs with *A. wattsi* relative to its perch position where it is alone. Figure 10.2 illustrates that *A. gingivinus* perches higher in the vegetation where it co-occurs with *A. wattsi* than where it is the only species present.

These distributional data suggest there is strong reciprocal present-day interspecific competition. The sea-level habitat from which *A. wattsi pogus* is absent appears to be suitable habitat because its relatives on other islands use this habitat. Hence, it may be that *A. gingivinus*, an animal of nearly the same size as *A. wattsi*, excludes *A. wattsi* from this habitat. On other islands, the relatives of *A. wattsi* co-occur with species much larger than *A. gingivinus*. And, in the other direction, *A. gingivinus* suffers both a lower abundance and an upwards shift in perch position in the presence of *A. wattsi*. However, this case, by itself, is not conclusive evidence of present-day interspecific competition.

THE THERMAL ALTERNATIVE

The distributional data are also consistent with an alternative hypothesis based on strongly differing habitat requirements for the species, without any present-day interaction between them. Perhaps, *A. gingivinus* requires sunny locations while *A. wattsi* requires shady locations. Since the species only co-occur in the habitat of higher elevations, where the air temperature is necessarily cooler, the shift of *A. gingivinus* upwards may represent its moving to places in the vegetation where there is more sun than near the ground.

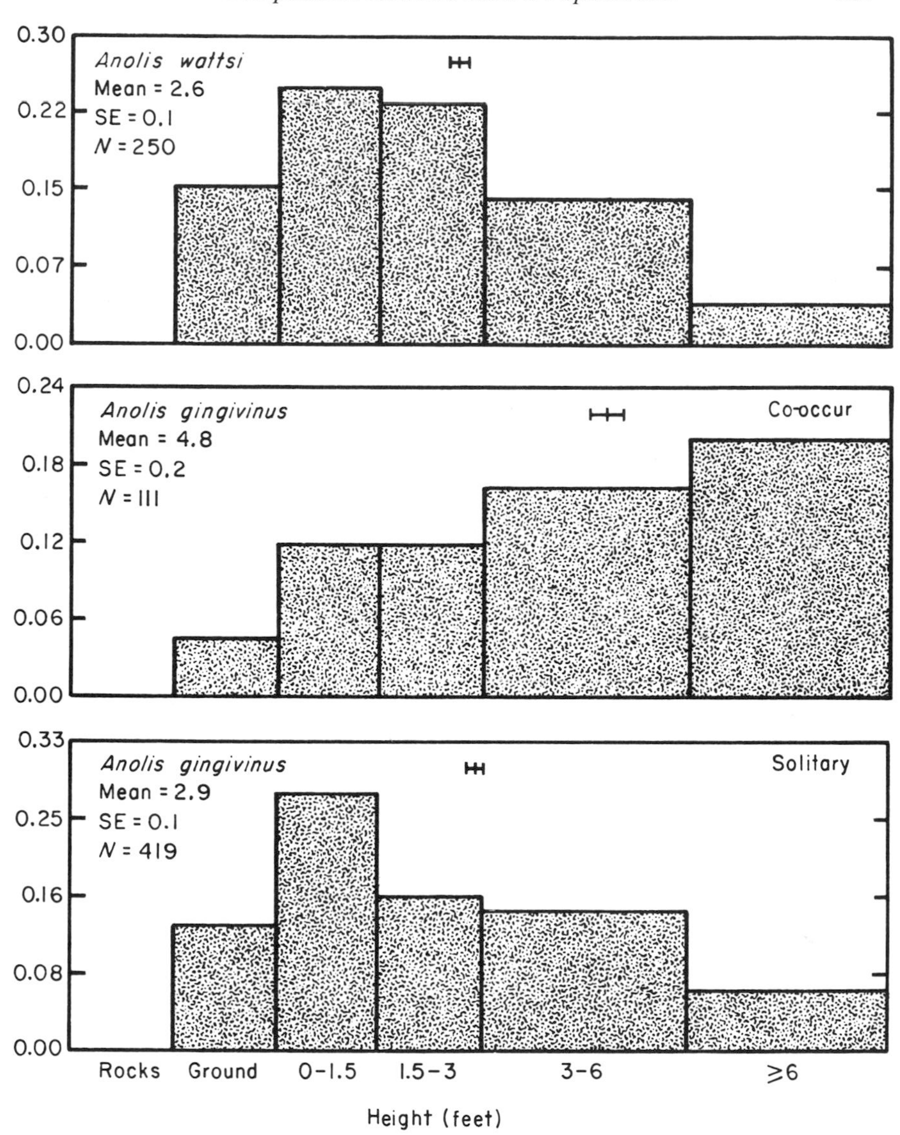

FIG. 10.2. Distribution of perch heights of the lizards on St Maarten. The horizontal axis is the perch height; the vertical axis is the fraction of observations. Notice that *A. gingivinus* shifts high in the crown of the vegetation where it co-occurs with *A. wattsi*.

And the low abundance of *A. gingivinus* in such habitat might be caused by the absence of sufficiently warm places there. Similarly the absence of *A. wattsi* from sea-level habitat may be caused by the absence of sufficiently cool perch locations in that habitat.

With this alternative in mind, we examined the thermal biology of the

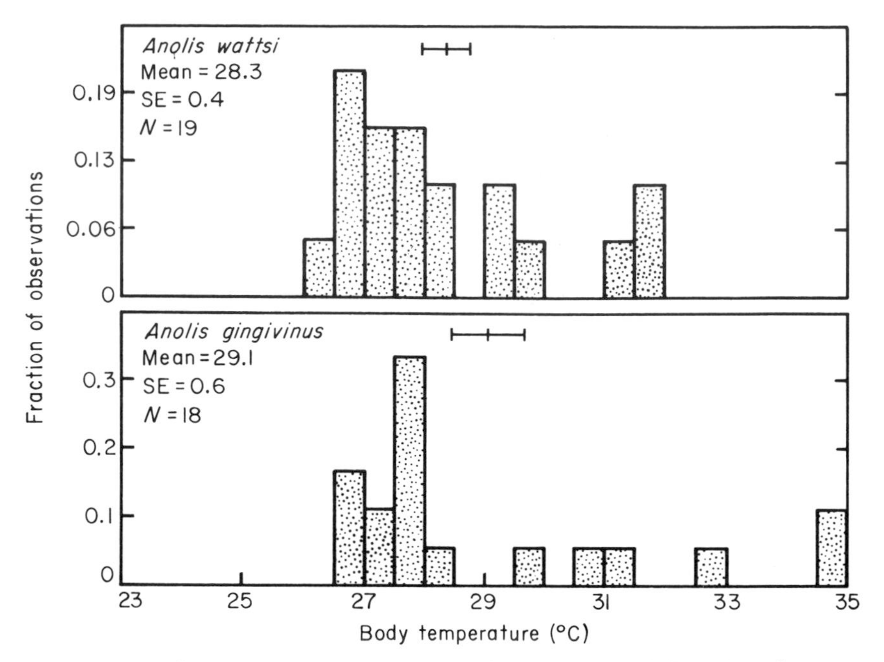

FIG. 10.3. Distribution of body temperatures for the lizards at Pic du Paradis on St Maarten. The body temperatures of the two species are not significantly different under field conditions, although there is a suggestion that *A. gingivinus* may occasionally attain higher temperatures than *A. wattsi* can tolerate.

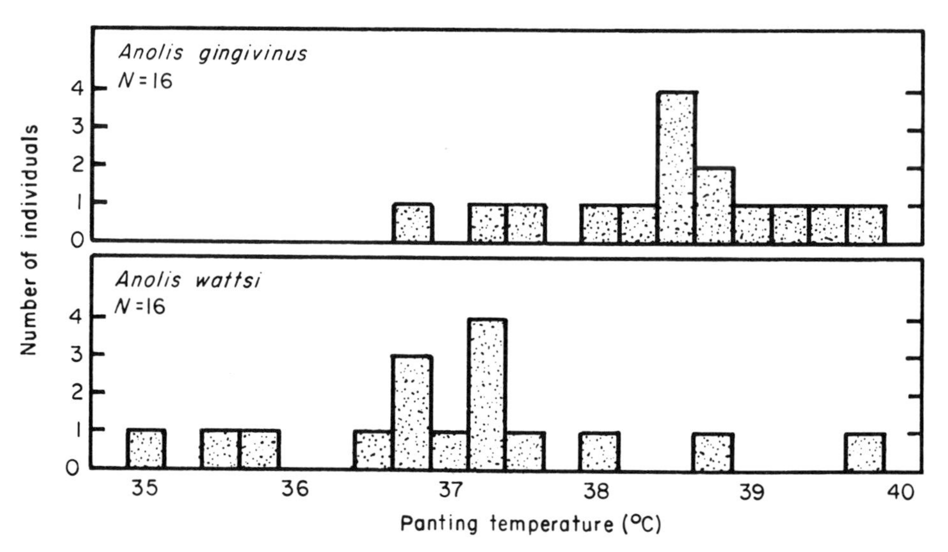

FIG. 10.4. Distribution of panting temperature for the lizards of St Maarten. *Anolis gingivinus* can generally withstand a higher body temperature before showing signs of heat stress than *A. wattsi*.

two species on St Maarten. We found that where they co-occur under field conditions, the lizards of the two species do *not* use perches with different micro-climates and their body temperatures are *not* significantly different. Figure 10.3 illustrates the distribution of body temperatures at Pic du Paradis.

However, we also found that the panting temperature of *A. gingivinus* is higher than that of *A. wattsi*, as illustrated in Fig. 10.4. This result suggests that *A. gingivinus* can tolerate higher temperatures than *A. wattsi* even though it does not seem to prefer a micro-climate any different from that used by *A. wattsi*.

Finally, there are *not* more hot micro-sites high in the woods of Pic du Paradis than there are near the ground. Figure 10.5 illustrates the distribution of the effective temperature of micro-sites in the Pic du Paradis woods throughout the day (Roughgarden, Porter & Heckel 1981). The essential point is that locations high in the woods are both sunny and have a higher wind speed. The higher light level tends to heat lizards, but the higher wind speed provides more convective cooling. Near the ground there is less light

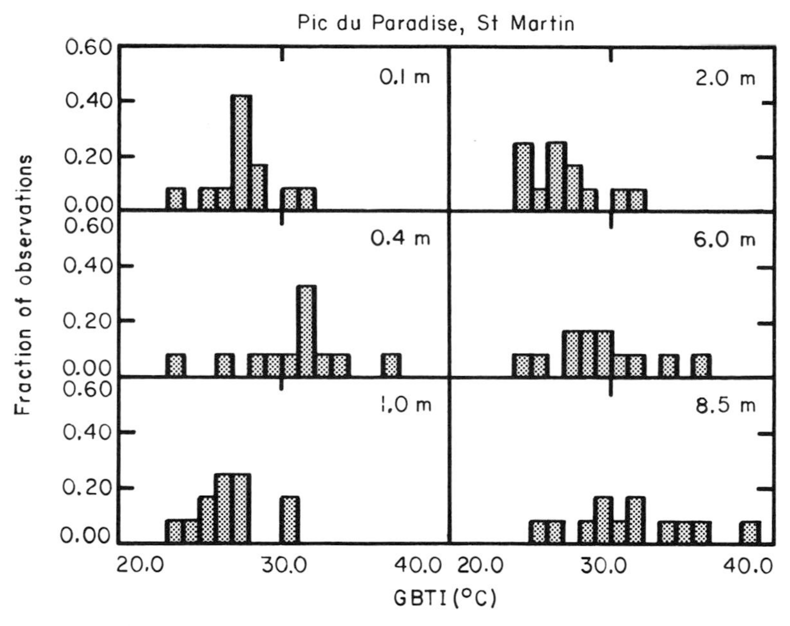

FIG. 10.5. Distribution of micro-climate throughout the day at various heights in the woods at Pic du Paradis, St Maarten. The GBTI stands for the Grey Body Temperature Index, an indicator of the effective temperature of a location. The GBTI is the temperature that an inanimate reference object would attain if allowed to come into thermal equilibrium at the place where a lizard was perched. Its calculation includes consideration of both convection and total incident radiation.

and less breeze. There are flecks of direct sun on the ground so that a limited amount of basking is possible there. In Fig. 10.5, the panel for 8.5 m is actually above the canopy in the open sun and wind, while the other panels refer to locations below the canopy.

Thus, the thermal biology of the species offers little support for the alternative hypothesis that the species distribution is directly caused by very restrictive habitat requirements. Since body temperature, and the effective temperatures of the perch micro-climates, are the same for both species in the field, there is no evidence for greatly differing requirements for the species. Also, the environment in the hills, where *A. gingivinus* perches above *A. wattsi*, does not have much vertical stratification in effective microclimate temperature and, hence, could not explain the perch height separation, even if the species did have greatly differing thermal requirements.

Nonetheless, *A. gingivinus* can withstand more heat stress than *A. wattsi*. This might diminish the daily activity time of *A. wattsi* relative to *A. gingivinus* in arid sea-level habitat. We return to this point in the discussion. The differing thermal biology of the species probably affects the rate constants of the interacting population dynamics of the species. This is the way that the comparative thermal biology of the species can influence community structure.

OTHER ENVIRONMENTAL FACTORS

There is no qualitative difference in the identity of the predators or of other non-competitive but biological factors in the hills of St Maarten relative to the sea-level habitat. The principal predator is probably the pearly-eyed thrasher (*Margarops fuscatus*). It is found throughout the island as are other predators like the sparrow hawk (*Falco sparverius*) and the introduced mongoose. Snakes are very rare.

We have also trapped insects at the lizard census sites since 1977. The same taxa of insects occur throughout St Maarten. Some sites always have more insects than others, but the kinds of insects invariably are the same throughout the island.

The topic where we are largely ignorant is how environmental factors influence the hatching rate for eggs.

INTRODUCTION TO AN OFFSHORE CAY

To obtain direct evidence concerning competition between the anoles of St Maarten, we began experiments to determine whether *A. gingivinus* has any effect on *A. wattsi*, as the absence of *A. wattsi* from sea-level habitat suggests. The obvious experiment would be to introduce individuals of

A. wattsi to sea-level habitat on St Maarten at sites where *A. gingivinus* was removed, and at sites where *A. gingivinus* was left undisturbed. However we thought that our results might somehow be compromised by migration of *A. wattsi* on St Maarten. As Table 10.2 shows, a site near the boundary of the range of *A. wattsi* can acquire lizards of this species in some years and lack them in others. Meanwhile, it was known that the offshore cays on the Anguilla bank all had resident populations of *A. gingivinus* and that these cays absolutely lacked *A. wattsi* (Lazell 1972). Hence, we devised an experimental plan based on introducing *A. wattsi* to an offshore cay that contained a resident population of *A. gingivinus*.

We selected a cay known as Anguillita that is west of Anguilla (Fig. 10.6). The cay is a limestone platform comprised of Pleistocene coral. The vegetation is primarily sea grape (*Cocaloba uvifera*), Manchaneel (*Hippomane mancinella*) and perennial grasses. The wind has shaped the vegetation into krumholtz. In three places the vegetation is tall, (4 m maximum height),

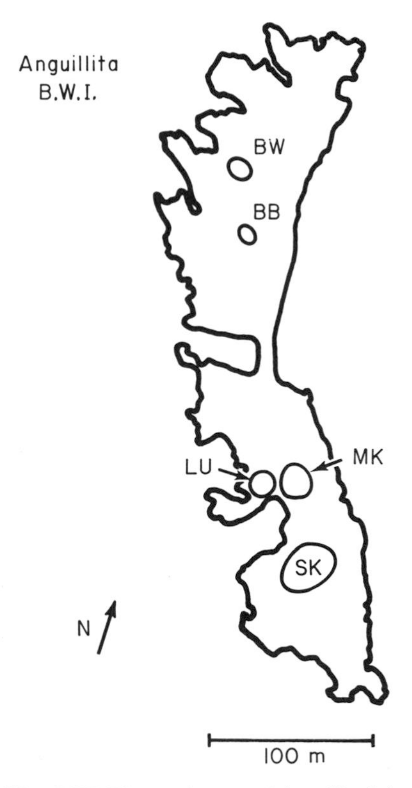

FIG. 10.6. Map of Anguillita, BWI. The cay is west of Anguilla. It is a limestone platform with an elevation of 15 m. above sea level.

TABLE 10.3. Introductions to Anguillita

Experiment		Date	*A. gingivinus*	*A. wattsi*
SK	Original	August 1979	196.3 (7.2)*	0
558.6 m^2	Start	August 1979	196.3 (7.2)	103
	End	May 1980	204.6 (9.3)	14.5 (7.0)
MK	Original	March 1980	115.6 (8.9)	0
262.6 m^2	Start	March 1980	74.6 (8.9)	55
	End	October 1980	116.4 (5.4)	15.5 (1.7)
SK_2	Original	March 1981	194.1 (14.5)	10.3 (7.6)
558.6 m^2	Start	March 1981	92.1 (14.5)	110.3 (7.6)
	End	October 1981	153.5 (3.3)	37.3 (2.4)
LU	Original	March 1981	103.9 (5.1)	0
295.9 m^2	Start	March 1981	103.9 (5.1)	48
	End	October 1981	127.2 (5.5)	7.0 (1.6)

*Standard error of estimate given in parentheses

forming groves separated by low scrub and grass. These groves are the sites of the experiments. *Anolis gingivinus* inhabits the space between the groves in very low abundance, and is common in the groves themselves.

Table 10.3 presents the results. The largest grove is SK. It contained about 200 *A. gingivinus* initially. We introduced 103 individually marked *A. wattsi* in an equal sex ratio, while leaving the residents undisturbed. After 8 months about fourteen animals remained. Also, we estimated that fifteen were there in October 1980 and the number had dropped to ten by March of 1981. All the surviving *A. wattsi* had assumed territories at the periphery of the grove where the vegetation height was 1 m or less. Thus, introduced animals are capable of surviving for almost 2 years in the environment of the cay. The initial 6-month survivorship is low. Meanwhile, the abundance of *A. gingivinus* remains virtually unchanged.

We calculate a 6-month survivorship as follows: First, the monthly survivorship is

$$l = (N(\Delta t)/N_0)^{1/\Delta t},$$

where N_0 is the initial number, $N(\Delta t)$ is the number after Δt months, and Δt is the elapsed time. For the first experiment at SK, the elapsed time is 8 months, so the monthly survivorship is 0.783. Then the 6-month survivorship is calculated by taking the sixth power of the monthly survivorship, yielding 0.230.

The next experiment was at the site, MK. Here we removed about 40% of the resident *A. gingivinus* before introducing the *A. wattsi*. We obtained

TABLE 10.4. Effect of *A. gingivinus* on survivorship of *A. wattsi pogus* on Anguillita

Experiment	Six-month *survivorship* of *A. wattsi*	*A. gingivinus* removed (%)
SK_1	0.230	0
LU	0.193	0
MK	0.337	35.5
SK_2	0.396	52.6

about twice the survivorship of *A. wattsi* here as compared with the experiment at SK. Moreover, the *A. wattsi* here assumed territories in the centre of the grove as well as at its periphery. Also, a hatchling of *A. wattsi* was observed. Meanwhile *A. gingivinus* from the surrounding habitat migrated into MK and restored its abundance to almost exactly its previous value.

To repeat these experiments we removed about 50% of the resident *A. gingivinus* from SK and again introduced *A. wattsi*. As with the experiment at MK, we observed about twice the survivorship of the first experiment at SK. Moreover, the introduced *A. wattsi* this time did establish territories in the centre of the grove.

We also repeated the experiment of introducing *A. wattsi* to a site where the resident *A. gingivinus* were left undisturbed. The site is LU. Again we obtained very low survivorship and territories were established by *A. wattsi* only at the periphery of the grove.

The 6-month summary statistics are recorded in Table 10.4. These experiments on the cay establish that *A. wattsi* adults can survive for long periods (at least 2 years) in the physical environment of sea-level habitat. They also establish a strong negative interaction of *A. gingivinus* against *A. wattsi*. The fact that *A. wattsi*'s territories are restricted to peripheral habitat when *A. gingivinus* is undisturbed, but occur in the centre of the groves when *A. gingivinus* is partially removed, suggests that interspecific territoriality is an important mechanism in the competition. Interspecific territorial behaviour between these species is readily observed although it has never been formally documented in the literature.

DENSITY EXPERIMENTS USING ENCLOSURES ON ST MAARTEN

The studies on the cay made it immediately clear that a sustained programme of density manipulation experiments would be impossible without a method for both confining lizards within an experimental site, and preventing the entry of lizards from outside the experimental area. We developed an enclosure technique and used it to determine whether *A. wattsi* has an effect on *A. gingivinus*; this is the reciprocal interaction to that detected in the cay experiments discussed above.

The enclosure design is illustrated in Fig. 10.7. An area 12 × 12 m area is fenced. The fence has a polypropylene overhang that cannot be crossed by

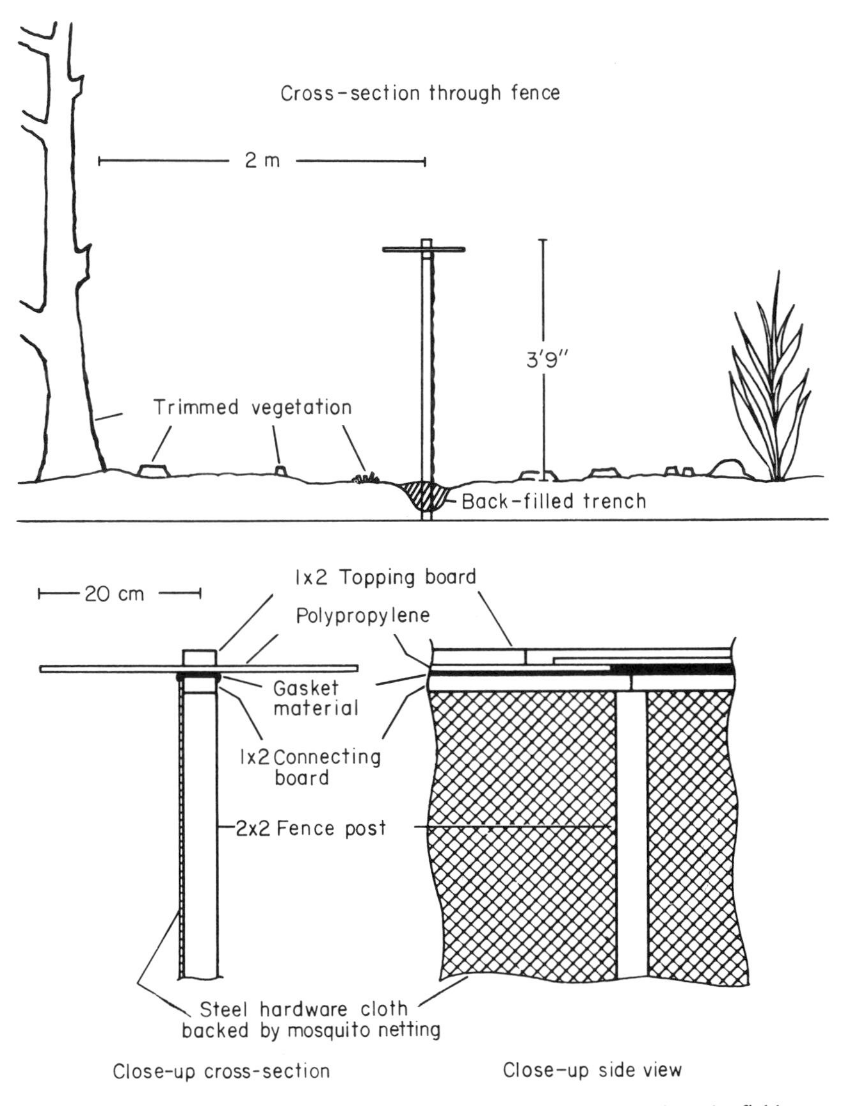

FIG. 10.7. Design for enclosures used to contain fixed densities of lizards under field conditions.

lizards. The fence is in a trench that is backfilled to prevent burrowing. A corridor is cut through the canopy of the vegetation to prevent lizards from moving in and out along branches.

Four enclosures were built on St Maarten near Pic du Paradis during November, December and January of 1980–81. The anoles in the enclosures were removed, and the enclosures were stocked with animals caught from the area outside the enclosures. Two enclosures were stocked with sixty *A. gingivinus* each. The other two were each stocked with sixty *A. gingivinus* plus 100 *A. wattsi*. The lizards were individually marked by toe clipping and the initial size distribution and sex ratio were matched among all the experiments. At monthly intervals from January until May 1981, *A. gingivinus* individuals were noosed and their length and weights were measured. At the end of the experiment most of the *A. gingivinus* individuals were collected. The specimens were analysed at the laboratory for stomach contents and reproductive condition.

The results appear in Table 10.5. Notice the entry for the amount of food found in the stomachs of the animals. *Anolis gingivinus* from the enclosures without *A. wattsi* have more food in their stomachs than those from the enclosures where *A. wattsi* is present. Probably this is responsible for most of the other entries in the table. *Anolis gingivinus* from the enclosures without *A. wattsi* grow faster than those from the enclosures with *A. wattsi*. And of great importance, the female *A. gingivinus* from enclosures without *A. wattsi* have more eggs in them, on the average, than those from enclosures with *A. wattsi*. Furthermore, the perch position of *A. gingivinus* shifts up in the

TABLE 10.5. Effect of *A. wattsi pogus* on *A. gingivinus* in Enclosure Experiments

	A. gingivinus alone				*A. gingivinus* and *A. wattsi*			
♂♂ Growth rate (mm day^{-1})	0.112	(0.008)	0.100	(0.007)	0.070	(0.005)	0.054	(0.006)
♀♀ Growth rate (mm day^{-1})	0.102	(0.005)	0.102	(0.006)	0.057	(0.004)	0.055	(0.005)
Egg volume/♀ (mm^3)	200.3	(23.4)	139.3	(21.0)	70.5	(17.8)	45.5	(11.5)
Stomach content volume, (mm^3)	81.7	(21.9)	54.3	(11.4)	27.6	(8.8)	27.0	(8.1)
Perch height, (m)	0.38	(0.03)	0.49	(0.05)	0.89	(0.06)	0.87	(0.06)
Prey size classified by volume, (mm^3)	3.6	(0.9)	3.9	(0.9)	1.9	(0.7)	1.6	(0.5)

*Standard error of estimate given in parentheses.
Each column is the result of one experiment.

presence of *A. wattsi*, as originally observed in the distributional data of 1977. Moreover, *A. gingivinus* takes, on the average, smaller prey when in the presence of *A. wattsi*.

These results suggest exploitative competition. *Anolis* growth rates depend on food availability (Stamps & Tanaka 1981). The reduced food level in the stomachs of *A. gingivinus* from the enclosures where *A. wattsi* was present is thus the probable cause of their lower growth rates.

Thus, the entries in Table 10.5 demonstrate a very strong negative effect of *A. wattsi* on *A. gingivinus*, and implicate exploitative competition as part of the mechanism for the competition.

DISCUSSION

The results consist of three distinct lines of evidence for the existence of strong present-day competition between the *Anolis* lizard populations of St Maarten. The distribution of the species on St Maarten indicates both an effect of *A. gingivinus* against *A. wattsi* (the restricted range of *A. wattsi*) and also an effect of *A. wattsi* against *A. gingivinus* (the lower abundance and perch shift of *A. gingivinus* where both species co-occur). The introduction experiments to the offshore cay of Anguillita demonstrate lower survival of *A. wattsi* in the presence of *A. gingivinus*. The enclosure experiments on St Maarten demonstrate lower growth and fecundity of *A. gingivinus* in the presence of *A. wattsi*.

We conjecture that present-day competition produces the community structure of anoles on St Maarten in the following way. The parameters of the population dynamics of competition vary from place to place depending primarily on elevation as related to heat and water stress. Using the terminology of the Lotka–Volterra competition equations, the carrying capacity of a site for *A. wattsi* and *A. gingivinus*, K_w and K_g, varies with elevation, as do the competition coefficients of *A. gingivinus* against *A. wattsi* and vice versa, $\alpha_{w,g}$ and $\alpha_{g,w}$. The evidence suggests that, at sea-level, *A. wattsi* is competitively excluded by *A. gingivinus* by present-day competition. If so, the Lotka–Volterra competition equations require that $\alpha_{w,g} > K_w/K_g$ at such locations. Also, the coexistence of these species in the hills requires that $\alpha_{w,g} < K_w/K_g$ at such places. Hence, along a transect from the hills to sea-level, the ratio K_w/K_g progressively drops and $\alpha_{w,g}$ rises. We conjecture that the species border for *A. wattsi* occurs approximately at the contour where $\alpha_{w,g}$ equals K_w/K_g.

An immediate significance to these studies is their ability to refute assertions that competition theory is a theory about nothing (e.g. Wiens 1977, Birch 1979, Connell 1980). It remains to be seen, of course, whether competition theory offers a useful mathematical representation of the competitive process. But it is clear that interspecific competition does occur, and that competition theory is a theory about a real process in nature.

The more lasting significance of these studies lies in the demonstration of strong present-day competition in a *co-evolved* system. Both populations on St Maarten are native to the Anguilla Bank (Anguilla, St Maarten, and St Barths), and are found nowhere else. It is widely believed that the co-evolution of competitors brings a reduction, even elimination of interspecific competition through niche divergence. Our findings of strong present-day competition on St Maarten challenges this belief.

On the other islands in the eastern Caribbean with two species, the separation of body sizes is much greater than on St Maarten. Studies on nearby St Eustatius show that there is virtually no present-day competition from the smaller lizard against the larger lizard (Pacala & Roughgarden 1982) or from the larger lizard against the smaller lizard (J. Rummel & J. Roughgarden, unpublished). According to traditional beliefs about the co-evolution of competitors, St Eustatius is a later stage in the process of niche divergence than is St Maarten.

However, there is no evidence, or reason to believe that the lizards of St Eustatius are older than those of St Maarten. To the contrary, the placement of the Anguilla Bank as the closest Lesser Antillean Bank to the Puerto Rico Bank has led Gorman & Kim (1976) to consider the Anguilla Bank as a stepping stone into the eastern Caribbean. If so, the St Maarten anoles are ancestral to islands like St Eustatius; and the species on St Maarten have been together for a longer time, not a shorter time, than those on St Eustatius.

A possible explanation for the biogeography of anoles in the eastern Caribbean is provided by mathematical theory for the coevolution of competitors (Roughgarden *et al.* 1983). The central idea is that there is a turnover of species on islands caused by an extinction during the co-evolution of competitors, followed by a re-invasion of the island after the extinction occurs. The extension results from co-evolution with asymmetrical competition. It is presumed that a large lizard has a larger competition coefficient against a small lizard than the reciprocal coefficient.

There are several phases to the cycle. First, a lizard species that has no other *Anolis* competitor evolves to a characteristic size, as observed on most of the islands with one species. The carrying capacity of an island is assumed to be a bell-shaped function of the logarithm of body size. The body size to which a solitary species evolves is the size corresponding to the highest carrying capacity. This 'solitary size' is the size that yields the highest net catch of insect food relative to a lizard's energy requirements.

Next, the island is invaded by a larger species. A species that has the solitary size, or smaller, cannot invade because of competition from the resident. In the model, the size of the invader is the size that produces the highest rate of increase in a propagule that has landed on an island where a resident is already established.

Because of the assumed asymmetry in the competition, only a larger lizard can successfully invade. It must come from outside the system. This large species presumably came from Puerto Rico, a complex island whose fauna includes large forms. It may have come during the Pleistocene when the sea level was as much as 100 m lower than it is today. But even then, the dispersers must have rafted across the sea, because there has never been a land-bridge connection between the Puerto Rico Bank and the small islands in the eastern Caribbean south of the Virgin Islands.

The history of introduced anoles in Bermuda may exemplify these assumptions about the invasion process (Wingate 1965). An anole from Jamaica was introduced that quickly spread across the island. Then a much larger anole from Antigua was introduced that now co-occurs with the first anole in forest habitat. Finally a third anole, the size of the first, was introduced from Barbados; it has not spread and exists only in small enclaves near the ocean.

After the invasion, the two species begin to co-evolve as competitors. The resident evolves a smaller body size to avoid competition from the invader. The invader also eventually evolves a smaller body size because it entered with a body size above that which yields the heighest carrying capacity. Indeed, the invader should shift to a smaller size because the resident has a lower abundance than when the invasion began, and also the resident is evolving away from its original size, thus vacating the niche space it originally occupied. Saint Eustatius is possibly at this stage.

As the resident continues its evolution of a smaller body size, the invader's body size continues to approach the characteristic size of a solitary species, as observed on St Maarten.

But during this co-evolution, the resident is eventually driven to extinction by competitive exclusion. When the extinction of the smaller species occurs, the invader has not attained the solitary size; it is still larger than the solitary size. This condition is observed on Marie Galante, the only island with one species whose size is not equal to the solitary size.

Finally, the invader completes its convergence to the solitary size and the island awaits another invasion.

Thus, on St Maarten we are witnessing intense interspecific competition as a possible prelude to the eventual extinction of *A. wattsi* from the entire island. At some time since the Pleistocene, *A. wattsi* had the size of a solitary species. Then St Maarten was invaded by an *A. gingivinus* population with a large body size. The *A. gingivinus* spread across the entire island and begin to evolve towards the solitary size. As *A. wattsi* responded by evolving to be smaller than the solitary size, its range contracted so that it is now concentrated in the central hills of St Maarten. The *coup de grace* may be a series of drought years during which the central hills of St Maarten experience weather similar to what is normal in xeric sea-level habitat. In these conditions, the extinction of *A. wattsi* from St Maarten as a result of competitive exclusion by *A. gingivinus* is a real possibility.

ACKNOWLEDGEMENTS

We thank the Fleming family of Saint Maarten for allowing our use of their land for long-term studies on St Maarten. We thank Carl and Vikki Leisegang of the *Vikki Too* for their seamanship and hospitality during the study in Anguillita. We also thank Stephen Adolph, David Heckel, and Warren Porter, for their help during this research. Finally, we gratefully acknowledge the support of the National Science Foundation.

REFERENCES

Adolph S.C. & Roughgarden J. (1983) Foraging by Passerine birds and *Anolis* lizards on St Eustatius (Neth. Antilles): Implications for interclass competition, and predation. *Oecologia*, **56**, 313–317.

Birch L.C. (1979) The effect of species of animals which share common resources on one another's distribution and abundance. *Population Ecology*, *Fortschr Zool*, **25**, 197–221.

Connell J.H. (1980) Diversity and the co-evolution of competitiors, or the ghost of competition past. *Oikos*, **35**, 131–138.

Connor E. & Simberloff D. (1979) The assembly of species communities, chance or competition? *Ecology*, **60**, 1132–1140.

Gorman G. & Kim Y.S. (1976) *Anolis* lizards of the Eastern Caribbean: A case study in evolution II: Genetic relationship and genetic variation of the *bimaculatus* group. *Systematic Zoology*, **25**, 62–77.

Heckel D.G. & Roughgarden J. (1979) A technique for estimating the size of lizard populations. *Ecology*, **60**, 969–975.

Lazell J. (1972) The Anoles (Sauria, Iguanidae) of the Lesser Antilles. Bull. Museum Comparative Zool. (Harvard University, Cambridge, Mass.) **143**, 1–115.

Pacala S. & Roughgarden J. (1982) Relation between resource partitioning and interspecific competition in two-species insular *Anolis* lizard communities. *Science*, **217**, 444–446.

Pregill G. & Olson S. (1981) Zoogeography of West Indian Vertebrates in relation to Pleistocene climatic cycles, *Annual Revue of Ecology and Systematics*, **12**, 75–98.

Roughgarden J., Porter W. & Heckel D. (1981) Resource partitioning of space and its relationship to body temperature in *Anolis* lizard populations. *Oecologia*, **50**, 256–264.

Roughgarden J., Heckel D. & Fuentes E.R. (1983) Co-evolutionary theory and the biogeography and community structure of *Anolis*. In *Lizard Ecology: Studies of a Model Organism*. (Ed. by R.B. Huey E.R. Pianka & T.W. Schoener) Harvard University Press, Cambridge, Mass. pp. 371–410.

Schoener T.W. (1983) Field experiments on interspecific competition. *American Naturalist*, **122**, 240–285.

Stamps J. & Tanaka S.K. (1981) The influence of food and water on growth rates in a tropical lizard *(Anolis aeneus)*. *Ecology*, **62**, 33–40.

Strong Jr D. Szyska L. & Simberloff D. (1979) Tests of community wide character displacement against null hypotheses. *Evolution*, **34**, 332–341.

Terborgh J. & Faaborg J. (1980) Saturation of bird communities in the West Indies. *American Naturalist*, **116**, 178–195.

Wiens J.A. (1977) On competition and variable environments. *American Science*, **65**, 590–597.

Wingate D. (1965) Terrestrial herpetofauna of Bermuda. *Herpetologica*, **21**, 202–218.

11. THE EVOLUTION OF MUTUALISM

JOHN VANDERMEER
Division of Biological Sciences, University of Michigan,
Ann Arbor, Michigan 48109, USA

INTRODUCTION

Though the word mutualism was first used in 1875 and attracted great attention from naturalists in the late nineteenth century, it was not until 1902 that the first attempt to treat the subject theoretically appeared, the famous *Mutual Aid: a Factor in Evolution* by Peter Kropotkin. But little attention was paid to the well-known anarchist Kropotkin—his undisguised contempt for social Darwinism was not to gain him a great deal of favour in those times—and with the exception of Kostitzin's theoretical work (which was equally ignored), theoretical treatment of mutualism lagged far behind that of competition and predation (cf. Risch & Boucher 1976; Boucher, James & Keeler 1982).

Almost any ecology textbook includes substantial treatments of competition and predation, but frequently neglects to mention mutualism, or at bext gives it only a superficial treatment. Yet of the three, mutualism is certainly the most commonly observed in nature. Pollination, nodulated legumes, mycorrhizal fungi, zooxanthellae, and a host of other spectacular mutualisms are ready observable in every corner of the globe. Yet the less frequently observed predation or the almost never observed competition have received most of the attention.

As to why this is the case, Risch & Boucher (1976) have suggested that a socio-political climate ripe for competitive or negative interactions may have subconsciously influenced the way in which ecologists formulate research problems in the first place. R. May (personal communication) feels that the most elementary demographic models one can construct for population interactions give interesting results for competition and predation but, at first glance, rather silly results emerge for mutualism. Consequently ecologists have sunk their teeth into the first two, but have not really pursued the latter until recently.

Regardless of the reason, it is clearly the case that a theoretical framework for mutualism has only recently come into its own. With the exception of brief treatments by Whittaker (1975), May (1974, 1976), and Christiansen & Fenchel (1977), the simple analysis of Vandermeer & Boucher (1978) was the first attempt to develop a systematic theory for demographic change under mutualism. Since that work several others have taken the problem to task (Goh 1979; Post, Travis & De Angelis 1981; Addicott 1978; Heithaus, Culver & Beattie 1980), and presently theoretical developments in mutualism seem to be gaining ground rapidly.

As May has noted, the simple Lotka–Volterra equations give quite interesting qualitative results: predator–prey pairs will oscillate and competitors will either coexist or one will exclude the other (the so-called competitive exclusion principle). Yet the same equations, when cast in a mutualist form predict that the populations will both grow beyond their carrying capacity, increasing forever towards infinity. This is the silly result. Vandermeer & Boucher (1978) were able to make sense out of such simple equations by utilizing two tricks; first, allowing populations to be either obligate or facultative mutualists, second by establishing a criterion of persistance rather than mathematical stability for coexistence. A summary of their results is presented in Table 11.1. Using this formulation the possible outcomes of mutualistic interactions are somewhat more complicated than those of predation or competition, but nevertheless are sufficiently regular to offer a useful summary. Briefly, if both mutualists are facultative, both will persist indefinitely, but if both are obligate either they will persist together or both go extinct, depending upon the strength of the mutualism. Finally, if one is obligate and the other faculative either the obligate species will go extinct or both will coexist, again depending upon the strength of the mutualism.

TABLE 11.1 Summary of biological conditions of mutualism (after Vandermeer & Boucher 1978).

Formal mathematical condition of the equilibrium point	Facultative	Obligate	Mixed (one facultative) (one obligate)
Stable	Coexistence	Extinction of both	Indeterminate (extinction)* Extinction of obligate
Unstable	Coexistence	Indeterminate (persistence)†	Indeterminate (extinction)

*Indeterminate (extinction) means that either the obligate mutual becomes extinct or both persist together.

†Indeterminate (persistence) means that either both species become extinct or both persist together indefinitely, depending on conditions.

With this theoretical demographic base in mind, it seems appropriate to explore potential evolutionary patterns.

ORIGINS OF DIRECT MUTUALISMS

It is a difficult subject. On the one hand we would like to develop a general framework for evolutionary origins, but on the other hand an evident diversity of origins creates an apparently unsurmountable task. Mutualisms, ubiquitous as they are in nature, seem to arise from many different ancestral situations, as briefly discussed in what follows.

Undoubtedly a large number of mutualisms have no interesting evolutionary origins in the first place. Koptur (1979) found that the *Vicia*, introduced from Europe, provides extrafloral nectar for the ant *Iridomyrmex*, introduced from Argentina. The ant then provides some degree of protection against herbivores and receives the nectar in return, a common form of mutualism, especially in the tropics (Bentley 1976). Thus two organisms from totally different parts of the world come together to form what looks like an obviously coevolved mutualism in their non-native California. I suspect many mutualisms, especially those subtle ones that have as of yet escaped our attention, are of this form.

Probably an important origin for several types of mutualisms is through a neutralistic association (Odum 1963)—neutralism to commensalism to mutualism. For example, imagine ants foraging on a spiny shrub, neither the ants nor the shrub gaining or losing anything from this accidental association. The ants begin nesting in the occasionally hollow rotten spine, spending more time on the shrub and eating more local insects, some of which might have eaten leaves of the plant had they not been eaten first by the ants. It does not take much imagination to continue the story and eventually come up with the popular ant–acacia mutualism of Central America (Janzen 1966).

A third point of origin is through a host–parasite relationship. Such an origin has been fully documented by Jeon & Jeon (1976) in which a normally virulent bacterium gradually lost its virulence. The host, *Amoeba proteus*, subsequently became dependent upon the bacterium. Rarely are documented cases of evolutionary origins so clear, but numerous other mutualisms seem to have this origin, including mycorrhizal fungi (Harley 1959), root nodule bacteria (Burns & Hardy 1975), coral and zooxanthellae (Taylor 1974) and lichens (Hale 1974). Roughgarden's (1975) cost–benefit model specifically applies to this evolutionary pathway.

A fourth possible point of origin is through ingestion, clearly the most likely route for gut symbionts of all sorts (Howard 1967).

A discussion of the origins of mutualisms ought really to include a consideration of breakdowns. If we postulate that a species pair can evolve from competition to neutralism to mutualism, for example, we could just as easily postulate the reverse course. The well-known example of ants eating the homopterans they usually mutualistically associate with is possibly an example of the initiation of a breakdown (Way 1963). Frequently an additional species may associate with a mutualistic pair in a hyperparasitic fashion, such as nectar-robbing. Finally, one of the species may be left behind after an emigration. The breakdown of the *Cecropia-Azteca* association (Janzen 1973) on the island of Puerto Rico is an excellent example. In this example the plant has even lost the mullearian bodies which normally provide food for the *Azteca* associates.

As discussed earlier, approaching the evolutionary origin of mutualism from a generalized point of view is quite difficult owing to the apparent diversity of origins. The best and most recent attempt to deal with this problem is that of Wilson (1980), as described by John Maynard Smith elsewhere in this volume.

EVOLUTIONARY PATTERNS OF DIRECT MUTUALISM

It may be reasonable to postulate a negative correlation between evolutionary changes in K, the carrying capacity, and α, the mutualism coefficient. In fact the classical notions of r and K selection are somewhat contrary to intuition here. I do not mean to imply that a dichotomy of r-selection and k-selection actually exists in nature. Rather, to the extent that at least one pattern of natural selection may be viewed as a continuum from r-selection characteristics to k-selection characteristics, it makes sense to analyse the ends of that continuum as a dichotomy. Normally α and K are expected to respond in the same way to ordinary selection pressures (e.g. Pianka 1970, 1972). But such an expectation is based upon α being a competition coefficient. A population near its carrying capacity or near its equilibrium point (as defined by the competition coefficients) is likely to respond quite differently than a population which is periodically reduced to a very small number—the former of course is thought to be K-selected, the latter r-selected.

But the situation is somewhat different if α represents a mutualism coefficient. For example, in Fig. 11.1 I have plotted a typical example of two facultative mutualists. A population mix starting at point P_0—the condition for 'r-selection'—will clearly favour those genotypes of X_1 corresponding to the isocline labelled B. On the other hand a population mix starting at point P_1—the condition for K-selection—will clearly favour those genotypes of X_1 corresponding to the isocline labelled A. Thus, under mutualism

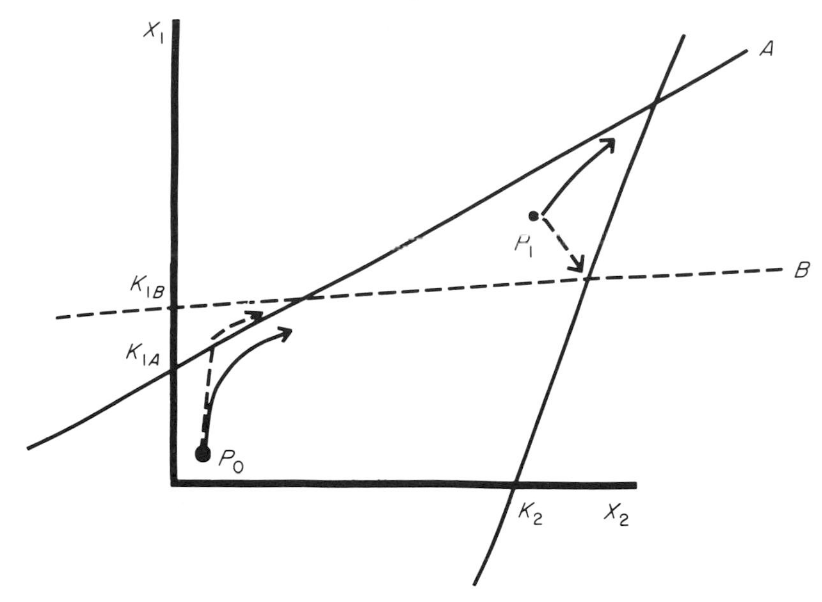

FIG. 11.1. Illustration of the evolutionary sequence in an r-selected population (near point P_0) and a K-selected population (near point P_1).

the rather surprising result is that r-selection leads to increased values of K and decreased values of α, whereas K-selection leads to decreased values of K and increased values of α. The above analysis is predicated upon the assumption that α and K are inversely correlated. A somewhat weaker yet qualitatively similar result emerges if that assumption is relaxed. With reference to Fig. 11.1 the obvious conclusion is that r-selection favours increasing K (see point P_0) and K-selection favours increasing mutualism (see point P_1). The corollaries (r-selection favouring decreasing mutualism and K-selection favouring decreasing K) derive from the assumed negative correlation of α and K. Relaxing that assumption weakens the result, but its general character remains.

If we take the above analysis to an extreme we are led to the interesting result that K-selection in a facultative mutualist system will lead to obligate mutualism, that is, as selective pressure drives K ever lower, eventually K approaches zero (or negative) and an obligate mutualist system results. We thus expect that facultative mutualists will be found most commonly amongst r-selected species and obligate mutualists amongst K-selected species.

The only data of which I am aware that might be brought to bear on this prediction is that of Janos (1980) on rain-forest plants. Through a series of

TABLE 11.2. Percentage of rain forest plants that are facultative or obligate mutualists with mycorrhizal fungi, as a function of their successional position. Pioneers, being the most *r*-selected, are all facultative. Primary forest species, being the most *K*-selected, are almost all obligate. Numbers in parenthesis indicate number of plant species. (data from Janos 1980).

	Facultative (%)	Obligate (%)
Pioneers	100(5)	0(0)
Second Growth	17(1)	83(5)
Primary	11(1)	89(8)

ingenious experiments Janos was able to determine whether a plant species was an obligate or facultative mutualist with respect to mycorrhizal fungi. If we look at only those plants he classified as pioneers or primary forest species we find that all five pioneers were facultative while eight of the nine primary forest species were obligate. These data correspond remarkably well with the prediction, if we presume that a pioneer is usually subjected to an *r*-selected environment and a primary forest species to a *K*-selected environment. Nevertheless, if we examine what Janos calls second-growth species, the prediction does not hold up nearly so strongly. Only 17% of the second-growth species were facultative, as compared to 11% of the primary forest species. While the percentages are in the predicted direction, the results are not all that spectacular. It could be argued that second-growth species are not subjected to *r*-selection environments nearly as strongly as are pioneers, thus explaining in a general qualitative way the results. Indeed, as is so frequently the case in evolutionary ecology, a patchwork of *post hoc* explanation can preserve almost any prediction. Rather than engage in such exercise, I ask the reader to digest the data in Table 11.2. While weak, the trend is certainly in the right direction. D.P. Janos (personal communication) has also noted that a truly accurate assessment must take account of different levels of soil fertility, a control which was not included in his 1980 study.

INDIRECT MUTUALISMS

Although the existence of indirect effects operating in communities has long been well known (e.g. Clements 1928, p. 5), recently they have been pinpointed as theoretically important in community structure (Lawlor 1979; Levins 1975; Levine 1976; Vandermeer 1980; Boucher *et al.* 1982). An example from nature is that of Yih (1982) who showed how in an intercropping situation two crops could combine to form positive relationships with one another through a kind of 'pre-emption' of the niche space that would normally be occupied by weeds in a monocultural situation. Yih's results showed that a positive association could be formed (by no means necessarily) from the underlying structure of the association. The basic system is illustrated in Fig. 11.2, where C_i represents the ith species, with $i = 3$ symbolizing the weeds. While the basic structure is motivated by the agro-ecosystem example, it is worthwhile to suggest that such triplets exist within the structure of other natural communities. Such is one of the basic structures analysed by Levine (1976), in which he showed that species 2 will have a positive effect on species 1 if,

$$a_{13}a_{32} - a_{12} > 1 \tag{1}$$

and species 1 will have a positive effect on species 2 if,

$$a_{23}a_{31} - a_{21} > 1 \tag{2}$$

in which case mutualism will exist if both relations 1 and 2 are satisfied. This is perhaps the simplest manner in which an indirect mutualism can be formed. What might be the evolutionary implications of such a structure?

It seems unlikely to suggest any direct coevolutionary changes to result from such an indirect association. Rather, several indirect evolutionary pressures are likely to be involved. Firstly, the existence of the mutualism is a consequence of particular combinations of interaction coefficients, the values of which change as a result of evolutionary pressures arising from

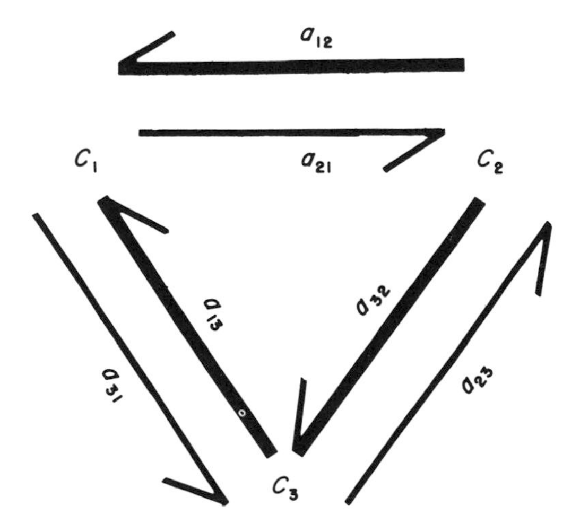

FIG. 11.2. Three species competitive situation where C_3 represents the weed community and C_1 and C_2 are two crop types. (From Yih 1982). The coefficients involved in generating a positive effect of species 2 on species 1 are highlighted (see eqn 1).

outside of the system. As the coefficients evolve, the propensity to form an indirect mutualism changes. For example, as the competitive ability of species 2 against species 1 increases, it becomes increasingly less likely that species 1 and 2 will from an indirect mutualism (see eqns 1 and 2). But secondly, this sets up what may be an additional evolutionary pressure. If, through evolutionary change resulting from natural selection originating outside the system, a move towards less mutualism results, the consequent negative consequences—implied by the loss of mutualism—could form a countervailing pressure. This countervailing pressure could in turn lead to the reverse evolutionary change, assuming that the appropriate population structure existed so as to allow for some evolutionary mechanism (individual, kin, interdemic, or trait-group selection) to operate, and that the appropriate genetic basis was in place. What emerges is a picture of counterbalancing evolutionary forces which suggest a limit on the evolution of competitive ability, as further analysed below.

A slightly more complicated structure which results in similar consequences is that of Levine (1976). Levine's equations (originally proposed by MacArthur 1972, for the purpose of studying competition) are summarized in Fig. 11.3. If the resources in the system of Fig. 11.3 interact, the net effect on the consumers can be that they are effectively mutualists. A further analysis of Levine's equations (Vandermeer 1980) indicates what parameters must be varied to generate indirect mutualism. Of particular interest are the parameters which specify the intensity of the positive effects of the resources on the consumers, the parameters that specify the consumers' eating habits. As the consumers become generalized in their diets (as their consumption

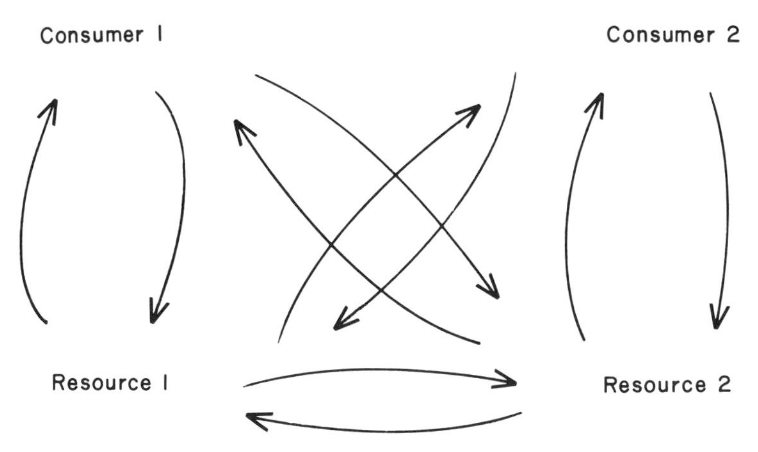

FIG. 11.3. Diagrammatic representation of MacArthur's consumer–resource system as analysed by Levine (1976).

of the non-favourite food approaches that of the favourite food), indirect mutualism is more likely to emerge (cf. Vandermeer 1980, for details).

Since the degree of specialization is a determinant of the presence or absence of indirect mutualism, it follows that those evolutionary mechanisms involved in niche breadth (or niche width, or degree of specialization) will also be involved in indirect mutualism. A great deal has been written about the evolution of the niche (e.g. Roughgarden 1972). It is not my intent to summarize that literature, rather only to note that certain evolutionary pressures lead to specialization, others to generalization. Those pressures leading to increased generalization also lead to indirect mutualism. While indirect mutualism will thus arise from evolutionary change, its appearance is itself the indirect result of evolutionary pressures applied elsewhere. In this case, indirect mutualism evolves indirectly from wholly independent evolutionary forces.

But the evolutionary pressures leading to specialization suggest something quite different with respect to indirect mutualism. If we begin with two highly generalized consumers, where their food sources are competing with one another, we are likely dealing with a case of indirect mutualism. If we then exert selective pressure that would ordinary result in an evolution towards specialization (for the consumers), as the populations change to become more specialized they experience an indirect selective pressure by way of one another. That is, as they become more specialized they move closer to a position of negative rather than positive interaction. Thus, again, the fact of indirect mutualism may exert a counter-evolutionary force.

A totally different form of indirect mutualism is suggested by the experiments of Boucher & Espinoza (1982). In an intercrop of corn and cowpea they discovered that the utilization of nitrogen by the growing corn resulted in nitrogen-deficient soil to which the cowpeas responded by extensive nodulating activity. Presumably a net gain in nitrogen for both species thus resulted from what was initially a competitive interaction.

This situation is similar in basic form to several other reported indirect mutualisms. For example, the well-known effect of lowered herbivore loads in intercrops (Bach 1980; Risch 1981; Tahvanainen & Root 1972) represents a similar situation in that the basic environment of one crop is modified by the presence of a second crop. Messina's recent example of the protection of goldenrod from two beetle herbivores by the ants that tend membracids on the goldenrod (Messina 1981) is a similar situation. The addition of the membracids causes an initial negative effect on the goldenrod. But the ants which tend the membracids effects control over two of the major herbivores. Again, a second species (the membracid) creates an environmental modification (by bringing the ants along).

The above examples suggest that we might usefully categorize indirect mutualisms into two categories, first, those that derive directly from species interactions and second, those in which one species effectively modifies the environment of the other. The sort of demographic models which might be applied to either case could be quite distinct. But the feature seemingly common to both situations is that there is an evolutionary pressure that results from the indirect mutualism and that may act as a countervailing force in evolution.

We can model the entire process by breaking down the two fitness components. Suppose w_1 is the fitness component which results from the direct selective pressure operating to increase specialization. Suppose w_2 is the fitness resulting from the indirect effect of decreasing mutualism. Then, according to the hypothesis,

$$\frac{\partial w_1}{\partial x} > 0$$

$$\frac{\partial w_2}{\partial x} < 0$$

where x is the characteristic changing as a result of external evolutionary pressure (e.g. competitive ability or degree of specialization). We now suppose that evolution operates through a linear combination of the two fitness components. Thus we may write the adaptive function as,

$$A = pw_1 + (1 - p)w_2 \qquad (3)$$

wherein we idealize the evolutionary process as maximizing A. We can represent all possible values of w_1 and w_2 as a set (more or less equivalent to the 'fitness set', Levins 1975) and solve for the largest value of A, within the constraints stipulated by eqn 3. The fitness set will be concave if,

$$\frac{\partial^2 w_i}{\partial x^2} > 0 \quad \text{for} \quad i = 1, 2; \quad \text{for all } x$$

and convex otherwise. The obvious result is that with a concave fitness set one or the other selective pressures will tend to dominate, while with a convex set it is likely that a balance of evolutionary forces will exist, the direct selective pressure pushing towards specialization and the indirect pressure pushing in the opposite direction. In this way we obtain the qualitative situation where, under some circumstances, indirect mutualism acts as a constraint on evolution. I should note in passing that there is no guarantee that selection will operate so as to maximize A of eqn 1. This caveat is especially important in the case of the concave fitness set. While a concave fitness set in no way suggests a balance of forces, it may be indeterminate which component of fitness dominates. That is, there may be a saddle point, to one side of which the direct evolutionary pressure dominates to the other side of which the indirect pressure dominates.

REFERENCES

Addicott J.F. (1978) Competition for mutualists: aphids and ants: *Canadian Journal of Zoology*, **56**, 2093–2096.

Bach C. (1980) Effects of plant density and diversity on the population dynamics of a specialist herbivore, the striped cucumber beetle, *Acalymma* vittata (Fab). *Ecology*, **61**, 1515–1530.

Bentley B. (1976) Plants bearing extrafloral nectaries and the associated ant community: interhabitat differences in the reduction of herbivore damage. *Ecology*, **54**, 815–820.

Boucher D.H. & Espinoza J. (1983) Increasing mutualism through competition: nodulation in a maize-bean mixture. *Tropical Agriculture* (Trinidad) (in press).

Boucher D.H., James S. & Keeler K.H. (1982) The ecology of mutualism. *Annual Review of Ecology and Systematics*, **13**, 315–347.

Burns R.C. & Hardy R.W.F. (1975) *Nitrogen Fixation in Bacteria and Higher Plants*. Springer, New York.

Christiansen F.B. & Fenchel T.M. (1977) *Theories of Populations in Biological Communities*. Springer, New York.

Clements F.E. (1928) *Plant Succession and Indicators*. H.W. Wilson & Co., New York.

Dawkins R. (1976) *The Selfish Gene*. Oxford University Press, New York.

Feder H.M. (1966) Cleaning symbiosis in the marine environment. *Symbiosis*, Vol. 1 (Ed. by S.M. Henry) Academic Press, New York.

Gilbert L.E. (1975) Ecological consequences of a co-evolved mutualism between butterflies and plants. *Coevolution of Animals and Plants*, (Ed. by L.E. Gilbert & P.H. Raven) University of Texas Press, Austir.

Goh B.S. (1979) Stability in models of mutualism. *American Naturalist*, **113**, 261–275.

Hale M.E., Jr. (1974) *The Biology of Lichens*, 2nd ed. Elsevier, New York.

Harley J.L. (1959) *The Biology of Mycorrhiza*. Leonard Hill Ltd, London.

Heithaus E.R., Culver D.C. & Beattie A.J. (1980) Models of some ant–plant mutualisms. *American Naturalist*, **116**, 347–361.

Howard B.H. (1967) Intestinal micro-organisms of ruminants and other vertebrates. *Symbiosis*, vol. 2 (Ed. by S.M. Henry). Academic Press, New York.

Janos D.P. (1980) Vesicular-arbuscular mycorrhizae affect lowland tropical rain forest plant growth. *Ecology*, **61**, 151–162.

Janzen D.H. (1966) Coevolution of mutualism between ants and acacias in America. *Evolution*, **20**, 249–275.

Janzen D.H. (1973) The disolution of mutualism between *Cecropia* and its *Azteca* ants. *Biotropica*, **5**, 15–28.

Jeon K.W. & Jeon M.S. (1976) Endosymbiosis in amoebae: Recently established endosymbionts have become required cytoplasmic components. *Journal of Cellular Physiology*, **89**, 337–344.

Koptur S. (1979) Facultative mutualism between weedy vetches bearing extrafloral nectaries and weedy ants in California. *American Journal of Botany*, **66**, 1016–1020.

Lawlor L.R. (1979) Direct and indirect effects of n-species competition. *Oecologia*, **43**, 355–364.

Levine S.H. (1976) Competitive interactions in ecosystems. *American Naturalist*, **110**, 903–910.

Levins R. (1975) Evolution in communities near equilibrium. *Ecology and Evolution of Communities*. (Ed. by M.L. Cody & J.M. Diamond). Belknap Press, Cambridge, Mass.

MacArthur R.A. (1972) *Geographical Ecology*. Harper and Row, New York.

May R. (1974) *Stability and Complexity in Model Ecosystems*, 2nd ed. Princeton University Press, Princeton.

May R. (1976) Models for Two Interacting Populations. (Ed. by R. May) *Theoretical Ecology: Principles and Applications*. Saunders, Philadelphia.

Messina F. (1981) Plant protection as a consequence of a ant-membracid mutualism: interactions on goldenrod (*Solidago* sp.). *Ecology*, **62**, 1433–1440.

Odum E.P. (1963) *Fundamentals of Ecology*. Saunders, Philadelphia.

Pianka E.R. (1970) On r-and K-selection. *American Naturalist*, **104**, 592–597.

Pianka E.R. (1972) *r* and *K*-selection or *b* and *d* selection? *American Naturalist*, **106**, 581–588.

Post W.M. III, Travis C.C. & DeAngelis D.L. (1981) Evolution of mutualisms between species. (Ed. by K.L. Cooke & S. Busenberg) *Differential Equations and Applications in Ecology, Epidemics, and Population Problems*. Academic Press, New York.

Risch S. & Boucher D. (1976) What ecologists look for. *Bulletin of the Ecological Society of America*, **57**, 8–9.

Risch S.J. (1981) Insect herbivore abundance in tropical monocultures and polycultures: an experimental test of two hypotheses. *Ecology*, **62**, 1325–1340.

Roughgarden J. (1972) Evolution of niche width. *American Naturalist*, **106**, 683–718.

Roughgarden J. (1975) Evolution of marine symbiosis: a simple cost benefit model. *Ecology*, **56**, 1201–1208.

Tahvanainen J.O. & Root R.B. (1972) The influence of vegetational diversity on the population ecology of a specialized herbivore, *Phyllotreta Cruciferae* (Coleoptera: Chrysomelidae). *Oecologia*, **10**, 321–346.

Taylor D.L. (1974) Symbiotic marine algae: taxonomy and biological fitness. *Symbiosis in the Sea*. (Ed. by W.B. Vernberg). University of South Carolina Press, Columbia, S.C.

Vandermeer J.H. (1973) Generalized models of two species interactions: a graphical analysis. *Ecology*, **54**, 809–818.

Vandermeer J.H. (1980) Indirect mutualism: variations on a theme by Stèphen Levine. *American Naturalist*, **116**, 441–448.

Vandermeer J.H. & Boucher D.H. (1978) Varieties of mutualistic interactions in populations models. *Journal of Theoretical Biology*, **74**, 549–558.

Way M. (1963) Mutualism between ants and honeydew—producing homoptera. *Annual Review of Entomology*, **8**, 307–343.

Whittaker R.H. (1975) *Communities and Ecosystems*, 2nd ed. Macmillan, New York.

Wilson D.S. (1980) *The Natural Selection of Populations and Communities*. Benjamin/Cummings Pub. Co., Menlo Park, Calif.

Yih K. (1982) Effects of weeds, intercropping, and mulch in the temperate zones and tropics—some ecological implications for low-thechnology agriculture. PhD dissertation, University of Michigan.

12. THE EVOLUTIONARY ECOLOGY OF PREDATION

JEREMY J.D. GREENWOOD
Department of Biological Sciences, University of Dundee, Dundee DD1 4HN

INTRODUCTION

The evolutionary ecology of feeding relationships has become a popular field of research. Much of the interest has centred on aspects of predation, about which an enormous body of literature has grown. I do not intend to review the whole field here: to do so within the available compass would entail a treatment so superficial as to be virtually worthless. Furthermore, in recent years there have been a number of reviews of the whole or of parts of the field, it has been the subject of entire symposia, and it has been summarized in various more general works (Curio 1976; Edmunds 1974; Kamil & Sargent 1981; Krebs & Davies 1978, 1981; Morse 1981), so another brief review could be no more than repetitive. Instead, I shall consider various consequences of a characteristic of predation that results in the evolutionary ecology of predation being rather different from that of other heterotrophic relationships and in which there has been a rapid increase in interest recently.

That characteristic feature is unpredictability. To provide a basis for this discussion, I shall first consider how predation may usefully be defined and then sketch in an evolutionary background, considering optimization principles and the evolutionary arms-race concept in particular.

WHAT IS PREDATION?

When members of one species regularly eat members of another, both the evolution and the ecology of each species will partly depend on that of the other. The effect on the evolution and population dynamics of the food species will depend on whether the feeding results in death of the affected individuals or merely in debilitation. The interaction will also be influenced by the numbers of the food species that an individual consumer utilizes. It is therefore convenient to consider feeding relationships as falling into four classes, following Williamson (1972). These are:

1 The consumer kills many whole organisms, though it may not eat the whole of each. Typical predators do this.

2 The consumer takes parts of several of many individual food organisms. Typical herbivores do this.

3 The consumer kills one whole food organism during its life. Larvae of insect parasitoids exemplify this.
4 The consumer takes part of only one organism during its life. Many parasites do this.

Like many classifications in biology, this one imposes divisions where there is continuity and overlap. A caterpillar on a tree belongs to class 4, unless it moves to a neighbouring tree, when it moves to class 2. Furthermore, a bird searching for worms in a lawn (class 1) is behaving similarly to a highly selective herbivore searching for a particular tissue of a particular species in a mass of vegetation (class 2). Yet the classes represent real clusters within the continuum of trophic relationships, witness of which is the fact that three chapters of this symposium volume correspond to three of them. Moreover, the characteristics by which we define the classes focus our attention on the fundamentals of the interactions, without the distraction of the particular features of the interaction between two particular species.

Class 1 is clearly the one to which our attention should be devoted at present. Some of the consumers in this class, typically the filter feeders, scarcely react to the individual members of their diet. I shall ignore these and consider as predators only those that in some sense treat their prey as individuals. Let us not forget that this includes a great variety of organisms, not only many of the animals is that feed on other animals but also most seed-eaters, the various small invertebrates that feed on whole individuals of microscopic fungi and algae, and the protistans that eat other protistans or prokaryotes. Because of the intense but ambiguous relationship between man and the larger carnivorous vertebrates, which is a study all of its own (cf. Eaton 1978), the lesser hunters are sometimes forgotten even by ecologists.

EVOLUTIONARY BACKGROUND

Optimization

My basic approach in this chapter will be that which has come to be known as 'optimal foraging theory'. I do not intend to discuss the validity of optimization models in ecology. That has been done before and, although much of the discussion has comprised tilting at windmills, I think that the issues are clear enough not to bear repeating. They have been carefully considered by Maynard Smith (1978). The very existence of this symposium attests to a general acceptance of the view that evolutionary theory has useful contributions to make to ecology. I believe that a major contribution is to provide a basis from which general principles may be deduced and I believe that optimality models are the means of doing so. I do not believe that the testing of such models involves a test of Darwinian theory but that the theory may be taken as established for our purposes, so that attention may be focused on particular features of individual models. In principle, optimization models do, of course, take the rather extreme view that organisms are perfect. That is not a prediction of Darwinism, which merely states that most of the attributes of organisms have evolved by natural selection. However, in practice, perfection is only assumed to occur within certain constraints. This is better Darwinism, though we must always be prepared to ask why an organism operates within particular constraints and not be too ready to believe that they simply reflect the yet-unbroken shackles of phylogenetic history. The history of evolutionary biology shows not only that such beliefs tend to inhibit further investigation but also that they are usually mistaken (Cain 1964, 1979).

Optimization theory is valuable to those biologists interested in the way in which evolution happens. It is also useful in ecology because it enables us to make predictions in areas where the available observations are so few or so complex that the principles are not already obvious. The predictions are thus wholly genuine. Lewontin (1979) has argued that testing principles predicted in this way requires so many experiments or observations that the generalizations would be obvious even in the absence of the underlying hypotheses. However, this ignores the way in which relevant observations tend to get submerged in the irrelevant in the absence of a guiding principle. Predictions from basic theories are especially valuable because their testing involves the collection of observations according to a defined programme. Such directed research tests principles far more efficiently than does research that is not directed towards such tests. Nowhere is this better shown than in the study of feeding behaviour, where investigations based on Darwinian predictions have resulted in a considerable increase in our understanding. We have come a long way since Buckland (1836) argued that the existence of carnivores was a manifestation of the benign dispensation of the Creator, since the quick and painless death afforded by predation lead to a greater aggregate of animal enjoyment than would occur if herbivores were condemned to die by slow starvation. Yet a major part of that advance has occurred only in the last 10% of the time that has elapsed since Buckland wrote, that is since the paper on optimal use of a patchy environment by MacArthur & Pianka (1966). As a result, while we can still acknowledge, and be fascinated by, the diversity of predator–prey interactions and agree with Leuthold (1977) that

> 'local circumstances in any one area may call for different patterns of antipredator measures in different species, or even in different populations of the same species',

we are also beginning to understand what circumstances produce what results, and why.

The fundamental optimization principle I shall use is that one expects organisms to behave in ways that maximize their fitness. I use the latter term in the population geneticists' sense of 'probable contribution to the next generation, relative to that of individuals of other genotypes', not only because of my upbringing in the school of ecological genetics rather than that of evolutionary ecology but also because it is the most precise and least ambiguous of the many senses in which the term has been used (cf. Dawkins 1982). Since the models I shall consider are concerned with single individuals, I can ignore the effects of individual behaviour on the fitness of relatives and what these mean for evolution, though these are relevant to certain aspects of predation, such as co-operative hunting and co-operative defence (Bertram 1978). I can also, for the same reason, ignore frequence-dependent and density-dependent effects, where the fitness of a genotype varies according to its frequency relative to that of other genotypes or to its absolute population density. Such effects are probably common and may have important evolutionary consequences (e.g. Greenwood & Elton, 1979; Clarke 1979). To introduce them here would, however, result mainly in complication and confusion.

Since fitness is a relative measure and may be frequency-dependent, it should be clear that the optimization principle I use applies to individuals and not populations. Even if natural selection always maximized 'mean fitness', which it does not (cf. Charlesworth in this volume), that parameter is merely the mean of a set of relative values, not an absolute property of the population (Cain & Sheppard 1954, 1956; Li 1955; Turner 1967). The only way in which any population property could be necessarily expected to be maximized in evolution is if evolution resulted mainly from group selection, which seems unlikely (Williams 1966; Maynard Smith 1976; Alexander & Borgia 1978).

ARMS-RACES BETWEEN PREDATORS AND PREY

It has been said (Popper 1978) that the publication of the *Origin* produced an astonishing change in the attitudes of scientists. In one important respect this is not, however, true. As Ruse (1982) has pointed out, the teleological interpretation of organic nature remained largely unchanged, though natural selection was substituted for God as the great designer.

That teleology persists today. Even though few, if any, evolutionary biologists would maintain that evolution is goal-directed, they find it convenient to use teleological analogy, such as Leuthold's 'call for' in the quotation above. This way of expression gives rise, of course, to a particular way of thinking, which is reinforced by speaking of the features of organisms as *ad*aptations *to* the environment. But organisms have not evolved in order to fit their environments: their evolution has been moulded *by* their environments, without any striving towards a goal.

Perhaps only by speaking of *ab*aptation *by* the environment, rather than *ad*aptation *to* the environment can we shed this teleological inheritance (Harper 1981). The need to do so is illustrated by the way in which thoroughgoing neo-Darwinians have been led into considering predator–prey evolution in group-selection terms by a teleological analogy.

Various authors have suggested that the co-evolution of predator and prey may usefully be considered as an evolutionary arms-race. Dawkins & Krebs (1979) write:

> 'Foxes and rabbits race against each other in two senses. When an individual fox chases an individual rabbit the race occurs on the time scale of behaviour. It is an individual race, like that between a particular submarine and the ship it is trying to sink. But there is another kind of racc, on a different time scale. Submarine designers learn from earlier failures. As technology progresses, later submarines are better equipped to detect and sink ships, and later-designed ships are better equipped to resist. This is an 'arms-race' and it occurs over a historical time scale. Similarly, over the evolutionary time scale the fox lineage may evolve improved adaptions for catching rabbits, and the rabbit lineage improved adaptations for escaping. Biologists often use the phrase 'arms-race' to describe this kind of evolutionary escalation of ever more refined mutual counter-adaptations.'

The concept of 'counter-adaptation' of prey to predators implies that natural selection promotes the evolution of features in prey that reduce the probability of success of their predators. But this is a mere side-effect, and not an inevitable one at that, of the result of natural selection, which is the evolution of features which increase the probability of survival of an individual, so that individuals with such features have a higher probability of escaping predation than those without them.

The importance of the subtle distinction may be illustrated by the story of the two trappers cowering in their tent in Greenland while a polar bear *Ursus maritimus* prowled outside. One, a native Greenlander, began to remove his boots. 'What are you doing?' whispered his Danish companion 'You know that even without your boots you can't run faster than a polar bear'. 'Yes' replied the Greenlander 'But without my boots *I can* run faster than you.'

The point is that prey do not race, in an evolutionary sense, against predators, and *vice versa*, but genotypes race against each other within each species. Only effective group selection could produce interspecific arms-races. Prey do not evolve anti-predator adaptations but anti-predation abaptations, which may not be the same thing at all.

A simple example will serve to show that so-called 'counter-adaptation' is not an inevitable side-effect of co-evolution. Many gregarious species clump together when approached by predators or when in places where predators are likely to attack them (Ainley 1972, 1974; Allen 1920; Altmann & Altmann 1970; Baerends & Baerends-van Roon 1950; Baskin 1974; Black 1970; Breder 1959; Bruns 1970; Crook 1960; Darling 1937; Eibl-Eibesfeldt 1962b; Galton 1883; Goss-Custard 1970; Hamilton 1971; Harding 1977; Horstmann 1953; Kitchen 1974; Klingauf & Segonca 1970; Kramer & Aeschbacher 1971; Kruuk 1964, 1972; Leuthold 1977; Lorenz 1963; Major 1977; Marshall 1965; McCullough 1969; Meinertzhagen 1959; Munro 1938; Nikolsky 1963; Potts 1970; Rudebeck 1950/1; Schaller 1967, 1972, 1977; Shanewise & Herman 1979; Spooner 1931; Tener 1965; Tinbergen 1951; Wursig & Wursig 1980). It seems reasonable to suppose that this is an abaptation by, among other things, predators selecting the prey that are closest to them, on the margins of flocks (Hamilton 1971). In some cases, it may be that clumping by the prey reduces the probability of success by the predators, as they are confused by the multiple targets milling before them or endangered by entering the maelstrom of fleeing bodies. But there are many cases where the clumping increases the predator's rate of success, because it can use methods of catching prey that are too imprecise to have much effect when used on isolated prey, or because it can take several prey in succession from a flock, or because prey individuals become more confused than the predator in the fleeing herd (Atwell 1954; Breder 1959; Bullis 1961; Crisler 1956; Hudleston 1958; Marshall 1965; Rich 1947). Some predators, indeed, appear to herd prey together, presumably for these very reasons (Bartholomew 1942; Bigelow & Schroeder 1953; Din & Eltringham 1974; Eibl-Eibesfeldt 1962a; Evans & Bastian 1969; Fink 1957; Hiatt & Brock 1948; Leatherwood 1975; Martinez & Klinghammer 1970; Morris & Dohl 1980; Saayman & Tayler 1973; Saayman, Tayler & Bower 1973; Serventy 1939; Springer 1957; Taverner 1926; Thomson 1964; Wursig & Wursig 1980). The highly integrated behaviour of the predators co-operating in such herding shows that it is an important feeding method, sufficiently productive for co-operative herding by cetaceans and men to have become established in several parts of the world (Busnel 1973; Fairholme 1856). There can be little doubt, therefore, that in many cases the clumping of prey, though an abaptation by predation, actually leads to greater success for the predators. Nothing could be further from the truth than 'counter-adaptation'.

Dawkins & Krebs (1979) went on to develop the argument that:

> 'The rabbit runs faster than the fox, because the rabbit is running for his life while the fox is only running for his dinner.'

They use this life–dinner principle to exemplify the conclusion that when the interaction of two species places stronger selection on one of the species than the other, then the evolutionary response of the strongly selected species is likely to be greater. While the general conclusion is undeniable, I have no doubt that the life–dinner principle is wrong as an expression of the general evolutionary interaction of predators and prey. It depends on the view that selection is weaker on foxes than rabbits, because failure in a chase means only slight loss of fitness for the fox, whereas failure to escape means complete loss of fitness for the rabbit. But this ignores the difference in encounter frequency between the two species. The relationships between successive trophic levels are such that, unless the predator is extremely inefficient, the number of encounters per unit time between predator and prey is usually much less for individual prey than for individual predators: to survive, foxes have to catch rabbits more often than rabbits have to escape foxes. Dawkins (1982) has dubbed this 'the rare enemy effect' and has accepted that the circumstances of particular cases will determine whether this effect or the life–dinner principle is the more powerful in determining the form of co-evolution.

Schaffer & Rosenzweig (1978) have considered the results of co-evolution of predators and prey in terms of its effects on the population dynamics of the two species. They concluded that there will be a steady state, at which the two species may be said to be evolving at the same rate, and which will usually entail an equilibrium of numbers, with the prey being sufficiently abundant that it is suffering considerably from resource depletion. This could be interpreted as the prey being well ahead in the so-called arms-race. However, their conclusion rests on the assumption that predation results in the evolution by the prey of features that decrease the absolute success rate of predators, and *vice versa*. Since this assumption is not universally valid, we cannot have much faith in the conclusions.

Consideration of population dynamics leads one to wonder whether the evolution of prey ever results in the extinction of predators, or *vice versa*. Theoretically, there is no reason why it should not. Nor, I believe, is there any a *priori* reason for supposing that prey evolution extinguishes predators more often than the converse. Turning to the data, it has been argued that the evolution of predatory crabs, teleosts, and snails in the late Mesozoic seas was directly responsible for the decline of endobyssate bivalves and articulate brachiopods on which some of them may have fed and that other extinctions may also have been predator-driven (Stanley 1979; Vermeij 1977).

Thus, in this sense also, there is no support for the broad idea that prey generally out-evolve predators.

UNPREDICTABILITY AS A CHARACTERISTIC OF PREDATION

The meaning of unpredictability

The predictability of a system is the extent to which its future state can be forecast from information obtained by observing its past and present states. Predictability depends not only on intrinsic properties of the system but also on how much observation has been made of its past state and how far ahead one is forecasting. In general, more past observations give more predictability, though the increase in predictability is usually less than proportional to the increase in the amount of observation and may, indeed, be limited. Also in general, but even less constantly, predictability is less for forecasts of longer range.

Since I define predictability in this operational fashion, its complement is not synonymous with strict randomness. A system may be completely determined, yet if it has no detectable structure to the observer it is unpredictable or 'perceived random' (Wimsatt 1980). To an organism, it is immaterial whether the world is truly subject to chance or whether it is merely Laplacean, so complex that it is impossible for an organism to gather enough information to make it completely predictable.

Daily life shows us that organisms' environments are all unpredictable to some extent, on the time-scale of the individual's lifetime and shorter. But it is also true that the degree of unpredictability on this time-scale is fairly constant over the time-scale of several generations. Since this is the time-scale of major evolutionary importance (Lewontin 1966), we may expect organisms to have attributes appropriate to the degree of unpredictability of their environments. This is, of course, the basis of considerations of r- and K-strategies. Its importance to the predator–prey relationship is that this relationship is characterized by greater unpredictability than other feeding relationships.

Unpredictability for predators

Consider a typical herbivore, a practitioner of the second class of feeding relationship in our four-fold classification. Presumably, it moves from plant to plant consuming part of each rather than consuming the whole of each because some parts are more nutritious or easier of access than others, so that it maximizes its overall rate of feeding by ignoring the less nutritious and less accessible parts. This will only be possible if new plants are easily available. Thus, for the typical herbivore, food-supplies are fairly predictable. This is even more true of typical parasites and parasitoids, except for the stage of finding a host, which may be cruelly unpredictable. The predator's search for food is usually less unpredictable than the parasite's search for a host but it is, of course, not just done once in a lifetime but many times. And it is normally more unpredictable than the herbivore's quest for food because the herbivore's technique of partial consumption demands a predictable supply of new food items. Hence, while it would be unwise to maintain that all predators have more unpredictable food-supplies than all other organisms, most predators do have more unpredictable food-supplies than most other organisms.

Unpredictability for prey

Being eaten by a predator is clearly an all-or-nothing affair. As a result, there will be a division of the prey population according to the loss of potential contribution to the next generation occasioned by the activities of predators (Fig. 12.1a). Of course, unsuccessful attacks by predators may reduce the reproductive output of the attacked individuals compared with those not attacked, while among those individuals that are successfully attacked there

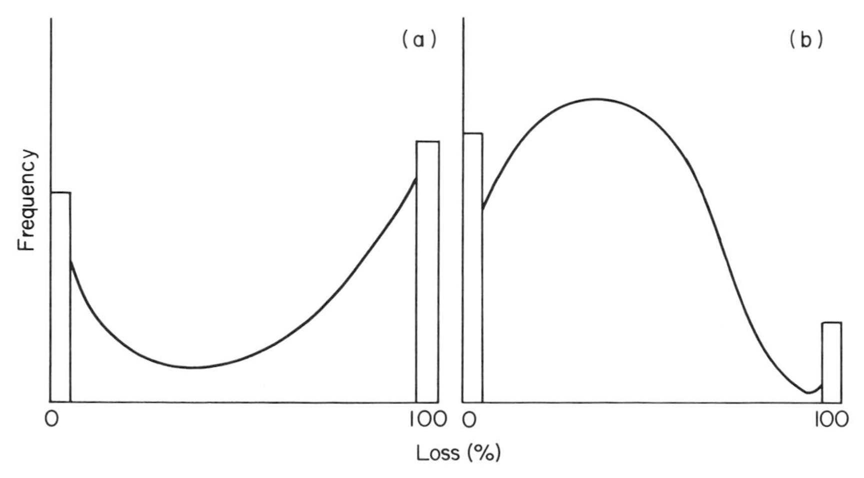

FIG. 12.1. Typical frequency distributions of the amount of fitness lost as the result of the activities of (a) predators and parasitoids and (b) herbivores and parasites.

will be some variation in the amount of reproductive output they attained before being killed, but the basic dichotomy remains. The same is true of the victims of parasitoids.

Parasites and herbivores have a much less dramatic effect (Fig. 12.1b). They attack a larger proportion of the victim population and levy a more even toll, though some members of the population may escape them entirely and others, especially if attacked early in life, may be killed.

Thus, although I shall concentrate on unpredictability as it effects predators, we should not loose sight of the complementary processes affecting prey.

Evolutionary effects of the unpredictability of predation

Unpredictable resources promote ecological generalization (Levins 1968). We should thus expect predators to be more general in their feeding habits than other consumers. They are (Williamson 1972).

We may also expect predators to be more capable than other consumers, particularly herbivores, of coping with variability of food availability, both in time and space. They should be capable of responding to such variation in ways that increase their feeding efficiency. Thus foraging becomes not just a matter of finding, catching, and ingesting food, but also a question of sampling to discover the nature and distribution of the currently available food-supply. I shall return to this below.

A third important aspect of unpredictability is that the unpredictability of the results of various courses of action may affect which of them is best. Alternatives with the same expected average outcome may differ in value, in terms of their effects on fitness, if they differ in the variance of outcome. Indeed, the importance of the differences in variance may be great enough that they outweigh differences in expectation, so that an alternative for which the expected outcome is less valuable may yet be superior to one for which the expected outcome is more valuable, because of differences in the variance of outcome. The effects of variance arise in various ways, which I shall consider in the next three sections.

MULTIPLICATIVE FITNESS COMPONENTS

Consider a small bird attempting to live through a so-called temperate winter. It loses considerable weight during the long, cold nights (Baldwin & Kendeigh 1938) and has to find food during the day at stupendous rates to compensate (Gibb 1960). Its probability of surviving a hard winter is not high (Cawthorne & Marchant 1980; Dobinson & Richards 1964; Marchant &

Hyde 1980). It is, of course, the product of the daily probabilities of survival. Unless these are the same for all days, their product will be less than their arithmetic mean raised to the appropriate power. Thus, if two places have the same arithmetic mean food-supply, the bird will do better to spend the winter in the one where that food-supply is more dependable. Indeed, it may do better to rely on a dependable food-supply than on a variable one, even when the latter has the higher arithmetic mean.

These considerations are important only for cases in which the survival rates over individual time periods are not close to unity. Fortunately, in most models, each individual unit is small enough for its effect on overall fitness to be minute, so the fundamentally multiplicative nature of successive contributions to fitness can be ignored (e.g. Sibly & McFarland 1976). This is, therefore, the least important aspect of unpredictability, though it should not be forgotten.

UNPREDICTABILITY AND DISPROPORTIONAL FITNESS

Fitness disproportional to resource harvest

The joint effect of unpredictability and disproportional fitness has been considered by Caraco (1980a, b) and by Real (1980a, b). They have pointed out that the fitness of an animal will usually increase at a rate that is less than proportional to any increase in the rate at which it is harvesting some resource (Fig. 12.2a). The fundamental reason for this is that the rate at

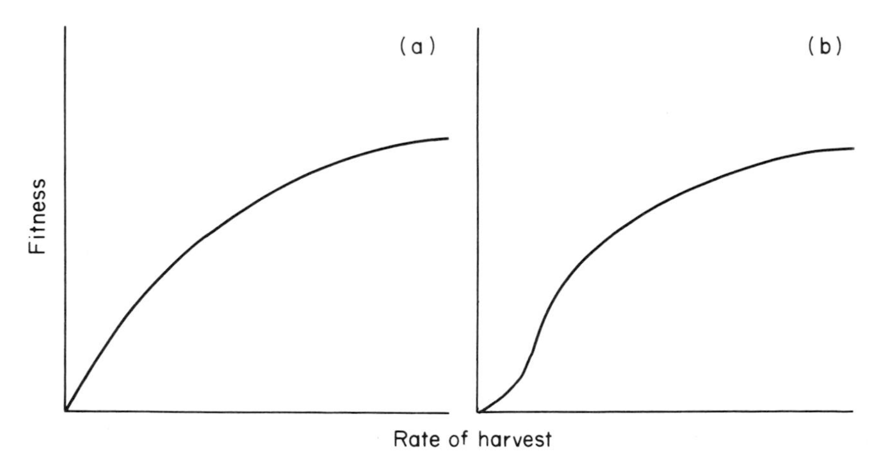

FIG. 12.2. (a) General relation between fitness and the rate at which an animal harvests food, showing the subproportionality of fitness. (b) The relation between fitness and rate of harvest when fitness is superproportional at very low rates of harvest.

which resources can be converted into units of survival or reproduction is limited by the animal's anatomical, physiological, and behavioural attributes. These will be most efficient at rates of resource harvest of about the average. (Note a deliberate vagueness here: the variance of harvest rates as well as its average over the last few generations will have moulded the animal's attributes.) Thus, because of the lower efficiency with which resources can be converted into units of survival or reproduction when they are harvested at above-average rates, fitness will be less than proportional to rate of resource harvest.

The role of unpredictability

Let us take a very simple model of the relationship between the '*harvest*' of survival or reproduction, F, and the harvest of resources per unit time, R:

$$F = bR^a \quad a < 1$$

Over n units of time, the mean rate of harvest of fitness is thus:

$$(\Sigma F_i)/n = b(\Sigma R_i^a)/n$$

Unless all the R_i values are the same, this is less than the fitness corresponding to the arithmetic mean harvest, of resource which is:

$$b(\Sigma R_i/n)^a$$

That the mean rate of harvest of fitness is less than the fitness corresponding to the mean harvest of resource is true not only for this model but for any case in which fitnesses is subproportional to harvest ('Jensen's inequality'). The extent to which the mean fitness falls below the fitness for the mean harvest depends on the variance of the rate of resource harvest and on the degree of subproportionality of the relationship between fitness and resource harvest.

Thus, if two foraging techniques produce the same mean yield of resources, that with the more unpredictable yield has a lower fitness associated with it (Caraco 1980b; Real 1980a). Indeed, if fitness is sufficiently subproportional to yield and if one foraging technique has a very unpredictable yield, that technique will be inferior to one that has a more predictable yield even when the mean yield of the more reliable technique is lower. Thus, Darwinian animals should adopt low-yield, low-risk techniques rather than high-yield, high-risk ones if the difference in yield is small compared with the difference in risk and if fitness is sufficiently subproportional. Caraco and his colleagues have carried out careful experiments with the seed-predators *Junco phaeonotus* and *J. hyemalis* in the laboratory, to determine their sensitivity to unpredictability of food-supplies (Caraco, Martindale & Whittam 1980; Caraco 1981). Though they did not define the relationship of fitness to rate of food intake precisely, they provided the birds with food at a mean rate greater than that sufficient to keep alive. Since the birds were non-breeding, this should have resulted in a subproportional relationship. In both species, when given the choice between two sources that delivered food at the same mean rates, the birds chose to feed at the source delivering the food at a constant rate rather than the one delivering it at a variable rate.

No experiments have yet been published in which mean and variance were simultaneously varied, to test the more extreme prediction that the source providing food with the higher mean may be rejected if it does so with sufficiently high variability.

Example: When should less profitable foods be ignored?

The behaviour of an animal faced with an array of prey of different profitabilities is 'the most throughly analysed problem in optimal foraging theory' (Pulliam 1981), having been solved independently 'no less than nine times' (Pyke, Pulliam & Charnov 1977). The solution is simply expressed. Consider a predator searching for prey of only one type. Let its rate of discovery of prey per unit time be d, the amount of time taken to catch and consume a discovered prey (the 'handling time') be h, and the net nutritive content of a prey (the excess of its gross content over the expenditure of resources involved in its capture, consumption, and digestion) be c. The mean time taken to discover a prey is $1/d$, so the mean rate of uptake of nutrients is:

$$c(h + 1/d) = dc/(1 + hd)$$

Now consider the same predator faced with a variety of types of prey, all randomly distributed. We rank the prey in order of profitability, such that for all i:

$$c_i/h_i > c_{i+1}/h_{i+1}$$

Then the predator will maximize its overall rate of uptake of nutrients if it eats the most profitable type of food ($i = 1$) whenever it encounters it but otherwise only eats a prey of the ith type when it encounters it if:

$$c_i/h_i > \left(\sum_{j=1}^{i-1} d_j c_j\right) \bigg/ \left(1 + \sum_{j=1}^{i-1} d_j c_j\right)$$

The criterion for including the ith type in the diet depends on expected rates of nutrient uptake. The left-hand side of the inequation is the rate of nutrient uptake in the near future if an encountered prey of type i is eaten. The right-hand side is the expected rate if the predator ignores that prey item and proceeds to search for more profitable items.

If fitness is directly proportional to net rate of harvest of nutrients, maximizing the mean rate of harvest of nutrients will maximize fitness. But this is not true with subproportional fitness, as can readily be seen in a model with only two prey types. If type 1 is the more profitable, the criterion for including type 2 in the diet, if net rate of harvest of nutrients is to be maximized reduces to:

$$1/d_1 \geq h_2 c_1/c_2 - h_1$$

However, if fitness is related to the ath power of nutrient content of the prey with a being less than unity, as in our simple model of subproportional fitness, then overall fitness will be maximized by using the criterion:

$$1/d_1 \geq h_2 (c_1/c_2)^a - h_1$$

Thus, under subproportional fitness either the rate of discovery of the more profitable type or the difference in profitability needs to be less than under proportional fitness if the less profitable type is to be included in the diet. This, however, is simply a result of subproportional fitness. What is the additional effect of uncertainty?

Imagine an animal that has just discovered a prey of the less profitable type. If it abandons it and continues to search, it may expect to find one of the more profitable type after $1/d_i$ seconds, perhaps having discovered, but ignored, others of the less profitable type while doing so. If this expectation is certain, then the above analysis is correct. If, however, it is not, we must consider the mean of the fitnesses corresponding to various searching times, rather than the fitness of the mean searching time. Under subproportional fitness, the high rates of harvest obtained by the lucky chance of finding profitable prey quickly have insufficient effect on fitness to balance the low rates obtained by the unlucky chance of having to search for a very long time before finding prey. This may markedly change the criterion for specializing on the more profitable type rather than generalizing.

The simple form of the optimal diet model, ignoring subproportional fitness and unpredictability, has been tested against both observational and experimental data, in both the field and the laboratory. In general, the model's predictions have been confirmed: animals do tend to ignore less profitable foods when more profitable ones are abundant; they are more likely to do so when the difference in profitability is larger; and the rate of discovery of the less profitable prey does not affect their choice (for a critical discussion of the evidence, see Krebs 1979). Experimental psychologists and some behavioural ecologists have used concurrent variable-ratio schedules to study feeding behaviour (Krebs, Kacelnik & Taylor 1978; Staddon 1980). In such experiments, two or more sources of food are available to the animal. It has to work at each by some action such as pressing a bar and the sources differ in the proportion of presses that produce a reward. The rate of obtaining rewards is maximized by using only that source yielding rewards at the highest rate and this is what animals usually do, after a sampling period in which they discover which source is the best.

Food abundance and satiation

The optimal diet model predicts specialization on the more profitable prey when they are abundant. Many studies have confirmed that selectivity increases as the more profitable prey, or all prey, become more abundant (Krebs 1979). However, as Orians (1981) has pointed out, with particular reference to subproportionality of fitness imposed by an upper limit on the rate of digestion, the difference in fitness value of two forms of prey when food is abundant may be too small to make up for the cost of discriminating between them. This could lead to satiated predators being less selective than somewhat less well-fed ones.

Even ignoring the cost of discriminating prey types, one could argue that subproportional fitness will decrease selectivity when the predator is close to satiation. It will then probably be true that the coefficient a in our simple model of subproportional fitness will become very small, so that the criterion for allowing less profitable prey into the diet will be lessened. This may explain why laboratory mice *Mus musculus* failed to select the more profitable seeds available when alternative food was supplied *ad libitum* (Emlen & Emlen 1975), for the mice were not only probably in an environment in which the benefit of selectivity was small but probably had an evolutionary history of such environments.

In contrast to the Emlen's mice, people at the end of a heavy meal will readily eat small quantities of sweet food but not bulky food, bicoloured antbirds (*Gymnopithys bicolor*) that have been feeding with swarms of army ants become more selective as they become more satiated (Willis 1967), and mantids *Hierodula crassa* with plenty of food in their guts will only stalk flies that are close to them (Charnov 1976a). The classic explanation of such observations is that the animals are using the fullness of the gut as a measure of food abundance. This may be correct but other explanations are possible. In an animal with only limited space remaining in the gut, the future rate of uptake of nutrients may be higher if bulky items are ignored than if they are eaten, when they will occupy space that could be occupied by more concentrated food: this may not apply if the gut is less close to its capacity.

This interpretation is supported by the observation that antbirds spend longer removing the less digestable parts of their prey when they were approaching satiation. The mantid observations cannot be explained in this way, since the preference shown was not between foods of different sorts, but in that case there may have been a change in cost or time of hunting when the predator was heavy with food: this would increase the ratio of the net yield of nearby flies to that of distant flies, so making selectivity more likely. Thus satiation may increase selectivity by changing the relative benefits or costs of alternative prey types. (The suggestion of Charnov [1976a] that his results can be explained by assuming that there is a risk of predation or other accident when hunting, which is greater for more distant flies, seems incorrect. Such a cost is independent of degree of satiation and therefore cannot alter selectivity according to degree of satiation.)

To predators that take both animal and plant food in particular, a balanced diet may be more valuable than a uniform diet (cf. Belovsky 1978, Pulliam 1975, Rapport 1981, Rapport & Turner 1975; Westoby 1974 for balanced diet theory; Kagel *et al.* 1980 and Real 1980a, b for the general theory of balance of behavioural spectra). Figure 12.3 represents a simple balanced diet model for two foods: fitness isopleths are concave lines, because fitness is higher for a given total food consumption when the diet is mixed

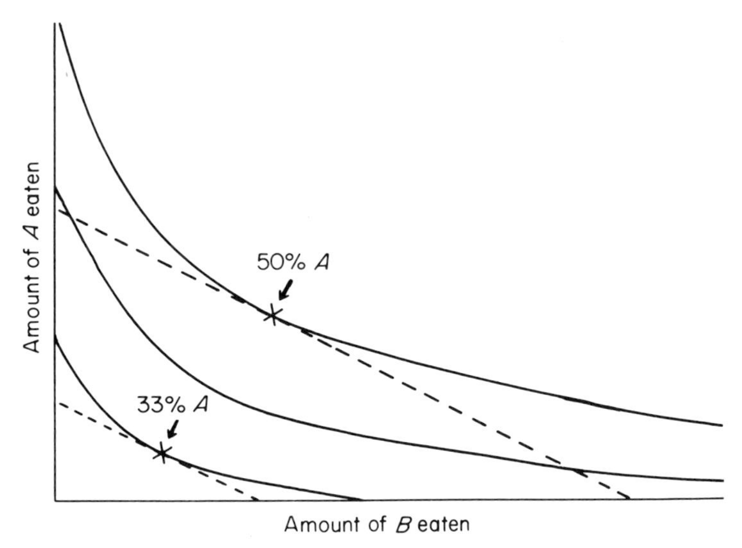

FIG. 12.3. A balanced diet model for two sorts of food. Continuous lines are fitness isopleths, fitness increasing towards the top right. Pecked lines are foraging potential isopleths. Crosses indicate optimal diets for given foraging potentials. Note that a change in foraging potential alters the proportions in the optimal diet.

than when it is unvaried; isopleths of foraging potential are straight lines, their slope departing from 45° if one prey is more valuable than the other; the optimal diet for a given foraging potential is the point at which the potential isopleth is tangential to the highest fitness isopleth it meets. The foraging potential is determined by both foraging effort and total prey availability. Thus the optimal dietary mix will usually change as total prey availability changes, even if the proportions of the two prey available are constant. The exact form of the fitness surface will determine whether the optimal diet is more or less selective at the higher level of prey availability: the general theory of balanced diets makes no prediction about the direction of change.

Finally, the theory of balanced behavioural spectra predicts that a satiated animal should turn from feeding to some other activity. For a predator, for whom knowledge of the form, availability, and distribution of prey must constantly be updated, that other activity may be sampling the available food spectrum. The significance of this is not as nutrient intake but as information intake, but the observer will usually register only a decrease in selectivity of the diet.

In summary, optimal diet models, especially those incorporating subproportional fitness, predict both greater and lesser selectivity at higher levels of food abundance and predator satiation. No general correlation is to be expected.

Unpredictability and starvation

Both Garton (1979) and Orians (1981) have maintained that if a food can provide nutrients only at a rate less than the rate at which they are being used, then that food should never be eaten. This is wrong. If that food provides nutrients more efficiently than any other currently available and if the net nutrient level of the body is depleted less rapidly by eating than by not eating, then the food should be eaten: slow starvation is better than quick starvation. Thus woodpigeons *Columba palumbus* starve when feeding only on *Brassica* because they cannot digest it fast enough (Kenward & Sibly 1977) but they are a menace to agriculture because, amongst other things, they feed avidly on *Brassica* when other food is not available. Similarly, Bullfinches *P. pyrrhula* may eat buds that are insufficiently nutritious, when other foods are unavailable (Newton 1964). In both cases, some birds live long enough to be able to recover when better foods appear.

We have so far assumed that fitness is related subproportionately to rate of food intake (Fig. 12.2a). But if an animal is starving, the converse may be true because a small amount of food cannot prevent starvation but a slightly larger amount can. Thus, as Schaffer (1978) pointed out in a different context, the overall form of the relationship between fitness and resources availability is likely to be sigmoid (Fig. 12.2b). Hence, when food is very scarce, Jensen's inequality is reversed, so a starving animal may do better to adopt the more unpredictable of two feeding techniques or of two foods: it has nothing to lose if the expected return on the predictable food-supply is insufficient to prevent starvation but it may be lucky enough to obtain a high yield of the unpredictable food-supply (Caraco 1980b). Starving animals should take risks.

Confirmation of this prediction was obtained in the experiments with *Junco phaeonotus* and *J. hyemalis* (Caraco *et al.* 1980; Caraco 1981b). When the mean rate of reward was insufficient to meet the birds' daily needs, so that fitness was zero unless the bird attained a rate of feeding well above the mean, the birds preferred the unpredictable food-source to the constant one. Experimental psychologists have generally found that laboratory rats *Rattus norvegicus* and pigeons *Columba livia* prefer variable to constant rewards (Davison 1969; Fantinor 1967; Herrnstein 1964; Killeen 1968; Levanthal *et al.* 1959; Pubols 1962). Since the experimental animals were maintained at body-weights less than they had when fed *ad libitum*, their physiological condition was like that of a starving animal. Thus their preference for variable rewards is to be expected. It is interesting that in experiments with larger mean food rewards, the rats of Levanthal *et al.* (1959) became indifferent to whether the reward was constant or variable: presumably the larger reward had raised their nutritional status to a level that no longer corresponded to probable starvation.

Caraco (1981) performed a third series of experiments with *J. hyemalis*, in which the mean rate of food reward was exactly sufficient to supply daily needs. Since the birds were non-breeding and in captivity, so that food ingested in excess of requirements or time saved by feeding rapidly for a short time could not be used for other purposes, the relationship of fitness to rate of feeding was probably more or less linear. Thus, if the variance of rate of feeding was sufficiently small that all values experienced fell within this linear region, the birds should have been indifferent to whether a food-source was constant or unpredictable. However, if the variance was greater, so that some values fell outside the region in which fitness was directly proportional to rate of feeding, this would not be true. Depending on whether the subportionality of the upper part of the curve was more or less important than the superproportionality of the lower part, the birds should have preferred or eschewed the constant source. In fact, the birds were indifferent when the variance of the non-constant source was small but preferred the constant source when the variance of the other was large (Caraco 1981).

Stephens (1981), commenting on the experiments on *J. phaeonotus*, showed that the preference for the predictable source when expected rewards exceeded requirements and *vice versa* would be predicted if the birds were minimizing the probability of starvation. But this is not the Darwinian criterion and only appears to apply because, in the model used, fitness is related directly to the probability of not starving. Its inappropriateness is empirically demonstrated by the results of the experiment in which *J. hyemalis* were given food at a mean rate just sufficient to balance requirements. Since the constant food-source gave no probability of starving but even a slightly variable one gave some probability of starving, birds that behaved in a way that minimized the probability of starvation should always have preferred the constant food-source. As we have seen, they did not, but were indifferent when the variance of the non-constant source was small.

Caraco (1980b) has also proposed that the probability of starvation is important in its own right. He suggested that when the animal is experiencing good conditions, with a low probability of starvation, then an improvement in conditions should lead to it becoming less averse to taking risks, since it is unlikely to starve even if it does so. Since fitness is more than a matter of avoiding starvation, especially when starvation is unlikely, this is a mistaken argument. It is the form of the curve relating fitness to feeding rate that determines the expected degree of risk-aversiveness.

OPTIMAL PATCH USE: QUITTING WHILE AHEAD

The marginal value theorem

The importance of 'quitting while ahead' arises in the context of exploiting a patchily distributed prey. Charnov's marginal value theorem provides a general solution to the problem of how patches should be exploited (Charnov 1976b; Parker & Stuart 1976 for a more general discussion of the same principle).

Suppose the patches are identical in terms of amount of food each contains and the ease with which that food may be harvested. Suppose that the amount of food is reduced by the feeding activity of the predator, so that its rate of harvest is reduced. If the food resources are non-renewable, the gross harvest will asymptotically approach a limit set by the initial level of resources, so that the net yield, allowing for the cost of foraging in the patch, will actually decline if the animal spends a long time in the patch (Fig. 12.4a). If the food resources are renewable, the effect may be less extreme—merely a reduction in the rate of harvest below its initial value (Fig. 12.4b). The best time, t^* for the animal to quit a patch is the time at which the instantaneous

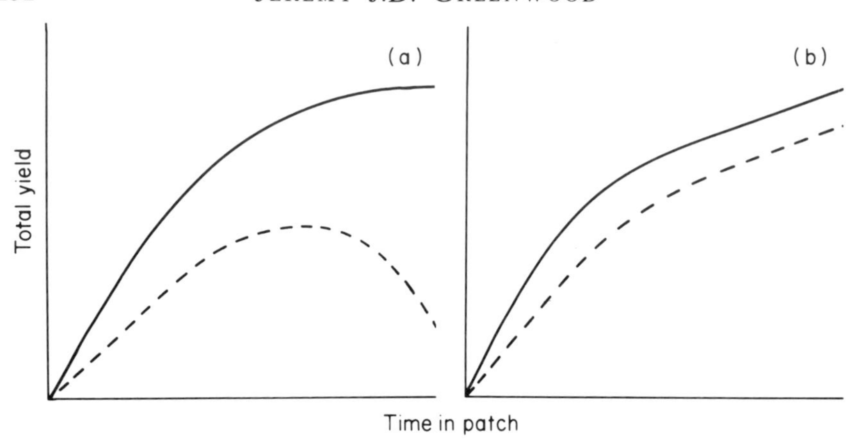

FIG. 12.4. Relationship between time spent in a patch, gross amount of food obtained (continuous line) and net amount obtained (gross amount minus cost of foraging) dashed line), for (a) non-renewable food resources and (b) resources that can be renewed fast enough to balance the rate of harvest.

rate of resource-harvesting is equal to the net rate that it will obtain in the next $T+t^*$ seconds by leaving the patch, taking T seconds to find another patch, and spending t^* seconds feeding in that other patch. This is shown graphically in Fig. 12.5. At time $-T$ the animal has begun to search for a patch, using nutrients at a rate $-k$ so that when it finds a patch at time zero it has incurred a 'searching cost' of $-kT$. The tangent to the curve of net yield from $-T$ has a slope corresponding to the highest net rate of harvest

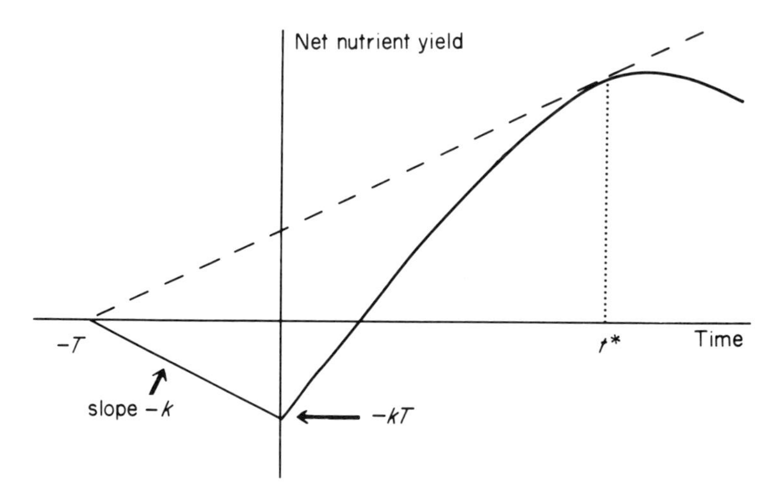

FIG. 12.5. Graphical solution to the problem of optimal use of patches. For explanation see text.

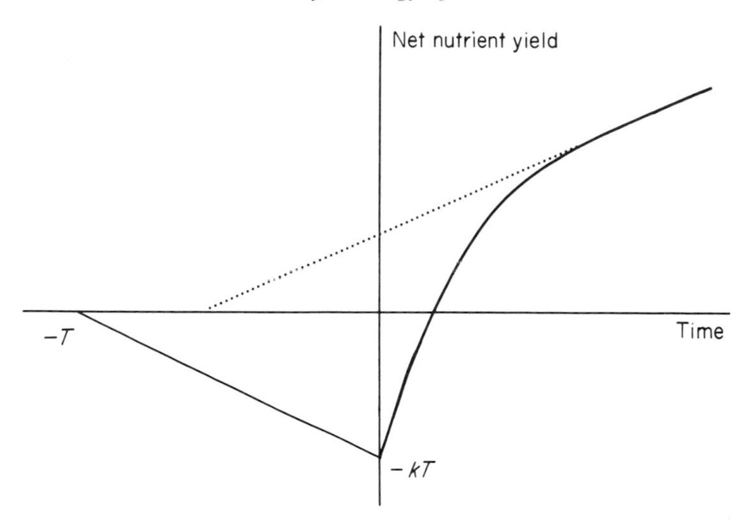

FIG. 12.6. Showing that if rate of net yield stabilizes, rather than declines, no line from $-T$ may be tangential to the curve of net yield. The optimal time to spend in the patch is infinite.

that the animal can achieve, given the values of $-T$ and of $-k$ and the form of the curve of net yield. Lines from $-T$ that intercept the curve of net yield other than tangentially, corresponding to quitting times longer or shorter than t^* have lesser slopes, implying lower overall rates of harvest. Figure 12.5 corresponds to a net yield that eventually declines with time, as in Fig. 12.4a. If the rate of net yield reaches a stable value, as in Fig. 12.4b, the same sort of result may follow. However, if the transit time between patches and the stable rate of net yield are high, no line from $-T$ can ever be tangential to the curve of net yield: the optimal solution is to stay in the patch for ever (Fig. 12.6).

Patch use and subproportional fitness

Before considering the principle of quitting while ahead, I shall briefly consider effects of subproportional fitness on the optimal use of patches. In Charnov's (1976b) original formulation, the model was applied to an array of patches of diverse content. In this case, the graphical solution involves plotting a curve of net yield that is the mean rate of harvest via the usual tangent (Fig. 12.7). The optimal quitting time for a particular patch is found by drawing the highest line parallel to this tangent that is itself tangential to the curve of the yield for that patch.

Once again, this model is only true if yield is measured in units that are

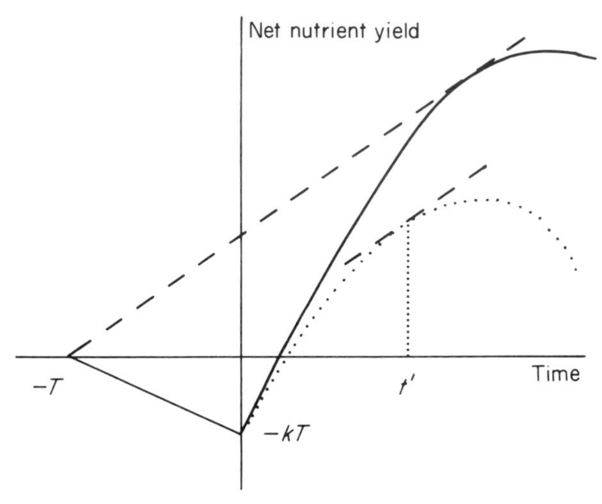

FIG. 12.7. Method of estimating the optimal quitting time for a particular patch (t^*) from the mean curve of net yield (continuous line) and the curve of net yield for that patch (dotted line).

linearly related to fitness. If fitness is subproportional to food harvest, maximum fitness will be attained at an overall mean rate of harvest of food resources that is less than the maximum, to an extent dependent on the extent to which fitness is subproportional to harvest and on the between-patch variance in food availability. One implication of this is that a predator should prefer prey species that are uniformly distributed to those that are unevenly distributed—if its foraging behaviour is adequately described by this model.

Perhaps one should point out here a deficiency in the original model that arises from uneven distribution, though not from subproportional fitness. If the variance in content of patches is sufficiently high, some patches will contain so little food that the optimal quitting time is zero. Unless the predator can avoid such patches, and the model assumes that it cannot judge the quality of patches until it has arrived at them, this situation will lead to considerable waste of the time and energy involved in travelling to such patches. The overall maximum attainable rate of harvest will be lower than if food was more evenly distributed.

Even if there is no variation between patches in content, there is almost certain to be variation in the time taken to travel between patches. If transit costs are linearly related to transit time, as our graphs imply, this is irrelevant. However, they may well not be. It has been argued, for example, that the instantaneous probability of survival of an animal is a quadratic function of the discrepancy between its actual reserves of nutrients and the optimal level of such reserves (Sibly & McFarland 1976). If long transits are disproportionately more costly, then the greater the variance in transit times the more the mean cost of travel will exceed the cost of travelling for the mean time. Hence fitness will be maximized by staying longer in each patch than the time that would maximize fitness if transit times were equal. As a result, the maximum overall rate of harvest will be lower than when transit times are equal. Once again, unpredictability lowers fitness.

A more specific model

To illustrate the principle of quitting while ahead, it is best to formulate a more specific version of the model, simplified to its essentials. Let us assume that the energy costs of searching for patches and within patches are zero, that handling times are zero, and that the food resources are non-renewable: these are simplifying assumptions that do not affect the conclusions to which we shall come. If search within a patch is random and the prey are so numerous that a continuous approximation is valid, then the rate of harvest at time t is

$$rM \exp(-rt)$$

where r is the rate of discovery per unit prey available at any time and M is the initial total amount of food available. Thus the overall rate of harvest over the time interval $(T+t)$ is

$$M[1-\exp(-rt)]/(T+t)$$

From the more general model, this is maximized when t is such that this overall rate is equal to the instantaneous rate of harvest at that time:

$$M[1-\exp(-rt^*)]/(T+t^*) = rM \exp(-rt^*)$$

This maximum overall rate is

$$Mr/(1+rT+rt^*)$$

Hence both the maximum overall rate of harvest and the optimal quitting time depend on transit time.

Prey of finite size and number

Suppose that the total biomass of prey in the patches is still M but that there are fewer of them: we have few, big prey rather than the innumerable, infinitessimally small prey that we have so far implicitly assumed. Suppose, indeed, that we have the extreme case of a single prey per patch, of biomass

M. Once the predator has found the prey, there is nothing to be gained by staying in the patch—we assume, of course, as we have done throughout the discussion so far, that the predator is omniscient and knows that there is only one prey per patch. If the predator is searching randomly, the probability of encountering prey within the next time unit remains constant until the prey is actually found. Hence, if it is worthwhile starting to search for prey, it remains worthwhile to continue until the prey is found. Thus the best method of searching is to the continue searching each patch until the prey is found, then quit at once (Cowie & Krebs 1979). This method of searching will lead to a negative exponential distribution of quitting times, with a probability density of $\exp(-rt)$ and a mean quitting time of $1/r$. Thus the amount of food obtained in each patch, the mean time spent in patches, and the times spent in any particular patch are all independent of transit time, unlike the model for innumerable, infinitesimally small prey.

Since T is constant, the net time to find a prey $(T + t)$ has the same probability density function as that of t. Thus the mean overall rate of harvest is

$$M/(T + 1/r) = Mr/(1 + rT)$$

That this is greater than the corresponding rate for innumerable, infinitesimally small prey is clear from observing that it implies a mean time to catch a prey of biomass M is $(1/r + T)$, whereas the mean time to harvest the same biomass of infinitesimally small prey is $(1/r + T + t^*)$, even though the biomass density per patch is the same.

The reason for the difference is that, at the moment it catches a prey, the predator searching for a single prey experiences a sharp increase in its overall rate of harvest, often to a value greater than the corresponding rate at that time for an animal feeding on innumerable infinitesimally small prey. Since it can quit while it is ahead, it can use this to increase its overall mean rate of harvest.

The randomness of search is, in fact, an irrelevant feature of the model. Suppose that the animal searches each patch systematically. If the food is continuously and uniformly distributed, then the rate of harvest is constant until the whole patch has been searched. Thus, if the patch is worth exploiting at all, the animal should not quit until it has searched all of it. If this takes W seconds, the overall rate of harvest is

$$M/(T + W)$$

But if the food is present as one big prey per patch, the probability distribution of the time taken to find it is a uniform rectangular distribution over the domain zero to W. If the animal acts for the best, quitting as soon as it finds the prey, its mean overall rate of harvest is thus

$$M/(T + W/2)$$

Once again, the predator's greatest attainable rate of harvest is greater when there is one big prey per patch than when the prey are innumerable but infinitesimally small.

I have not analysed cases intermediate between the two extremes considered here. It may not prove easy to do so. The apparently simple solution implied by Cowie & Krebs (1979, Fig. 9.8) depends on the predator knowing, at the time it enters the patch, when it will encounter each prey: such omniscience seems unlikely. However, it would be surprising if intermediate cases did not give intermediate results. Thus the variability in rate of harvesting imposed by the discountinuous form of their food allows predators to harvest food from patches at an enhanced rate, because they can quit while ahead. Whether this effect or the countervailing effects of subproportional fitness and of fitness multiplicativity will be the more important will depend on the details of a particular case.

Through no experiments to test the effect of quitting while ahead have been published, the following report by Cowie & Krebs (1979) agrees with what one would expect if birds acted on that principle:

> 'Cowie presented six great tits (*Parus major*) with two different environments in which the total energy intake with respect to the time spent at a (food source) was identical, but in one environment the prey were four times as big and were encountered at intervals four times as long. Cowie found that in both environments the birds frequently left after capturing a prey item, but that by the end of the experiment the birds were doing so more frequently in the "large prey environment".'

SAMPLING

Abaptation by changing environments

Environmental variability promotes the evolution of generalized organisms (Levins 1968). However, it will often be true that in any particular environment a particular specialist is superior to a generalist. Even if this is not true, a generalist that uses the environmental resources in one particular way will be superior at a particular time to a generalist that uses the environment in a way that is superior when many generations are averaged. Thus, given environmental variability, one might expect that animals should be able to respond to their environment in ways that enhance their fitnesses, rather than adopting a constant pattern of behaviour, no matter how superior that pattern is to other constant patterns (Pulliam & Dunford 1980). Harley (1981) has formally demonstrated, indeed, that there exists an evolutionarily stable learning rule that is uninvadable by a mutant adopting learning rules with different properties and that results in the behaviour of the population approaching the evolutionarily stable strategy appropriate to the environmental conditions currently experienced. For the ESS to be reached requires, of course, that the learning shall be rapid in relation to the rate of change of the environment. Thus predators, faced with unpredictable food resources, may be expected to respond to variations in those resources, making use of any predictability that exists.

No organism is *tabula rasa*, on which the environment can write any message. Responses to the environment will depend on innate features, themselves determined by the circumstances under which recent evolution has occurred. Thus we may expect predators to show forms and degrees of learning appropriate to the unpredictability of their food-supplies—or, more exactly to that of the food-supplies of their recent ancestors. Memories should be abaptively short in predators with food-supplies that change quickly (Pulliam 1981), unless the changes are regular, in which case information gathered over a long period may allow that regularity to be used to reduce the unpredictability of food-supplies. If we think of learned responses as being decision processes, then innate features, determined by the environments of recent ancestors, will determine the prior probabilities for the Bayesian procedures used to arrive at probabilities on which the decisions are based (McNamara & Houston 1980).

While predators can respond directly to their prey, the availability and quality of prey may be partly predictable from environmental cues. If the relationship between prey and environment is constant from generation to generation, innate responses to the cues may evolve—the timing of breeding seasons in relation to food availability is an obvious example. Less constant correlations may better be learned, since they will differ from generation to generation.

I shall make no effort to review here the vast literature on sampling, even that part of it concerned only with learning in relation to food. Instead, I wish to make some general remarks about how we may expect predators to learn, largely through considering responses to prey patchiness.

Responding to prey patchiness

There is no doubt that most predators can discover where prey are most available and concentrate their hunting there (cf. Curio 1976; Krebs 1979; Kamil & Sargent 1981). Some problems of exploitation of patchily distributed prey correspond to the statistical model of the many-armed bandit: there are several patches, within each of which prey availability is constant but between which there are differences in prey availability; the predator has no certain knowledge of prey availability, though it may have prior probability distributions; it can be in only one patch at once; sampling and exploitation comprise the same process—searching for and catching prey; the total time or effort available, sampling and exploitation combined, is fixed. Clearly, the predator needs to sample all the patches if it is to judge which is likely to yield most prey and the more time it devotes to sampling, the more reliable its judgement will be. Equally clearly, the less time it spends sampling, the more time it can spend in the most productive patch, exploiting at the highest possible rate. The problem is to determine the best pattern of sampling and exploitation, i.e. the pattern that yields the greatest expected rate of feeding. This problem has been solved for the two-armed bandit (Groot 1970). The general solution is that the predator should sample the two patches equally and should then concentrate the remainder of its effort on exploiting the patch from which most prey was obtained during sampling. The amount of sampling should increase when there is a greater total amount of time available. The predator should also devote a greater proportion of its time to sampling when the patches are similar in quality than when they are dissimilar, both because more information is then required to achieve the same degree of certainty about which patch is superior and because the rate of feeding during sampling is not then markedly lower than that during exploitation of the better patch.

Great tits *Parus major* have been presented with this problem, in the form of two food-sources that delivered food probabilistically in relation to number of hops made on two corresponding perches (Krebs *et al.* 1978). The experiments allowed the birds a modal number of 150 hops in total, for sampling plus exploitation, and were performed at five different levels of difference in reward probability. The birds not only showed a fairly clear division of activity into sampling the two sources equally and exploiting the apparently better one but the mean number of hops made before the 'decision point' was close to that predicted by the optimal solution for the particular conditions used (Fig. 12.8). Further experiments, in which the total number of hops allowed was varied, again showed mean responses close to the optimum for each particular number (Houston, Kacelnik & McNamara 1982).

Harley (1981) has discussed a behavioural mechanism that would give results close to, though not identical with, the optimal solution for the two-armed bandit. Its results differ most significantly in predicting a gradual change from equal sampling to exploiting the preferred food-source, rather than a sudden switch. He has pointed out that such gradual changes appear to have occurred in experiments on laboratory rats *Rattus norvegicus*

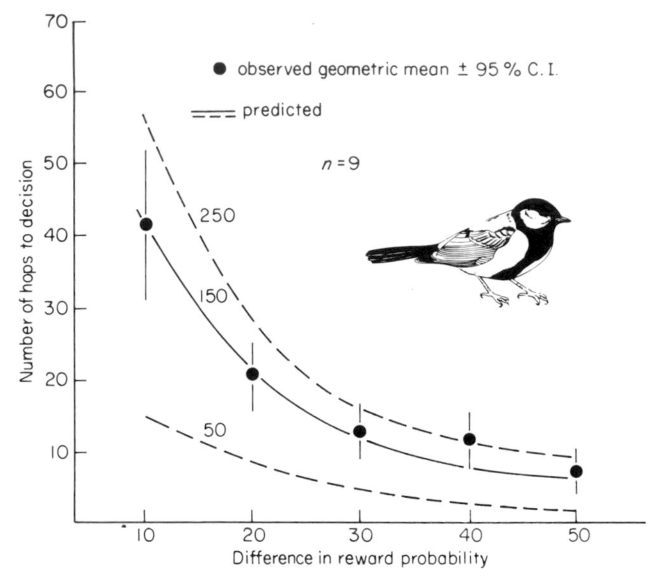

FIG. 12.8. The results of the experiment of Krebs, Kacelnik & Taylor (1978). The number of sampling hops made prior to the switch to exploitation is plotted against the difference in reward probability between the two food-sources (e.g. 30% differences means that one gave a reward on 40% of hops and the other on 10% of hops). The continuous line gives the predicted results, from the two-armed model with a total number of hops per feeding-bout equal to the model number in the experiments (150). The pecked lines are predictions for other hop-numbers. (Reprinted from Cowie & Krebs 1979).

(Roberts 1966) and on paradise fish *Macropodus opercularis* (Bush & Wilson 1956) and has suggested that the sudden switches made by tits were artefacts of the experimental design. However, while that design involved certain biases and unnatural features (though no more than did the experiments on rats and fish), it is difficult to see how these can have produced sudden switching as an artefact. Much more important is the fact that the birds differed from the rats and the fish in that they had been thoroughly trained to expect two important features of the two-armed bandit, that is constant reward rates and a constant length to each experiment (150 hops). In nature, reward rates may change quickly, especially if the predators' hunting changes prey availability, and the length of a foraging bout may vary greatly and unpredictably, depending on a diversity of possible interruptions. My intuition says that the best behaviour in such circumstances is probably to sample relatively briefly, then to begin exploiting the apparently better food-source but to continue sampling to some extent, both to check one's earlier assessment of the relative quality of the two food-sources and to monitor any changes. Animals that have an innate or trained expectation that patch quality or length of feeding-bouts will vary, will adopt such behaviour even when faced with a two-armed bandit and will thus show a gradually developing preference for the better food-source rather than a sudden switch. Furthermore the experiments on rats and fish extended over several days and the data were not presented in a way that shows that pattern of responses within days. It is possible that on each day the animals were first sampling and then suddenly switching to exploiting, like the tits. The gradual increase in concentration on the better food-source, seen over the whole of each experiment, could result from less sampling being done on successive days, as the animals accumulated experience of the environment.

I introduce these ideas not to explain away the difference between the rats and the fishes on the one hand and tits on the other but to emphasize the need to consider what an animals 'expectations' may be. An inappropriate response to a problem posed by the investigator may occur if that problem is not one to which its recent ancestors have been exposed. In particular, we need to consider how patch quality is likely to change with time in nature and how predators may be expected to respond to such changes. This is relevant to an interestingly suboptimal facet of the tits' behaviour in the experiments of Krebs *et al.*: after a run of unrewarded hops, they tended to switch perches more readily than would have been best. This was probably because the birds had the phylogenetic, if not the ontogenetic, experience that a drop in the rate at which food is being found is likely to indicate a decrease in its availability, though it may simply be a run of bad luck. Though ingenious in designing experiments, even Krebs and his colleagues have not devised a way of assuring birds that they really are being asked to play a true two-armed bandit! It would be interesting to know if animals abapted by or trained to environments in which patch quality changes relatively rapidly were quicker to respond to apparent changes in quality.

Confusion between the effects of chance and changes in food availability may be relevant to the interpretation of an experiment reported by Cowie & Krebs (1979). They allowed great tits to feed in a series of mediocre patches, then gave them a series of much better patches, and then gave them mediocre ones again. If the birds were using Charnov's marginal value theorem, they would have compared the rate of feeding in the first few good patches with a mean rate based largely on the preceding mediocre patches. As a result, they would have spent too long a time in the first few patches, only reducing to the optimal time after they had built up enough experience in the good patches for their expectations to be based on these (Fig. 12.9). The converse would have occurred on moving back to mediocre patches. Suppose, however, that they were using another criterion, remaining in each patch for a fixed time, determined by the mean prey availability experienced in recent

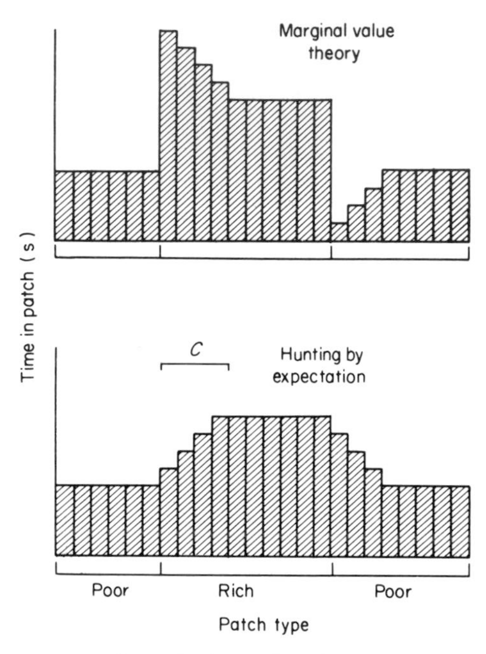

FIG. 12.9. The responses to a change in the quality of patches made by the marginal value theorem and 'hunting by expectation' models, with a hypothetical 'memory window' of fixed length. (Reprinted from Cowie & Krebs 1979).

patches. Providing the time is adjusted correctly, this can give maximum rates of feeding if all patches are identical: if they vary, it is a suboptimal behaviour. If the tits were using this method, labelled 'hunting by expectation' by Cowie & Krebs (1979), one would expect the time spent in each patch to change gradually and monotonically after changes in patch quality (Fig. 12.9). The experimental birds showed behaviour that was somewhat closer to the time-expectation model than to the marginal value model (Fig.12.10). Let us ignore the variability of the data, and their rather poor fit to either model, and consider instead whether, even if the results were clear-cut, we should conclude that the tits were using a suboptimal technique. They may not have been. The prediction of the upper part of Fig. 12.9 is based on a deterministic model, with the predator able to judge exactly the quality of the patch in which it finds itself. But the tits were faced with real life, stochastic in nature, and had to estimate the quality of the patches. An apparent sudden increase in patch quality may, in these circumstances, be no more than a run of good luck. Further experience would make the latter possibility less and less likely, causing the bird to upgrade its estimate of currently available patch quality, that is to increase the length of time spent in each patch gradually.

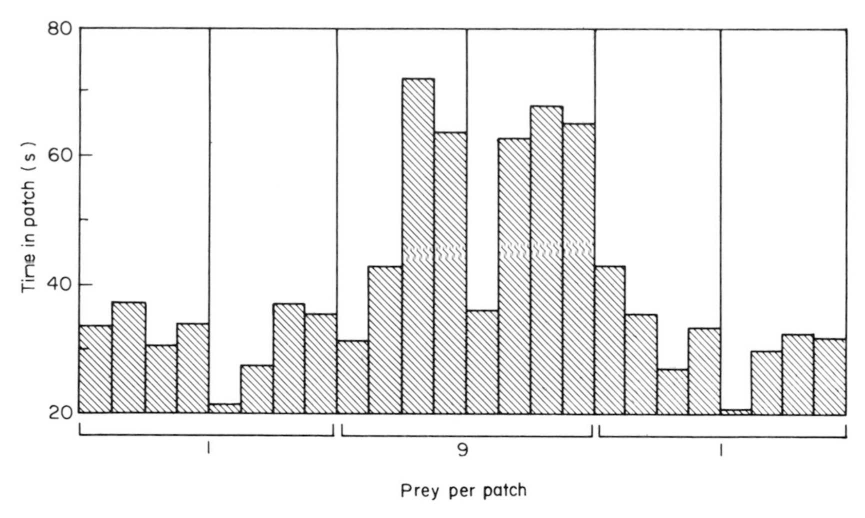

FIG. 12.10. The mean responses of six great tits which were subjected to two trials in poor patches, two in rich patches, and two in poor, in that order. The data for each trial are divided into quartiles. (Reprinted from Cowie & Krebs 1979).

Patches that decline in quality with exploitation

If the quality of a patch declines as the predator catches prey in it, because, for example, the eaten prey are not replaced, the predator's problems may become even more difficult. If it catches many prey in the first few minutes after entering the patch, this may be a sign that the patch is a good one and that the predator should stay, or it may be a sign that the predator has already caught most of the prey in the patch and that it should now move on to look for a fresh patch. This problem has been considered by Green (1980) in terms of a model that is probably of fairly general application. The more essential assumptions are that the prey are discrete, not uniformly distributed within patches, and that there may be systematic differences between patches in number of prey, that is differences greater than would be expected from the binomial within-patch distribution. The predator is assumed to know the parameters of the prey distribution, but not the number of prey in a particular patch. It is assumed to search each patch systematically. The solution that maximizes the number of prey caught per unit time, including both between-patch transit time and within-patch time, demands that the predator leaves a patch when the number of prey caught in relation to the amount of the patch so far searched indicate that the predator will probably do better to quit, and search for another patch, than to spend more time in the patch. As in the two-armed bandit model, a backward-recurrent dynamic programming method can be used to obtain the optimal solution for a given parameter set.

Green finds that a predator adopting the optimal solution will usually do markedly better than a 'naive' predator that assumes that the patches are identical and so searches the whole of each. However, if the variation between patches in prey number arises only from the within-patch variance, there is no advantage in sampling and assessment: the naive predator does just as well. (This presumably happens because in such a situation, the probability of future success in a patch is uncorrelated with the level of success already attained in that patch: the sampling provides no information). If, in contrast, there is marked between-patch variance in prey number, sampling may be very valuable. Furthermore, the rate of feeding attainable by an optimal predator is greater in such a variable environment, presumably because the greater variation allows the predator to spend more time exploiting good patches and less sampling poor ones.

Green points out that, for any given set of prey-distribution parameters, there is a set of searching times $t(k)$ at which the predator should leave the patch if it had observed only k prey by that time. The predator needs only to 'know' these numbers to show the optimal behaviour: it does not need to be continuously computing the optimal solution. However, this is only true if the predator knows the parameter values of the prey distribution. This will rarely be true. In real life, the predator will merely have a Batesian expectation of these parameters, which may itself need to be adjusted in the light of experience. This obviously demands more of the predator—and of the theoretical biologist.

Optimal solutions or suboptimal approximations

It has been argued not only that the increase in brain size found in many mammalian lineages is the result of the ability to process information being valuable to both predators and prey, and that there has been a co-evolutionary positive feedback at work, but also that carnivore brains have always tended to be larger than those of herbivores (Jerison 1973; Gould 1978; cf. Radinsky 1978). The latter has been taken by Dawkins & Krebs (1979) as an example of an evolutionary arms-race, with the predators always ahead (an unremarked violation of the life–dinner principle!). It may, alternatively or also, be a result of the more complex problems that predators face, by virtue of their more unpredictable food-supplies. These problems become most obviously complex when optimality theories require predators to be capable of solving sampling problems. The dynamic programming solution of the two-armed bandit, for example, requires large amounts of computing.

This has lead some people to argue that these problems are too difficult or, at least, too costly for animals to solve. For example, a 'parasite might be better off learning to avoid superparasitism than learning to use Lagrange multipliers' (Oaten 1977). This may be true, though we should not forget that a problem which seems complex in abstract formulation and when solved by digital computing may be easily soluble with an analogue device, especially if the range over which a solution is required is restricted, as it generally will be for predators' foraging problems. S.F. Hubbard (personal communication) tells me that an accurate numerical solution to the two-armed bandit requires much less computing than the exact solution.

The feeling that true solutions to sampling problems are too costly has led to the search for simple models or rules-of-thumb that give approximate solutions (Harley 1981; Houston *et al.* 1981; Jones 1975, 1976; McNamara & Houston 1980; Ollason 1980; Waage 1979). Usually, however, these simple models do not provide such good approximations as the animals attain or fail to mimic their behaviour satisfactorily (Harley 1981; Houston *et al.* 1981). Furthermore, and in principle more important, the degree of approximation regarded as satisfactory to a working statistician or behavioural ecologist may be less than satisfactory in evolutionary terms. Selective differences must be very small to be effectively negligible.

The apparent difficulty of attaining optimal solutions must not lead us to abandon optimality models as a basis for investigation. The most fundamental reason for this is that optimality models and rules-of-thumb are concerned with quite different things: optimality models are concerned with the selective factors leading to the origin and maintenance of phenotypic traits in an evolutionary lineage, rules-of-thumb with the proximate mechanisms responsible for the development of these traits in individuals. Less fundamentally but more practically, models of proximate mechanisms rather than ultimate factors are, by their very nature, less general: a mechanistic model of how an insect parasitoid exploits host patches can have little relevance to how a bird finds patches of seeds, whereas general optimality models can be identical for the two species and provide general insights into the interactions between species. This is not to deny the value of studying proximate mechanisms in their own right but simply to argue that optimality models are more useful in providing general principles.

Sampling and disproportional fitness

I have hinted earlier at the importance of considering the relationship between sampling and disproportional fitness. The sampling models so far considered by evolutionary ecologists ignore disproportional fitness. Since sampling and exploitation differ in riskiness, this needs to be corrected. However, while it is easy to see how to introduce non-linear relationships between fitness and feeding rates into models with a restricted time-scale, such as the two-armed bandit, it is less easy to model the value of sampling at one time for feeding in the more distant future. As I have pointed out, a satiated animal will usually enhance its fitness by adopting some activity other than feeding and that activity may be sampling—which looks like feeding to us but is really information-gathering.

Sampling and learning by prey

Ungulates modify their use of particular habitats or their daily activity patterns according to the frequency with which they are disturbed by potential predators (e.g. Chapman & Chapman 1975; Douglas 1971; Geist 1971). How much experience should they accumulate before avoiding a habitat at a particular time of day? How much re-sampling should they carry out to monitor changes in its dangerousness? These questions are particularly intriguing in that the cost of using a dangerous habitat may be extreme, i.e. death. Furthermore, the animal cannot directly estimate the probability of that particular cost being incurred since, by the very nature of the cost, the animal has either never experienced it or is no longer in a position to benefit from the experience. Similar considerations apply to other aspects of predator avoidance. They may give rise to some interesting theoretical and experimental work in future. A particularly interesting question, in view of the difficulties faced by individual animals in assessing risk of predation, is whether social learning is particularly important in learning about predators. This has received rather little attention in recent discussions of predation and anti-predation defences, which is surprising in view of the dramatic nature of some social responses to predators, such as mobbing, which can give rise to cultural transmission of predator recognition (Altmann 1956; Curio 1978; Curio, Ernst & Vieth 1978a, b; Frankenberg 1981; Rasa 1981; Vieth, Curio & Ernst 1980).

CONCLUDING REMARKS

This paper has contained two messages. The first is that, if we are to use evolutionary theory as a basis for some ecological investigations, we must use the right theory. Just as our optimality arguments must be based on on individual selection rather than ideas of species' benefit, so we must beware of misleading teleological analogies. Predation does not result in the evolution of anti-predator defences but in the evolution of features that reduce the probability of an individual being eaten relative to that of its fellows being eaten.

The second message is that the unpredictable nature of predation may have profound effects. Our models must take this unpredictability into account if we are to understand the evolutionary and ecological relationships between predators and their prey.

ACKNOWLEDGEMENTS

I thank Dr S.F. Hubbard for useful discussions and criticism of an earlier draft; Mr J.P. Johnston for drawing the figures; Mrs M. Ramsay and Mrs E. Patterson for typing; Dr R.M. Dicker and members of the Dundee University Computing Centre for computing advice; the staff of the libraries of Dundee University, the Scottish Ornithologists' Club, and the Zoological Society of London for much assistance with literature; Dr Hubbard and Professor P.S. Corbet for taking over my teaching during the time in which this paper was being prepared; and many colleagues around the world for answering my simple-minded queries and for sending reprints and pre-prints.

REFERENCES

Ainley D.G. (1972) Flocking in Adélie penguins. *Ibis*, **114**, 388–90.

Ainley D.G. (1974) The comfort behaviour of Adélie and other penguins. *Behaviour*, **50**, 16–51.

Alexander R.D. & Borgia G. (1978) Group selection, altruism, and the levels of organization of life. *Annual Review of Ecology and Systematics*, **9**, 449–74.

Allen W.E. (1920) Behavior of loon and sardines. *Ecology*, **1**, 309–10.

Altmann S.A. (1956) Avian mobbing behaviour and predator recognition. *Condor*, **58**, 241–53.

Altmann S.A. & Altmann J. (1970) *Baboon ecology: African field research*, Chicago University Press, Chicago.

Atwell R.I.G. (1954) Crocodiles feeding on weaver birds. *Ibis*, **96**, 485–6.

Baerends G.P. & Baerends-van Roon J.M. (1950) An introduction to the study of the ethology of cichild fishes. *Behaviour*, Suppl. 1, 1–243.

Baldwin S.P. & Kendeigh S.C. (1938) Variations in the weight of birds. *Auk*, **55**, 416–67.

Bartholomew G.A. (1942) The fishing activities of double-crested cormorants of San Francisco Bay. *Condor*, **44**, 13–21.

Baskin L.M. (1974) Management of ungulate herds in relation to domestication. In *The Behaviour of ungulates and its Relation to Management*, (Ed. by V. Geist & F. Walther), Vol. 2, pp. 530–41. IUCN Publications New Series No. 24. International Union for the Conservation of Nature and Natural Resources, Morges, Switzerland.

Belovsky G.E. (1978) Diet optimization in a generalist herbivore: the moose. *Theoretical Population Biology*, **14**, 105–34.

Bertram B.C.R. (1978) Living in groups: predators and prey. In *Behavioural Ecology. An Evolutionary Approach* (Ed. by J.R. Krebs & N.B. Davies), pp. 64–96. Blackwell Scientific Publications, Oxford.

Bigelow H.B. & Schroeder W.C. (1953) Fishes of the Gulf of Maine *Fisheries Bulletin of the US Fish and Wildlife Service U.S.*, **53**, 1–577.

Black J.H. (1970) A possible stimulus for the formation of some aggregations in tadpoles of *Scaphiosus bombifrons. Proceedings of the Oklahoma Academy of Science*, **49**, 13–14.

Breder C.M. (1959) Studies on social groupings in fishes. *Bulletin of the American Museum and natural History*, **117**, 397–481.

Bruns E.H. (1970) Winter predation of golden eagles and coyotes on pronghorn antelopes. *Canadian Field Naturalist*, **84**, 301–304.

Buckland W. (1836) *The Bridgewater Treatises on the Power Wisdom and Goodness of God as manifested in the Creation. Treatise VI. Geology and Mineralogy considered with reference to Natural Theology*. William Pickering, London.

Bullis H.R. (1961) Observations on the feeding behaviour of white-tip sharks on schooling fishes. *Ecology*, **42**, 194–5.

Bush R.B. & Wilson T.R. (1956) Two-choice behavior of paradise fish. *Journal of Experimental Psychology*, **51**, 315–22.

Busnel R.G. (1973) Symbiotic relationship between man and dolphins. *Transactions New York Academy of Sciences, Series II*, **35**, 112–31.

Cain A.J. (1964) The perfection of animals. In *Viewpoints in Biology*, (Ed. by J.D. Carthy & C.L. Duddington), Vol. 3, pp. 36–63. Butterworths, London.

Cain A.J. (1979) Introduction to general discussion on the evolution of adaptation by natural selection. *Proceedings of the Royal Society of London B*, **205**, 599–604.

Cain A.J. & Sheppard P.M. (1954) The theory of adaptive polymorphism. *American Naturalist*, **88**, 321–6.

Cain A.J. & Sheppard P.M. (1956) Adaptive and selective value. *American Naturalist*, **90**, 202–3.

Caraco T. (1980a) On habitat selection in a multiattribute stochastic environment. *Evolutionary Theory*, **5**, 127–33.

Caraco T. (1980b) On foraging time allocation in a stochastic environment. *Ecology*, **61**, 119–28.

Caraco T. (1981) Energy budgets, risk and foraging preferences in dark-eyed juncos *(Junco hyemalis)*. *Behavioural Ecology and Sociobiology*, **8**, 213–7.

Caraco T., Martindale S. & Whittam T.S. (1980) An empirical demonstration of risk-sensitive foraging preferences. *Animal Behaviour*, **28**, 820–30.

Cawthorne R.A. & Marchant J.H. (1980) The effects of the 1978/79 winter on British bird populations. *Bird Study*, **27**, 163–72.

Chapman D. & Chapman N. (1975) *Fallow Deer. Their History, Distribution and Biology*. Terence Dalton, Lavenham, Suffolk.

Charnov E.L. (1976a) Optimal foraging: attack strategy of a mantid. *American Naturalist*, **110**, 141–51.

Charnov E.L. (1976b) Optimal foraging, the marginal value theorem. *Theoretical Population B*, **9**, 129–36.

Clarke B.C. (1979) The evolution of genetic diversity. *Proceedings of the Royal Society of London Biology*, **250**, 453–74.

Cowie R.J. & Krebs J.R. (1979) Optimal foraging in patchy environments. In *Population Dynamics* (Ed. by R.M. Anderson, B.D. Turner & L.R. Taylor), pp. 183–205. 20th Symposium of the British Ecological Society, London 1978. Blackwell Scientific Publications, Oxford.

Crisler L. (1956) Observations of wolves hunting caribou. *Journal of Mammalogy*, **37**, 337–46.

Crook J.H. (1960) Studies on the social behaviour of *Quelea q. quelea* (Linn.) in French West Africa. *Behaviour*, **16**, 1–55.

Curio E. (1976) *The Ethology of Predation*. (*Zoophysiology and Ecology*, Vol. 7.) Springer, Berlin.

Curio E. (1978) The adaptive significance of avian mobbing. Telenomic hypotheses and predictions. *Zeitschrift für Tierpsychologie*, **48**, 175–183.

Curio E., Ernst U. & Vieth W. (1978b) The adaptive significance of avian mobbing. II. Cultural transmission of enemy recognition in blackbirds: effectiveness and some constraints. *Zeitschrift für Tierpsychologie*, **48**, 184–202.

Curio E., Ernst U. & Vieth W. (1978a) Cultural transmission of enemy recognition: one function of mobbing. *Science*, **202**, 899–901.

Darling F.F.(1937) *A Herd of Red Deer*. Oxford University Press, Oxford.

Davison M. (1969) Preferences for mixed-interval *versus* fixed-interval schedules. *Journal for the Experimental Analysis of Behavior*, **12**, 247–52.

Dawkins R. (1982) *The Extended Phennotype. The Gene as the Unit of Selection*. W.H. Freeman, Oxford.

Dawkins R. & Krebs J.R. (1979) Arms races between and within species. *Proceedings of the Royal Society of London B*, **205**, 489–511.

Din N.A. & Eltringham S.K. (1974) Ecological separation between white and pink-backed pelecans in the Ruwenzori National Park, Uganda. *Ibis*, **116**, 28–43.

Dobinson H.M. & Richards A.J. (1964) The effects of the severe winter of 1962/63 on birds in Britain. *British Birds*, **57**, 373–434.

Douglas M. (1971) Behaviour responses of red deer and chamois to cessation of hunting. *New Zealand Journal of Science*, **14**, 507–18.

Eaton R.L. (1978) The evolution of trophy hunting. *Carnivore*, **1**, 110–21.

Edmunds M. (1974) *Defence in Animals. A Survey of Anti-predator Defences*. Longman, Harlow.

Eibl-Eibesfeldt I. (1962a) Freiwasserbeobachtungen zur Deutung des Schwarmver-haltens verschiedener Fische. *Zeitschrift für Tierpsychologie*, **19**, 165–82.

Eibl-Eibesfeldt I. (1962b) Die Verhaltenstwicklung des Krallenfroches (*Xenopus laevis*) und des Scheibenxunglers (*Discoglossus pictus*) unter besonderer Berucksichtigung der Beutefanghandlungen. *Zeitschrift für Tierpsychology*, **19**, 385–93.

Emlen J.M. & Elmen M.G.R. (1975) Optimal choice in diet: test of a hypothesis. *American Naturalist*, **109**, 427–36.

Evans W.E. & Bastian J. (1969) Marine mammals: social and ecological problems. In *Biology of Marine Mammals* (Ed. by H.T. Andersen), pp. 425–75. Academic Press, New York.

Fairholme J.K.E. (1856) The blacks of Moreton Bay and the porpoises. *Proceedings of the Zoological Society of London*, **24**, 353–4.

Fantinor D.J. (1967) Preference for fixed- versus mixed-ratio schedules. *Journal for the Experimental Analysis of Behavior*, **10**, 35–44.

Fink B.D. (1957) Observation of porpoise predation on a school of pacific sardines. *California Fish and Game* **45**, 216–7.

Frankenberg E. (1981) The adaptive significance of avian mobbing. IV. 'Alerting others' and 'Perception advertisement' in Blackbirds facing an owl. *Zeitschrift für Tierpsychologie*, **55**, 97–118.

Galton F. (1883) *Inquiries into Human Faculty and its Development*. Macmillan, London.

Garton E.O. (1979) Implioations of optimal foraging theory for insectivorous forest birds. In *The Role of Insectivorous Birds in Forest Ecosystems* (Ed. by J.G. Dickson, R.N. Conner, R.R. Fleet, J.C. Kroll & J.A. Jackson), pp. 107–118. Academic Press, London.

Geist V. (1971b) A behavioural approach to the management of wild ungulates. In *The Scientific Management of Animal and Plant Communities for Conservation* (Ed. by E. Duffey & A.S. Watt), pp. 413–24. Blackwell Scientific Publications, Oxford.

Gibb J.A. (1960) Populations of tits and goldcrests and their food supply in pine plantations. *Ibis*, **102**, 163–208.

Goss-Custard J.D. (1970) Feeding dispersion in some overwintering wading birds. In *Social Behaviour in Birds and Mammals: Essays on the Social Ethology of Animals and Man* (Ed. by J.H. Crook), pp. 3–65. Academic Press, London.

Gould S.J. (1978) *Ever since Darwin. Reflections in Natural History.* W.W. Norton, New York.

Green R.F. (1980) Bayesian birds: a simple example of Daten's stochastic model of optimal foraging. *Theoretical Population Biology*, **18**, 244–56.

Greenwood J.J.D. & Elton R.A. (1979) Analysing experiments on frequency-dependent selection by predators. *Journal of Animal Ecology*, **48**, 721–37.

Groot M.H. de (1970) *Optimal Statistical Decisions.* McGraw-Hill, New York.

Hamilton W.D. (1971) Geometry for the selfish herd. *Journal of theoretical Biology*, **31**, 295–311.

Harding R.S.O. (1977) Patterns of movement in open country baboons. *American Journal of Physical Anthropology*, **47**, 349–54.

Harley C.B. (1981) Learning the evolutionarily stable strategy. *Journal Theoretical Biology*, **89**, 611–33.

Harper J.L. (1981) After description. In *The Plant Community as a Working Mechanism* (Ed. by E.I. Newman), pp. 11. Blackwell Scientific Publications, Oxford.

Herrnstein R.J. (1964) Aperiodicity as a factor in choice. *Journal of the experimental Analysis of Behavior*, 7, 179–82.

Hiatt R.W. & Brock V.E. (1948) On the herding of prey and the schooling of the black skipjack *Futhynnus vaito* Kishinonye. *Pacific Science*, **2**, 297–8.

Horstmann E. (1953) Form und Struktur von Starenschwarmen. *Zoologischer Anzeiger*, suppl. 17 (*Verhandlungen der Deutschen Zoologischen Geselleschaft*, 1952), 153–9.

Houston A.I., Kacelnik A. & McNamara J.M. (1982) Some learning rules for acquiring information. In *Functional Ontogeny* (Ed. by D.J. McFarland), pp.

Hudleston J.A. (1958) Some notes on the effects of bird predators on hopper bands of the desert locust (*Schistocerca gregaria* Forskal.) *Entomologists monthly Magazine*, **94**, 210–4.

Jerison H.J. (1973) *Evolution of the Brain and Intelligence.* Academic Press, New York.

Jones P.W. (1975) The two-armed bandit. *Biometrika*, **62**, 523–4.

Jones P.W. (1976) Some results for the two-armed bandit problem. *Mathematische Operationsforschung und Statistik*, **7**, 471–5.

Kagel J.H., Battalio R.C., Green L. & Rachlin H. (1980) Consumer demand theory applied to choice behavior of rats. In *Limits to Action. The Allocation of Individual Behavior.* (Ed. by J.E.R. Staddon), pp. 237–67. Academic Press, New York.

Kamil A.C. & Sargent T.D. (1981) In *Foraging Behavior. Ecological, Ethological and Psychological Approaches.* Garland STPM Press, New York.

Kenward R.E. & Sibly R.M. (1977) A woodpigeon (*Columba palumbus*) feeding preference explained by a digestive bottleneck. *Journal of Applied Ecology*, **14**, 815–26.

Killeen P. (1968) On the measurement of reinforcement frequency in the study of preference. *Journal of the experimental Analysis of Behavior*, **11**, 263–9.

Kitchen D.W. (1974) The social behaviour and ecology of the pronghorn. *Wildlife Monographs*, **38**, 1–96.

Klingauf F. & Segonca C. (1970) Koloniebildung von Rohrenblattlausen (Aphididae) unter Feindeinwirkung. *Entomophaga*, **15**, 359–77.

Kramer A. & Aeschbacher A. (1971) Zum Fluchtverhalten des Steinwildes (*Capra ibex*) im Oberengadin, Schweiz. *Säugetierkundlich Mitteilungen*, **19**, 164–71.

Krebs J.R. (1979) Foraging strategies and their social significance. In *Handbook of Behavioural Neurobiology*, vol. 3 (Ed. by P. Marler & J.G. Vandenbergh), pp. 225–270. Plenum Press, New York.

Krebs, J.R. & Davies N.B. (1978) *Behavioural Ecology. An Evolutionary Approach.* Blackwell Scientific Publications, Oxford.

Krebs J.R. & Davies N.B. (1981) *An Introduction to Behavioural Ecology.* Blackwell Scientific Publications, Oxford.

Krebs J.R., Kacelnik A., & Taylor P. (1978) Test of optimal sampling by foraging great tits. *Nature*, **275**, 27–31

Kruuk H. (1964) Predator and anti-predator behaviour of the black-headed gull (*Larus ridibundus* L.). *Behaviour*, Suppl. **11**, 1–127.

Kruuk H. (1972) *The Spotted Hyena*. Chicago University Press, Chicago.

Leatherwood S. (1975) Some obeservations of feeding behavior of bottlenosed dolphins (*Tursiops truncatus*) in the northern Gulf of Mexico and (*Tursiops* cf. *T. gilli*) off southern California, Baja California and Nayarit, Mexico. *Marine Fisheries Review*, **37(9)**, 10–16.

Leuthold W. (1977) *African Ungulates: A Comparative Review of their Ethology and Behavioural Ecology*. Springer, Berlin.

Levanthal A.M., Morrell R.F., Morgan E.J. & Perkins C.C. (1959) The relation between mean reward and mean reinforcement. *Journal for Experimental Psychology*, **57**, 284–7.

Levins R. (1968) *Evolution in Changing Environments. Some Theoretical Explorations*. Princeton University Press, Princeton.

Lewontin R.C. (1966) Is nature probable or capricious? *Bioscience*, **16**, 25–7.

Lewontin R.C. (1979) Fitness, survival, and optimality. In *Analysis of Ecological Systems*. (Ed. by D.J. Horn, G.R. Stairs & R.D. Mitchell), pp. 3–21. Ohio State University Press, Columbus.

Li C.C. (1955) The stability of an equilibrium and the average fitness of a population. *American Naturalist*, **89**, 281–295.

Lorenz K. (1963) *On Aggression*. Methuen, London.

MacArthur R.H. & Pianka E. (1966) On optimal use of a patchy environment. *American Naturalist*, **100**, 603–9.

Major P.F. (1977) Predator-prey interactions in schooling fishes during periods of twilight: a study of the silverside *Pranosus insularum* in Hawaii. *Fishery Bulletin*, **75**, 415–26.

Marchant J.H. & Hyde P.A. (1980) Population changes for waterways birds, 1978–79. *Bird Study*, **27**, 179–182.

Marshall N.B. (1965) *The Life of Fishes*. Weidenfeld and Nicolson, London.

Martinez D.R. & Klinghammer E. (1970) The behaviour of the whale *Orcinus orca*: a review of the literature. *Zeitschrift für Tierpsychologie*, **27**, 828–39.

Maynard Smith J. (1976) Group selection. *Quarterly Review of Biology*, **51**, 277–83.

Maynard Smith J. (1978) Optimization theory in evolution. *Annual Review of Ecology and Systematics*, **9**, 31–56.

McCullough D.R. (1969) The tule elk: its history, behaviour, and ecology. *University of California Publications in Zoology*, **88**, 1–209.

McNamara J. & Houston A. (1980) The application of statistical decision theory to animal behaviour. *Journal of Theoretical Biology*, **85**, 673–90.

Meinertzhagen R. (1959) *Pirates and Predators*. Oliver and Boyd, London.

Morse D.H. (1981) *Behavioral Mechanisms in Ecology*. Harvard University Press, Cambridge, Mass.

Munro J.A. (1938) The Northern Bald Eagle in British Columbia. *Wilson Bulletin*, **50**, 28–35.

Newton I. (1964) Bud-eating by bullfinches in relation to the natural food-supply. *Journal of Applied Ecology*, **1**, 265–79.

Nickolsky G.V. (1963) *The Ecology of Fishes* Academic Press, London.

Norris K.S. & Dohl T.P. (1980) The structure and functions of cetacean schools. In *Cetacean Behavior: Mechanisms and Functions* (Ed. by L.M. Herman), pp. 211–261. John Wiley, New York.

Oaten A. (1977) Optimal foraging in patches: a case for stochasticity. *Theoretical Population Biology*, **12**, 263–85.

Ollason J.G. (1980) Learning to forage-optimally? *Theoretical Population Biology*, **18**, 44–56.

Orians G.H. (1981) Foraging behavior and the evolution of discriminatory abilities. In *Foraging Behavior. Ecological, Ethological and Psychological Approaches* (Ed. by A.C. Kamil & T.D. Sargent), pp. 389–405. Garland STPM Press, New York.

Parker G.A. & Stuart R.A. (1976) Animal behavior as a strategy optimizer: evolution of resource assessment strategies and optimal emigration thresholds. *American Naturalist*, **110**, 1055–76.

Popper K. (1978) Natural selection and the evolution of mind. *Dialectica*, **32**, 339–55.

Potts G.W. (1970) The schooling ethology of *Lutianus monostigma* (Pisces) in the shallow reef environment of Aldabra. *Journal of Zoology*, **161**, 223–35.

Pubols B.H. (1962) Constant vs. variable delay of reinforcement. *Journal of Comparative and Physiological Psychology*, **55**, 52–6.

Pulliam H.R. (1975) Diet optimization with nutrient constraints. *American Naturalist*, **109**, 765–8.

Pulliam H.R. (1981) Learning to forage optimally. In *Foraging Behavior. Ecological, Ethological and Psychological Approaches* (Ed. by A.C. Kamil & T.D. Sargent), pp. 379–88. Garland STPM Press, New York.

Pulliam H.R. & Dunford C. (1980) *Programmed to Learn. An Essay on the Evolution of Culture.* Columbia University Press, New York.

Pyke G.H., Pulliam H.R. & Charnov E.R. (1977) Optimal foraging: a selective review of theory and tests. *Quarterly Review of Biology*, **52**, 137–54.

Radinsky L. (1978) Evolution of brainsize in carnivores and ungulates. *American Naturalist*, **112**, 815–31.

Rapport D.J. (1981) Foraging behavior of *Stentor coeruleus*: a microeconomic approach. In *Foraging Behavior. Ecological, Ethological and Psychological Approaches* (Ed. by A.C. Kamil & T.D. Sargent), pp. 77–93. Garland STPM Press, New York.

Rapport D.J. & Turner J.E. (1975) Feeding rates and population growth. *Ecology*, **56**, 942–9.

Rasa O.A.E. (1981) Raptor recognition: an interspecific tradition? *Naturwissenschaften*, **68**, 151–2.

Real L.A. (1980a) On uncertainty and the law of diminishing returns in evolution and behaviour. In *Limits to Action. The Allocation of Individual Behaviour.* (Ed. by J.E.R. Staddon), pp. 37–64. Academic Press, New York.

Real L.A. (1980b) Fitness uncertainty and the role of diversification in evolution and behaviour. *American Naturalist*, **115**, 623–38.

Rich W.H. (1947) The swordfish and swordfishery of New England. *Proceedings of the Portland Society of Natural History*, **4(2)**, 5–102.

Roberts W.A. (1966) Learning and motivation in the immature rat. *American Journal of Psychology*, **79**, 3–24.

Rudebeck G. (1950/1) The choice of prey and modes of hunting of predatory birds with special reference to their predatory effect. *Oikos*, **2**, 65–88, **3**, 200–31.

Ruse M. (1982) Teleology redux. In *Scientific Philosophy Today* (Ed. by J. Agassi & R.S. Cohen), pp. 299–309. D. Reidel, Dordrecht.

Saayman G.S. & Tayler C.K. (1973) Social organization of inshore dolphins (*Tursiops aduncus* and *Sousa*) in the Indian Ocean. *Journal of Mammalogy*, **54**, 993–6.

Saayman G.S., Tayler C.K. & Bower D. (1973) Diurnal activity cycles in captive and free-ranging Indian Ocean bottlenose dolphins (*Tursiops aduncus* Ehrenburg). *Behaviour*, **44**, 212–33.

Schaffer W.M. (1978) A note on the theory of reciprocal altruism. *American Naturalist*, **112**, 250–3.

Schaffer W.M. & Rosenzweig M.L. (1978) Homage to the Red Queen. I. Coevolution of predators and their victims. *Theoretical Population Biology*, **14**, 135–57.

Schaller G.B. (1967) *The Deer and the Tiger.* University of Chicago Press, Chicago.

Schaller G.B. (1972) *The Serengeti Lion.* Chicago University Press, Chicago and London.

Schaller G.B. (1977) *Mountain Monarchs: Wild Sheep and Goats of the Himalaya.* Chicago University Press, Chicago.

Serventy D.L. (1939) Notes on cormorants. *Emu*, **38**, 357–71.

Shanewise S. & Herman S.G. (1979) Flocking behavior in wintering Dunlin. In *Shorebirds in Marine Environments* (Ed. by F.A. Pitelka), p. 191. Cooper Ornithological Society Studies in Avian Biology No. 2. The Cooper Orn. Society.

Sibly R.M. & McFarland D.J. (1976) On the fitness of behaviour sequences. *American Naturalist*, **110**, 601–17.

Spooner G.M. (1931) Some observations on schooling in fish. *Journal of the Marine Biological Association of the U.K.*, **17**, 421–48.

Springer S. (1957) Some observations on the behaviour of schools of fishes in the Gulf of Mexico and adjacent waters. *Ecology*, **38**, 166–71.

Staddon J.E.R. (1980) Optimality analyses of operant behavior and their relation to optimal foraging. In *Limits to Action. The Allocation of Individual Behavior*. (Ed. by J.E.R. Staddon), pp. 101–41. Academic Press, New York.

Stanley S.M. (1979) *Macroevolution. Pattern and process*. W.H. Freeman, San Franciso.

Stephens D.W. (1981) The logic of risk-sensitive foraging strategies. *Animal Behaviour*, **29**, 628–9.

Taverner P.A. (1926) *Birds of western Canada*. Victoria Memorial Museum Bulletin No. 41. Biological Series No. 10. King's Printer, Ottawa.

Tener J.S. (1965) *Muskoxen in Canada*. Queens Printer, Ottawa.

Thomson A.L. (1964) Pelican. In *A New Dictionary of Birds* (Ed. by A.L. Thomson), pp. 607–8. Nelson, London.

Tinbergen N. (1951) *The Study of Instinct*. Oxford University Press, Oxford.

Turner J.R.G. (1967) Mean fitness and the equilibria in multilocus polymorphisms. *Proceedings of the Royal Society of London B*, **169**, 31–58.

Vermeij G.J. (1977) The Mesozoic marine revolution: evidence from snails, predators and grazers. *Paleobiology*, **3**, 245–58.

Vieth W., Curio E. & Ernst U. (1980) The adaptive significance of avian mobbing. III. Cultural transmission of enemy recognition in blackbirds: cross-species tutoring and properties of learning. *Animal Behaviour*, **28**, 1217–29.

Waage J.K. (1979) Foraging for patchily-distributed hosts by the parasitoid, *Nemeritis canescens*. *Journal of Animal Ecology*, 48, 353–71.

Westoby M. (1974) An analysis of diet selection by large generalist herbivores. *American Naturalist*, **108**, 290–304.

Williams G.C. (1966) *Adaptation and Natural Selection. A Critique of Some Current Evolutionary Thought*. Princeton University Press, Princeton.

Williamson M. (1972) *The Analysis of Biological Populations*. Edward Arnold, London.

Willis E.O. (1967) The behaviour of bicolored antibirds. *University of California Publications in Zoology*, **79**, 1–132.

Wimsatt W.C. (1980) Randomness and perceived-randomness in evolutionary biology. *Synthése*, **43**, 287–329.

Würsig B. & Würsig M. (1980) Behavior and ecology of dusky dolphins, *Ladenorhynchus obscurus*, in the South Atlantic. *Fisheries Biology*, **77**, 871–90.

13. THE GENETICS OF HOST–PARASITE INTERACTION

JOHN A. BARRETT
Department of Genetics, University of Liverpool, P.O. Box 147, Liverpool L69 3BX

INTRODUCTION

Population dynamics and population genetics are inextricably interwoven in any real population of organisms. It is a matter of historical accident that these two fields of study have developed separately in the distinct biological disciplines of ecology and genetics. This chapter is an attempt to draw together some of the ideas which have developed about the genetics of host–parasite interactions to show some of the range of genetic mechanisms which are involved in host–parasite interactions and to argue that the genetic interactions between host and parasites are important in the ecology of these systems. In this chapter I shall not discuss properties of host–parasite systems which appear to be the outcome of evolutionary processes and which, it can be argued, are 'adaptations', for example, lateral flattening in fleas, but more the evidence of genetic variation and evolutionary processes in host–parasite systems. I shall not consider in any detail explicit models of the population dynamics of host–parasite systems which have been developed by small groups of ecologists and parasitologists over recent years (cf. Anderson 1974, 1976; Price 1980; Gilpin, 1975; Hirsch, 1977). This chapter is not aimed at biologists already working with host–parasite systems but more at geneticists and ecologists with only a passing aquaintance with these associations.

Evolutionary ecology is the study of dynamic processes and as Harper (1977) has pointed out, the only way to study such systems is to perturb them. Whilst there may be moral or ethical problems in disturbing 'natural' ecosystems, in order to explore the underlying dynamic processes, this problem is avoided in the fields from which I shall draw most of my examples since most of them represent gross disturbances of ecosystems by man's activities; in many cases the purpose of the initial work has been to determine what form of intervention in the ecosystem would be most advantageous to man's well-being. Indeed, man's activities have left us with a rich heritage of 'natural' experiments which have only recently begun to be appreciated by evolutionary biologists interested in the dynamic properties of populations (e.g. Bishop & Cook 1981). However, the fact that much of this evidence is a direct consequence of man's activities produces two major problems:

1 The disturbance of ecosystems may be so great, for example, agriculture, that there may be no easy way to extend the conclusions to natural ecosystems.

2 The data have often been collected for other purposes and consequently information may be less than complete.

It would be almost impossible to attempt a comprehensive survey of all the host–parasite interactions that have been shown to contain a genetic component in a short chapter and so this essay is to a large extent a personal selection. But I hope that the references are sufficient for an entry into the literature for most host–parasite systems.

HOST–PARASITE DYNAMICS

Part of Darwin's original argument in the development of his ideas about the origin of species was that since more offspring are produced than can possibly survive, natural selection would occur. Fisher (1930) pointed out that this argument was fallacious, since the 'excess' reproductive output envisaged by Darwin was with respect to some imaginary world in which mortality was absent. He further argued that '... if there is any mortality whatever ...., either through inorganic causes, or by reason of predators or parasites, it necessarily follows that young must be produced in excess of the parental numbers'. Consequently, the reproductive 'excess' is itself a product of natural selection. The establishment of a parasitic association in an already 'stable' host population, will either lead to the extinction of the host population or the establishment of a new equilibrium host population size since all the parasite will do is to impose a further source of mortality on the host population (Lotka 1924; Maynard Smith 1974; cf. Anderson 1976). But these conclusions are based exclusively on the dynamic processes of population growth and regulation; they ignore the possibility that variation may exist in either the host or parasite population with respect to the ability to form a parasitic association and hence to be subject to selection at the time of first encounter between host and parasite.

FIRST ENCOUNTER BETWEEN A HOST AND PARASITE SPECIES

At some time in the past, a potential host species and a potential parasite species must have come into contact for the first time. This initial contact may have been phylogenetically ancestral to extant host–parasite associations or between host and parasite species taxonomically identical to present day species. When such a new host–parasite association occurs, there are four possible outcomes.

1 Chance fluctuations in population numbers eliminate the host, the parasite or both.

2 The parasite eliminates the host.

3 The association persists but the host eventually evolves resistance/immunity, and the parasite becomes extinct with respect to this host.

4 An equilibrium between host and parasite is attained and the association persists.

Only the last of these outcomes will leave any evidence of the association ever having existed and it is on this type of association that our modern observations and interpretations are based (cf. Harper 1977). In other words, when host–parasite systems are examined, only a biased sample of those that have existed can be investigated.

On initial contact with a new host species, a parasite may induce symptoms of such a debilitating effect that survival or reproduction of the host is seriously impaired. Since the absolute fitness of a parasite is dependent on the survival of the host for sufficient time for the parasite to reproduce and transmit its offspring or propagules to other hosts, natural selection can act on inherited variation in the parasite via its effects on its host. If there is variation in the parasite population which reduces its debilitating effects on the host whilst at the same time not affecting the reproductive potential of the parasite, then natural selection will tend to increase the frequency of the parasite genotypes which produce reduced symptoms, ultimately, perhaps, producing almost symptomless infection. However, there may be pleiotropic effects of the genes controlling the reduction of symptoms and which impair the reproductive output of the parasite, especially if the symptoms are associated with the reproduction or the dissemination of propagules or offspring. In this case there will be a 'trade-off' between reproduction and severity of symptoms, culminating in a reduced but not total suppression of symptoms. However, what natural selection is acting upon, in this case, is *transmissibility* and the symptoms in the host and reproductive output are two factors contributing to transmissibility (Burnet & White 1972).

One of the principal problems in the investigation of host–parasite interactions, especially in the laboratory, is that often the only phenotypic differences that can be discerned between individuals in a parasite population are in characters such as infectivity, level of infection and severity of symptoms and the parameters associated with these characters. Consequently, if a change in one of these characters coincides with an evolutionary change in the parasite, there is a temptation to attribute the change to selection of that character. The phenomenon of 'attenuation' of virulence in parasites may involve such an interpretation. Reduced symptoms or severity of disease can be associated with longer host survival and consequently, perhaps, an extended period of parasite reproduction, permitting more new infections to be established before the original host ultimately dies. In this case, natural selection is acting on the character of transmissibility which manifests itself *in the characters we can measure* as an apparent 'attenuation' of virulence. Whether individual or group selection arguments are required to account for the evolution of 'attenuation' depends on the biology of both host and parasite, for example, mode of transmission, density of host, the possibility of multiple infection, number of infective particles or propagules produced per infection, etc. (cf. Lewontin 1970).

Perhaps the best–documented example of the evolution of 'attenuation' of virulence under natural conditions is that of myxoma virus in its interaction with the European rabbit *Oryctolagus cuniculus*. The social, economic and political history of rabbit control by myxomatosis is a fascinating story and has been extensively reviewed by Fenner & Ratcliffe (1965) and more recently by Mead-Briggs (1977) (cf. Pimentel in this volume). Myxoma virus is found as an endemic parasite in the South American rabbit *Sylvilagus brasiliensis* and in the Californian rabbit *Sylvilagus bachmani* in which it causes mild symptoms in a substantial proportion of the population. However, in the European rabbit, the virus produces a rapidly fatal systemic disease. Its possible use as a controlling agent was recognized earliei in this century and after may attempts it was successfully introduced into rabbit populations in Europe and Australia in the early 1950s (Fenner & Ratcliffe 1965; Mead-Briggs 1977). Primarily because of the significant damage caused in agriculture by rabbits, the progress of the disease in the field was closely monitored. After some time it was noticed that infected rabbits in the field were not succumbing to the disease as readily as previously and that isolates of the virus taken from the field and tested in a standardized laboratory test did not produce such severe symptoms and the infected rabbits survived longer (Fenner & Ratcliff 1965). Indeed, by careful analysis, it was found that up to five different levels of virulence could be distinguished between isolates of virus derived from field samples (Table 13.1). Although very 'attenuated' strains of virus could be identified, the virus populations in the field appeared to consist predominantly of virus strains with intermediate levels of virulence (Fenner & Ratcliffe 1965; Fenner 1965), both in Australia and the UK. Despite differences in the biology of the main vectors in Australia and the UK, similar outcomes were obtained and this suggests important similarities in the epidemiology of the disease. Firstly, on introduction the disease was rapidly fatal. Secondly, the vectors (mostly mosquitoes in Australia and the

TABLE 13.1. Frequency distributions of different virulence grades of myxoma virus recovered from the field in Great Britain after Fenner & Ratcliffe (1965).

	Virulence Grade					
	I	II	IIIA	IIIB	IV	V
Mean survival time of infected test rabbits (days)	<13	14–16	17–22	23–28	29–50	—
Date recovered from field						
1953–54	1.00	—	—	—	—	—
1962	0.041	0.176	0.388	0.248	0.144	0.005

rabbit flea *Spilopsyllus cuniculi* in the UK) are haemophagous and transmission is from a live infected animal to other animals. Thus an animal infected with a virus strain of low grade virulence will survive longer than an animal infected with a strain of higher virulence and thus remain attractive to vectors for longer. Although a single virus particle can, in theory, initiate an infection, the vectors appear to transmit many virus particles in a single bite (Fenner & Ratcliffe 1965). The vector can thus transmit a sample of the virus population present in the infective host. If mutation to different levels of virulence can occur in the infected host, the transmitted infective doses may contain a range of different virus genotypes. In other words, chance variations in the proportions of different virus genotypes due to the effects of sampling by the vector would manifest themselves as changes in the proportions of the different genotypes in the recipient hosts and hence in the length of time before the host animal succumbed. This would necessarily increase the chances of transmission of the virus infections consisting of populations containing genotypes with reduced virulence. This model was originally proposed by Lewontin (1970) and has recently been modelled by Levine & Pimentel (1981) and Anderson & May (in press). The process of chance variation produced by sampling effects would then enable attenuated strains to become established. Two further processes appear to be involved in the evolution of attenuation. First, highly attenuated strains appear to have lower rates of increase in the blood of the host, and second, the highly virulent virus strains kill the rabbit before the host immune system is fully mobilized, whereas infection with an attenuated virus strain which allows the host to survive longer, may allow sufficient time for the immune system to respond to the presence of the virus. Once hosts are able to survive infection and reproduce, resistance to infection could begin to evolve by natural selection. In rabbit populations which have been exposed to continual infection by myxoma virus, an increase in resistance has been observed when these rabbits have been tested with standard strains of the virus (Fenner & Ratcliffe, 1965).

Natural selection towards resistance in the host could then begin to select for increased virulence in the virus since virus strains with extended periods of infectivity would obviously be at a selective advantage. But, again, this evolutionary process would be by group selection between subpopulations of the virus. Recent evidence suggests that this is happening (Fenner & Myers 1979; Mead-Briggs 1977). It remains to be seen whether a 'climax' association of the type found in the *S. brasiliensis*/myxoma and *S. bachmani*/ myxoma systems will eventually evolve.

Newcastle disease of poultry is a modern disease caused by a virus. It was first identified in the Dutch East Indies in 1926 and in the UK later in the same year (Anon 1965; Shope 1964). It is a highly virulent contagious disease and slaughter of all infected flocks was instituted as a control measure in the UK and elsewhere. In Britain, this policy appeared to be succeeding until about 1950 when two new forms of the disease, the *acute* and *sub-acute* forms, appeared and began to displace the original, *peracute*, form. Isolates of virus strains causing acute and sub-acute forms of the disease are stable when cultured in the laboratory and hence appear to be genetically different (Siegmund 1979). The two new forms exhibit reduced symptoms relative to the peracute form in infected poultry and rapidly displaced the latter form. In its centenary report, the Animal Health Division of the British Ministry of Agriculture, Fisheries and Food concluded that (Anon 1965). '... the slaughter policy had succeded in stamping out the peracute form of the disease'.

This choice of words is revealing in that it does not consider the possibility that the evolution of the acute and subacute forms could have been brought about by the slaughter policy (Barrett 1981). The peracute form produces the classic symptoms of the disease very rapidly. The poultry man recognizes the symptoms and the whole flock is slaughtered, but not before transmission to other flocks has possibly occurred. Obviously, any virus variant which can reduce the symptoms of the disease whilst at the same time maintaining its transmissibility will obviously be at an evolutionary advantage. Although the mode of transmission, by contact or aerosol, is different from that of myxoma virus, it is easy to envisage a process similar to that which probably brought about attenuation in myxoma virus acting in a similar way in Newcastle disease.

These two examples provide some evidence that on initial contact between a host and a parasite species, evolution towards reduced virulence in the parasite can occur but the process which brings this about is selection for transmissibility and not 'to prevent the parasite eliminating the host' (e.g. Browning 1981; Dawkins 1982). The rabbit-myxoma example also provides some evidence that a host can evolve reduced susceptibility to counter infection by the parasite. The next section discusses the possible consequences of evolution after a host–parasite association has become established.

MICRO-EVOLUTIONARY CHANGE IN HOST–PARASITE SYSTEMS

For adaptive evolution to occur at all three criteria must be satisfied:
1 Variation exists within populations.
2 The variation is, at least partly, inherited.
3 The variation must affect the probability of leaving offspring in the next generation.

If all of these criteria are satisfied, then natural selection will necessarily occur. For evolution to occur in host–parasite systems, these criteria must be satisfied in both host and parasite populations.

Host variation

We have already seen that there may be reciprocal genetic changes in both host and parasite as each responds to the presence of the other in the case of the rabbit-myxoma virus interaction. However, at the moment, the formal genetics of the inheritance of resistance to myxoma infection has not been fully worked out. Similarly, although Hutt (1958) has been able to show that the resistance to Newcastle disease in White Leghorn fowl can be substantially increased by selective breeding, a formal analysis of the inheritance of this resistance has not been carried out. Indeed, the largest part of the evidence for the inheritance of resistance to parasites in domestic animals is in the form of either observations that breeds or strains differ in their susceptibility to diseases or that when selection for increased or decreased susceptibility to disease is applied, a response can usually be obtained. For example: White Leghorns are less susceptible to *Salmonella pullorum* than Rhode Island Reds but both breeds can be selected for improved resistance. Sheep breeds can differ quite substantially in their levels of infection by intestinal nematodes. (Hutt 1958, 1964). Sheep breeds have been shown to differ in their susceptibility to the scrapie agent and experimental breeding has produced lines of sheep with increased and decreased susceptibility from a common stock (Nessbaum *et al.*, 1975; Parry 1979a, b; Pattison 1974). So there is good evidence for the inheritance of resistance to parasitic infection in domestic animals and poultry, although in many cases the formal genetics have not been worked out (cf. Taylor & Muller 1976; Skamene, Kongshavn & Landy 1980).

Conversely, three of the better known and extensively studied genetic polymorphisms in man, sickle cell trait, glucose-6-phosphate dehydrogenase deficiency and thalassaemia, are associated with resistance to *falciparum* malaria. In the bacteria it is surprisingly easy to select for resistance to bacteriophage infection and an extensive literature has built up on the genetics and mechanism of phage resistance (cf. Lewin, 1977).

Parasite variation

Although it is possible to show the existence of inherited variation in pathogenicity in both plant and animal parasites, our knowledge of the formal genetics of this variation is almost completely confined to plant parasites. To a large extent this is due to the differences in approach to the control of parasitic infection in cultivated plants, man and domesticated animals (Barrett 1981). In both man and domesticated animals, the basic approach is to improve hygiene to prevent infection and medical intervention to deal with any infections which arise in spite of the preventative procedures (e.g. Halpin 1975). Consequently, any breakdown in control is by the evolution of tolerance to or evasion of the control measure by the parasite and the influence of the host in the evolution of the parasite is minimal. Any genetic variation in the ability of parasites to induce infection tends only to be noticed during experimental manipulations in the laboratory; for example, in the mammalian parasites *Trypanosoma cruzi* and *Trichomonas vaginalis*, the severity of infection induced in standard test animals is dependent on the strain of the parasite used and the strain of test animal (Levine 1973). On the other hand, in cultivated plants, the use of genetically controlled resistance is the primary defence against parasitic infection. Consequently, any breakdown of control due to any evolution of the ability to overcome this resistance and detected in the field can be directly attributed to a genetic component of the host–parasite interaction. An understanding of the genetics of pathogenicity in plant parasites is a major objective in controlling plant disease (Day 1974).

In 1917 and 1918 Stakman and his colleagues described the phenomenon of 'physiologic specialisation' in the interaction of stem rust, *Puccinia graminis*, and wheat (Stakman & Piemeisel 1917; Stakman, Parker & Piemeisel 1918; Stakman, Piemeisel & Levine 1918). Isolates of stem rust were taken from different varieties of wheat and then used to inoculate a range of different varieties. It was found that each isolate of rust always grew best on the variety from which it was originally isolated. Since then this observation has been made repeatedly on other plant host–parasite systems (cf. Day, 1974 for discussion). This phenomenon is not confined to fungal parasites; other parasites, for example, nematodes, can be shown to have "biotypes" adapted to different host varieties (Day 1974).

To some extent it is difficult to discuss genetic variation in host–parasite systems by considering the host or the parasite separately. The essence of the host–parasite system is that there is an interaction and it is the genetics of the interaction rather than of just one of the members of the association that should be studied. In the next section, the form of genetic interaction between host and parasite will be discussed. This discussion is inevitably biased towards plant host–parasite systems since it is these that have received the most attention.

Genetic interactions between hosts and parasites

Starting with the observation that resistance in flax to flax rust *Melampsora lini* differed between varieties and that physiologic specialization in the flax rust fungus also existed, Flor (1942, 1955, 1956) began a series of experiments designed to investigate the genetics of pathogenicity and resistance simultaneously. He found that not only was resistance in the host controlled by simple Mendelian factors but also that the ability of the fungus to attack different varieties was also controlled by major genes which segregated in an *F*2 (Table 13.2). He summarized his results in what has now become known as the *gene-for-gene* hypothesis, which states that for every gene controlling resistance in the host, there is a corresponding gene in the parasite controlling pathogenicity. Gene-for-gene systems have been demonstrated for many fungal diseases of crop plants, for example, powdery mildew of barley and

TABLE 13.2. Reaction of flax varieties *Ottawa* and *Bombay* to races 22 and 24 of flax rust (after Flor, 1942, 1955, 1956).

Flax Variety	Rust Race			
	22		24	
Bombay	Resistant		Susceptible	
Ottawa	Susceptible		Resistant	
Plants resistant to race:	22 and 24	22 only	24 only	Neither
Segregation in flax F2:	9	3	3	1
Rust isolates able to attack variety:	*Ottawa* and *Bombay*	*Ottawa* only	*Bombay* only	Neither
Segregation in rust F2:	1	3	3	9

stem rust of wheat (cf. Day 1974), although the number of diseases in which it has been unequivocally demonstrated is still small. Gene-for-gene systems are not confined to plant–fungus associations. In parts of North America, substantial damage is caused to wheat by the wheat hessian fly, *Mayetolia destructor*; wheat varieties resistant to hessian fly attack have been bred and resistance is controlled by major genes. However, in the field, these resistant varieties have had limited success since once the resistant varieties have been grown on a large scale, the hessian fly has evolved biotypes capable of attacking them. The control of the ability to attack the resistant varieties is on a gene-for-gene basis (Gallun, 1971). Despite its appealing simplicity, the gene-for-gene hypothesis has a number of conceptual problems, not least what exactly we mean by a gene. The simple locus-for-locus system with two alleles per locus originally described by Flor is no longer the general rule (Barrett 1983); systems can be found in which resistance is controlled by multiple alleles in the host but the corresponding pathogenicity in the parasite is controlled by different loci, for example, the *P* locus in the flax–flax rust system (Lawrence, Mayo & Shepherd 1982). However, the basic conceptual framework of both resistance and pathogenicity controlled by major genes is a useful one. Virtually all of the cases that have been described as exhibiting a gene-for-gene relationship have been in agricultural crops; there is only circumstantial evidence for their existence in natural systems. Indeed, there is a suspicion that gene-for-gene systems may be artefacts of man's agricultural activities (Day 1974; Browning 1981).

When plant breeders are looking for resistance to incorporate into their breeding programmes, their work is made easier if the effect is large and the character is easy to handle. This necessarily means that the resistance characters they are looking for will more often be controlled by one or few major genes. Polygenically controlled resistance will tend to be exploited only if no simpler system is available (Day, Barrett & Wolfe 1983). Ideally, a plant breeder is looking for complete resistance to the prevailing parasite population, so that when the new variety is released for general cultivation it is free from disease. A variety fulfilling this requirement will be grown on a larger and larger acreage as growers become aware of its benefits. But as it is grown on an increasingly larger scale it imposes a larger and larger selective force on the parasite population and will select out of it any genotypes capable of overcoming the resistance. Since the breeders try to select resistance characters which are outside the range of adaptation of the parasite population, it seems that only major mutations will take the parasite into the range of adaptation required by the new resistance, in a manner similar to the processes involved in the evolution of Batesian mimicry and industrial melanism in *Lepidoptera* (Fisher 1930; Sheppard 1975; Turner 1977). This is not to say that the new adaptation of the parasite could not be achieved by polygenic systems, but that it appears that in most cases it is achieved more easily by changes at one or a few loci. The new mutations do not have to be particularly good at attacking the new resistance, since there will be little competition from other genotypes. As the new mutant forms increase on the resistant variety, there will be selection between the different mutations which have taken the parasite into the zone of adaptation to the resistant variety and also selection of the genetic background to enhance the expression of these mutations with respect to their pathogenicity and also to reduce any deleterious pleiotropic effects of the mutation (Fisher 1930). Thus by the time that the resistance of the variety has 'broken down', and the parasite has become a serious problem, selection will have tended to eliminate all but one of the mutations which had originally been able to attack the new variety and also produced a co-adapted genetic background which enhances the expression of this mutation. Subsequent crossing to other parasite genotypes will reveal an apparent single gene determination of pathogenicity and the data will be interpreted as a gene-for-gene relationship. Whilst the interpretation offered here is speculative, it is based on more extensively understood systems in which organisms have evolved under fairly intense selective pressures, for example, insecticide resistance, Batesian mimicry and industrial melanism (Fisher 1930; Lees 1981; Sheppard 1975; Turner 1977; Wood, 1981; Wood & Bishop 1981). So it is possible that the gene-for-gene relationship is an artefact of agriculture generated by the *ad hoc* nature of evolution, rather than some basic molecular or physiological relationship between the genes involved in the interaction between host and parasite (Day *et al.* 1983).

Although the argument that the gene-for-gene relationship is a consequence of the cultivation of host plants is fairly compelling, there is evidence that major genes controlling resistance and pathogenicity do exist in natural populations. Plant breeders often obtain resistance genes for use in breeding programmes by screening wild relatives of crop plants; indeed, in Biffen's (1905, 1907) experiments on the inheritance of disease resistance, one of the lines he used in his barley crosses was derived from a wild *Hordeum spontaneum*. By testing material collected from wild populations of host plants with fungal isolates of known genotypes, it has been possible to show that in wild populations of the *Avena sterilis*/*Puccinia coronata* and *Hordeum spontaneum*/*Erysiphe graminis* systems, the plants exhibit disease reactions characteristic of the presence of major resistance genes and these phenotypes can be present at fairly high frequencies in the host populations (Dinoor 1977; Wahl *et al.* 1978). Similarly, by testing samples of fungal parasites from natural ecosystems on sets of varieties carrying known resistance genes (differential sets), it has been shown that there is variation in these fungus populations for the ability to attack these varieties and again the frequencies of these different phenotypes can be fairly high in some populations. The inference from these types of experiment is that major genes for resistance and pathogenicity are segregating in wild populations, but a formal genetic investigation of these interactions has not yet been carried out and it is not yet clear whether we are dealing with systems which are homologous with those found in agriculture. Indeed the use of cultivated varieties and fungal 'races' derived from agricultural crops may be introducing biasses into the interpretation of the data.

When fungal parasites are isolated from agricultural crops, it is often found that, although morphologically identical fungi can be isolated from different crop species, cross infection between the different host species is not possible, for example, powdery mildew from oats will not infect barley and similarly powdery mildew from cultivated barely does not attack oats. Where such host specialization can be demonstrated, the fungal species is often subdivided into *formae speciales* based on the host plant species from which they were isolated. In cultivated crops, this convenient taxonomic approach is usually unequivocal. However, if a fungal parasite is taken from, say, the wild relatives of cultivated crops, and tested on a range of different species from the same ecosystems as the wild species, the specificity of the parasites characteristic of the cultivated forms is often absent (for discussion cf. Wahl *et al.* 1978; Browning 1981). For example, Israeli isolates of powdery mildew, *Erysiphe graminis*, from *Hordeum murinum* have been shown to be able to attack seventeen species from thirteen genera and four tribes (Wahl *et al.* 1978). Moreover, there may be no taxonomic consistency in the range of hosts that can be attacked by a single parasite isolate, i.e. the host species attacked by a given parasite isolate is not confined to the host from which it was originally collected and members of the same genus or all the members of another genus but perhaps a couple of species from a number of different genera (Gerechter-Amitai 1973; Eshed & Dinoor, 1981). Moreover, some fungal parasites have different hosts at different stages of their life cycle. For example, *Puccinia graminis* can attack a wide range of grass species on which it reproduces asexually but the sexual stage of all forms is on the alternate host, (*Berberis* spp.). So the possibility exists that even though some specialization to different host species may be manifest in the asexual stage, hybridization between different forms is possible on the alternate host.

When parasites exhibit a wide host range, the density-dependent feedback mechanisms which underly many models of host–parasite systems may not operate. Under these conditions, the dynamics of a particular host–parasite combination may be strongly influenced by the composition of the ecological community to which the host–parasite system belongs. The importance of genetic factors in the interpretation of experiments with host–parasite systems has been emphasized by Richards who has investigated the inheritance of differential susceptibility in different populations of the snail *Biomphalaria glabrata* to *Schistosoma mansoni*. Moreover, lines of *Schistosoma mansoni* have been selected from a single population which differ in their ability to infect *Biomphalaria glabrata*.

Population genetics of host–parasite interactions

Even in his earlier writings, Haldane (1932) was convinced that parasites had played an important role in evolution, perhaps even being instrumental in the extinction of species. Later he was led, mainly thought studying blood groups, to considering the possibility that parasites could be a determining factor in the maintenance of polymorphism. His argument (1949, 1954) was that if parasites are successful because they cannot be distinguished antigenically from 'self' by a host, then the success of a single parasite genotype would be reduced in a host population which was variable for blood protein antigens. Hence, parasite populations would tend to evolve adaptations enabling them to attack the commoner host genotypes and the rarer host genotypes would be at a selective advantage. This would give rise to frequency-dependent selection on both host and parasite populations as each responded to the influence of the other. He also argued that similar evolutionary processes would act on the wheat/stem–rust interaction (cf. Pimentel 1961 and this volume).

The simplicity of the gene-for-gene hypothesis has facilitated the development of explicit genetic models of host–parasite co-evolution. Indeed, it is probable that even had the gene-for-gene hypothesis not existed, it would have been necessary for population geneticists to invent it in order to model host–parasite systems. By and large, despite differences in the assumptions about breeding systems and the biology of the hosts and parasites, these models have shown the same types of dynamic behaviour obtained in population dynamic models of host–parasite interactions (e.g. Clark 1976; Mode 1958, 1961; cf. Barrett 1983; Anderson & May 1982 for further references). Stable and unstable equilibria can be obtained in gene, genotype and phenotype frequencies and where unstable equilibria exist, stable limit cycles can be produced. However, most of these models have generally considered only genetic changes and neglected changes in population size. Some understanding of the evolutionary processes in host–parasite interactions has been achieved at the expense of neglecting the population dynamics of the system. The complete integration of epidemiological and genetic aspects of host–parasite systems has yet to be achieved although over recent years, some attempts have been made at combining both epidemiological and genetic parameters in models of special cases, (e.g. Barrett 1978, 1980; Østergaard 1983; Leonard 1977; Leonard & Czochor 1978; Sedcole 1978).

Although the 'boom and bust' cycle characteristic of variety useage in agriculturally developed countries does generate a curve similar to a limit cycle when disease incidence or severity is plotted against the area of the variety planted, there is no evidence of genetic cycling behaviour in natural ecosystems. Although reactions characteristic of major gene resistance or pathogenicity can be shown to exist in the wild relatives of cultivated cereals and their fungal diseases (see above), there is no data of any sort on the evolutionary dynamics in these wild ecosystems. The results of the mathematical models also suggest that, even if cycling behaviour did occur in wild populations, it is unlikely that with the techniques currently available, the changes could be detected through the 'ecological noise'.

In man, there is some circumstantial evidence that biochemical variation may be involved in host–parasite interactions. The major tissue antigen system *HLA* has four loci with an ever increasing number of alleles at each locus (Bodmer & Thomson 1977; Bodmer & Bodmer, 1978). In population studies, the following properties of the *HLA* polymorphism have been observed:

1 Different 'racial' groupings show different frequencies of some alleles.

2 Substantial linkage disequilibrium for some haplotype combinations exists in some populations.

3 There is a cline in the frequency of one haplotype from the Near East across to north-west Europe. Although this cline may reflect migration patterns in the past, calculations suggest that there has been sufficient time for linkage equilibrium to have been achieved since the last major migration.

4 As well as the implication of some *HLA* genotypes in some 'auto-immune' diseases, there also appear to be associations with susceptibility to some infectious diseases.

These data have suggested to Bodmer and his colleagues that the *HLA* system is subject to the type of selection predicted by Haldane (1932)—it is highly unlikely that the system evolved to make tissue transplants difficult!

One of the central problems in evolutionary genetics over the last 15 years has been to produce a plausible hypothesis to explain the high levels of genetic polymorphism that have been shown to exist at the biochemical level by the use of gel electrophoresis. Clark (1979) has speculated that if successful parasitic infection requires a certain level of 'fine tuning' by the parasite to the physiology and biochemistry of the host, then a high level of protein polymorphism in a host population would make it less likely that a single parasite genotype would be completely compatible with the majority of the host population. This elaboration of Haldane's argument would then produce frequency–dependent selection on each of the protein polymorphisms.

THE IMMUNE RESPONSE

Parasites which elicit an immune response in their hosts sow the seeds of their ultimate demise, because each host successfully infected is converted to an immune (resistant) form. If such a parasite exhibits no antigenic variation, and if it is to be maintained, it requires a constant source of susceptible hosts. Apart from the transient effects of passive immunity in offspring produced by the antibodies of their female parent, immunity is not inherited. So, each cohort of offspring will consist of a new population of susceptible individuals. Hence, if the reproduction rate of the host population is high enough, the parasite may be maintained by infection of successive cohorts of offspring. However, this can only occur in large populations since, in small populations, the recruitment of susceptible individuals may not be high enough to maintain the infection within the population. Measles and smallpox, which show no antigenic variation, exhibit this type of epidemiology (e.g. Black 1966) (for an ecological application cf. Anderson & May 1979; May & Anderson 1979).

On the other hand, if a parasite species does show antigenic variation, then immunity to one genotype will not necessarily confer immunity to others. Although one genotype may ultimately fail to be maintained in a population, it will be replaced by other genotypes, since for them, the host population is susceptible. Indeed, a characteristic of influenza epidemics caused by influenza *A* virus is that of successive waves of different antigenic types (Pereira 1980). The antigenic variation in influenza *A* virus has been interpreted as being caused in two ways; mutation of the viral coat (drifts) and major genetic rearrangement through hybridization with other closely related viruses (shifts) (Webster & Laver 1975). A range of antigenic types is characteristic of some highly contagious virus diseases, for example, polio and foot-and-mouth disease. Although the examples given here are of virus diseases, the immune system is involved in reactions to many forms of parasites, for example, helminths and nematodes.

In an earlier part of this chapter, I argued that natural selection will tend to act on the transmissibility of parasites. In parasites which elicit an immune response, it is obvious that there is a limited time for transmission to occur before the host's immune system overwhelms the parasite. On the other hand, we have seen that if a parasite is capable of producing antigenically different forms then re-infection of a host can occur by a different antigenic type.

Taking an anthropopmorphic view of the system, it would appear that a parasite which, having infected a host could spawn a succession of different antigenic types, could increase the time available for transmission to other hosts. The tryapanosome *Trypanosoma brucei* appears to have evolved such a system (Cross 1978; Williams, Young & Majiwa 1979). After initial infection, the host's immune system begins to manufacture antibody against the coat protein of the parasite. But then, at regular intervals, the parasite changes over to producing a different coat protein with different antigenic properties, to which the immune system must produce a new antibody. Although much progress has been made in investigating the molecular basis of this process, the details have not yet been fully worked out (Turner 1982). On being picked up by the vectors (*Glossina* spp.) of trypanosomiasis, the molecular mechanism controlling expression of this antigenic 'library' appears to be reset, so that, on infecting a new host, the sequence of coat protein expression starts at the beginning.

CONCLUSIONS

It is becoming more and more apparent that an approach fusing both genetics and ecology is essential to our understanding of host–parasite systems since evolutionary changes can modify the population dynamics of the systems and conversely changes in population structure and behaviour can have substantial effects on the co-evolution of host and parasite. However, at the present time, this synthesis is more or less confined to theoretical treatments and retrospective analyses of the better documented cases, for example, myxomatosis. What is lacking are data from field studies specifically designed to examine both ecological and genetic components of host–parasite systems. Whether such an ideal could be attained is problematical; even in plant pathology, where both epidemiological and genetic techniques are well developed, it has proved difficult to obtain complete analyses and unambiguous interpretations. What is certain, however, is that although the evolutionary ecology of host–parasite systems is still in its infancy, it has an exciting future.

REFERENCES

Anderson R.M. (1974) Mathematical models of host-helminth parasite interactions. *Ecological Stability* (Ed. by M.B. Usher & M.H. Williamson), pp. 43–69. Chapman & Hall, London.

Anderson R.M. (1976) Dynamic aspects of parasite population ecology. *Ecological Aspects of Parasitology* (Ed. by C.R. Kennedy), pp. 439–462. North Holland Publishing Company, Amsterdam.

Anderson R.M. (1979) The influence of parasitic infection on the dynamics of host population growth. *Population Dynamics* (Ed. by R.M. Anderson, B.D. Turner and L.R. Taylor), pp. 245–281. Blackwell Scienitific Publications, Oxford.

Anderson R.M. & May R.M. (1979) Population biology of infectious diseases, Part I. *Nature*, **280**, 361–367.

Anderson R.M. & May R.M. (1982) Co-evolution of hosts and parasites. *Parasitology*, **85**, 411–426.

Anon (1965) *Animal Health—a Centenary, 1865–1965*. HMSO, London.

Barrett J.A. (1978) A model of epidemic development in variety mixtures. *Plant Disease Epidemiology* (Ed. by P.R. Scott & A. Bainbridge), pp. 129–137. Blackwell Scientific Publications, Oxford.

Barrett J.A. (1980) Pathogen evolution in multilines and variety mixtures. *Zeitschrift fur Pflanzenkrankheiten und Pflanzenschutz*, **87**, 383–396.

Barrett J.A. (1981) The evolutionary consequences of monoculture. *Genetic Consequences of Man-Made Change* (Ed. by J.A. Bishop & L.M. Cook), pp. 209–248. Academic Press, London.

Barrett J.A. (1983) Plant–fungus symbioses. *Co-evolution* (Ed. by D.J. Futuyma, M. Slatkin, J. Roughgarden & B.R. Levin). Sinauer Associates Inc., Sunderland, Mass.

Biffen R.H. (1905) Mendel's Laws of inheritance and wheatbreeding. *Journal of Agricultural Science (Cambridge)*, **1**, 4–48.

Biffen R.H. (1907) Studies in the inheritance of disease resistance. *Journal of Agricultural Science (Cambridge)*, **2**, 109–128.

Bishop J.A. & Cook L.M. (1981) *Genetic Consequences of Man-Made Change*. Academic Press, London.

Black F.L. (1966) Measles endemicity in insular populations: critical community size and its evolutionary implication. *Journal of Theoretical Biology*, **11**, 207–211.

Bodmer W.F. & Thomson G. (1977) Population genetics and evolution of the HLA system. *HLA and Disease* (Ed. by J. Dausset & A. Svejgaard), pp. 280–295. Munksgaard, Copenhagen.

Bodmer W.F. & Bodmer J.G. (1978) Evolution and function of the HLA system. *British Medical Bulletin*, **34**, 309–316.

Browing J.A. (1981) The agroecosystem-natural ecosystem dichotomy and its impact on phytopathological concepts. *Pests, Pathogens and Weeds* (Ed. by J.M. Thresh), pp. 159–172. Pitman Books Ltd, London.

Burnet M. & White D.O. (1972) *Natural History of Infectious Disease*, 4th ed. Cambridge University Press, Cambridge.

Clark B. (1976) The ecological genetics of host-parasite relationships. *Genetic Aspects of Host–Parasite Relationships* (Ed. by A.E.R. Taylor and R.M. Muller), pp. 87–104. Blackwell Scientific Publications, Oxford.

Clark B. (1979) The evolution of genetic diversity. *Proceedings of the Royal Society, Series B*, **205**, 453–474.

Cross G.A.M. (1978) Antigenic variation in trypanosomes. *Proceedings of the Royal Society, Series B*, **202**, 55–72.

Darwin C. (1872) *The Origin of Species by Means of Natural Selection or the Preservation of Favoured Races in the Struggle for Life*. 6th ed. John Murray, London.

Dawkins R. (1982) *The Extended Phenotype*. W.H. Freeman, Oxford.

Day P.R. (1974) *Genetics of Host–Parasite Interaction*. W.H. Freeman, San Francisco.

Day P.R., Barrett J.A. & Wolfe M.S. (in press) The evolution of host-parasite interaction. In *Genetic Engineering in Plants. An Agricultural Perspective* (Ed. by T. Kosuge and C.P. Meredith), pp. 419–430. Plenum Press, New York.

Dinoor A. (1977) Oat crown rust in Israel. *Annals of the New York Academy of Sciences*, **287**, 357–366.

Dinoor A. (1981) Epidemics caused by fungal pathogens in wild and cultivated crop plants. *Pests, Pathogens and Weeds* (Ed. by J.M. Thresh), pp. 143–158. Pitman Books Ltd, London.

Eshed N. & Dinoor A. (1981) Genetics of pathogenicity in *Puccinia coronata*: the host range among grasses. *Phytopathology*, **71**, 156–163.

Fenner F. (1965) Myxoma virus and *Oryctolagus cuniculus*: two colonising species. *The Genetics of Colonizing Species* (Ed. by H.G. Baker and G.L. Stebbins), pp. 485–501. Academic Press, New York.

Fenner F. & Myers K. (1978) Myxoma virus and myxomatosis in retrospect: the first quarter century of a new disease. *Viruses and Environment* (Ed. by E. Kurstak & K. Maramorosch), pp. 539–570. Academic Press, New York.

Fenner F. & Ratcliffe F.N. (1965) *Myxomatosis*. Cambridge University Press, Cambridge.

Fisher R.A. (1930) *The Genetical Theory of Natural Selection*. Clarendon Press, Oxford.

Flor H.H. (1942) Inheritance of pathogenicity in *Melampsora lini*. *Phytopathology*, **32**, 653–669.

Flor H.H. (1955) Host–parasite interaction in flax rust—its genetic and other implications. *Phytopathology*, **45**, 680–685.

Flor H.H. (1956) The complementary genic systems in flax and flax rust. *Advances in Genetics*, **8**, 29–54.

Gallun R.L. (1977) Genetic basis of hessian fly epidemics. *Annals of the New York Academy of Sciences*, **287**, 223–229.

Gerechter-Amitai Z.K. (1973) Stem rust, *Puccinia graminis* Pers, on cultivated and wild grains in Israel. PhD. Thesis, Hebrew University, Jerusalem. (Data quoted by Browning, 1980. Genetic protective mechanisms of plant pathogen populations: their co-evolution and use in breeding for resistance. In *Biology and Breeding for Resistance to Arthropods and Pathogens of Cultivated Crops* (Ed. by M.K. Harris), pp. 52–75. Texas A and M University, College Station, Texas.)

Gilpin M.E. (1975) *Group Selection in Predator–Prey Communities. Monographs in Population Biology*, vol. 9. Princeton University Press, Princeton.

Haldane J.B.S. (1932) *The Causes of Evolution*. Longmans, London.

Haldane J.B.S. (1949) Disease and evolution. *La Ricerca Scientifica*, **19**, Suppl., 68–76.

Haldane J.B.S. (1954) The statics of evolution. *Evolution as a Process*. (Ed. by J. Huxley, A.C. Hardy & E.B. Ford), pp. 109–121. Allen & Unwin, London.

Halpin B. (1975) *Patterns of Animal Disease*. Ballière Tindall, London.

Harper J.L. (1977) *Population Biology of Plants*. Academic Press, London.

Hirsch R.P. (1977) Use of mathematical models in parasitology. *Regulation of Parasite Populations* (Ed. by, G.W. Esch), pp. 169–207. Academic Press, New York.

Hutt F.B. (1958) *Genetic Resistance to Disease in Domestic Animals*. Constable, London.

Hutt F.B. (1964) *Animal Genetics*. Ronald Press Co., New York.

Kimberlin R.H. (1979) Aetiology and genetic control of natural scrapie. *Nature*, **278**, 303–304.

Lawrence G.J., Mayo G.M.E. & Shephered K.W. (1982) Interactions between genes controlling pathogenicity in the flax rust fungus. *Phytopathology*, **71**, 12–19.

Lees D.R. (1981) Industrial melanism: genetic adaptation of animals to air pollution. *Genetic Consequences of Man-Made Change* (Ed. by J.A. Bishop and L.M. Cook), pp. 129–176. Academic Press, London.

Leonard K.J. (1977) Selection pressures and plant pathogens. *Annals of the New York Academy of Sciences*, **287**, 207–222.

Leonard K.J. & Czochor R.J. (1978) In response to 'selection pressures and plant pathogens: stability of equilibria'. *Phytopathology*, **68**, 971–973.

Levine N.D. (1973) *Protozoan Parasites of Domestic Animals and Man*. 2nd ed. Burgess Publishing Co., Minneapolis.

Levine S. & Pimentel D. (1981) Selection of intermediate rates of increase in parasite–host systems. *American Naturalist*, **117**, 308–315.

Lewin B. (1977) *Gene Expression, Volume 3: Plasmids and Phages.* John Wiley & Sons, New York.

Lewontin R.C. (1970) The units of selection. *Annual Review of Ecology and Systematics,* **1**, 1–18.

Lotka A.J. (1924) *Elements of Physical Biology.* Williams Wilkins Co. Inc. (republished (1956) *Elements of Mathematical Biology* Dover Publications Inc., New York).

May R.M. & Anderson R.M. (1979) Population biology of infectious disease, part II. *Nature,* **280**, 455–461.

Maynard Smith J. (1974) *Models in Ecology.* Cambridge University Press, Cambridge.

Mead-Briggs A.R. (1977) *The European rabbit, the European rabbit flea and myxomatosis.* (Ed. by T.H. Coaker), *Applied Biology,* vol. 2, pp. 183–261. Academic Press, London.

Mode C.J. (1958) A mathematical model for the co-evolution of obligate parasites and their hosts. *Evolution, Lancaster,* Pa., **12**, 158–165.

Mode C.J. (1961) A generalized model of a host–pathogen system. *Biometrics* **17**, 386–404.

Nessbaum R.E., Henderson W.M., Pattison I.H., Elcock N.V. & Davies D.C. (1975) The establishment of sheep flocks of predictable susceptibility to experimental scrapie. *Research in Veterinary Science,* **18**, 49–58.

Østergaard H. (1983) Predicting development of epidemics in cultivars mixtures. *Phytopathology,* **73**, 166–172.

Parry H.B. (1979a) Aetiology of natural scrapie. *Nature,* **280**, 12.

Parry H.B. (1979b) Elimination of natural scrapie in sheep by sire genotype selection. *Nature,* **277**, 126–129.

Pattison I.H. (1974) Scrapie in sheep selectively bred for high susceptibility. *Nature,* **248**, 594–595.

Pereira M.S. (1980) The effects of shifts and drifts on the epidemiology of influenza in man. *Philosophical Transactions of the Royal Society, Series B,* **288**, 423–432.

Pimentel D. (1961) Animal population regulation by the genetic feedback mechanisms. *American Naturalist,* **95**, 65–79.

Price P.W. (1980) *Evolutionary Biology of Parasites. Monographs in Population Biology,* vol. 15. Princeton University Press, Princeton.

Segal A., Manisterski J., Fischbek G. & Wahl I. (1980) How plant populations defend themselves in natural ecosystems. In *Plant Disease: An Advanced Treatise. Volume V—How Plants Defend Themselves* (Ed. by J.G.S. Horsfall & E.B. Cowling), pp. 75–102. Academic Press, New York.

Sedcole J.R. (1978) Selection pressures and plant pathogens: stability of equilibria. *Phytopathology,* **68**, 967–970.

Sheppard P.M. (1975) *Natural Selection and Heredity.* 4th ed. Hutchinson, London.

Shope R.E. (1964) Introduction: the birth of a new disease. *Newcastle Disease Virus* (Ed. by R.P. Hanson), pp. 3–22. University of Wisconsin Press, Madison.

Siegmund O.H. (1979) *The Merck Veterinary Manual.* Merck and Co. Inc., Rahway, New Jersey.

Skamene E., Kongshavn P.A.L. & Landy, M. (1980) *Genetic Control of Natural Resistance to Infection and Malignancy.* Academic Press, New York.

Stakman E.C. & Piemeisel F.J. (1917) Biologic forms of Puccina graminis on cereals and grasses. *Journal of Agricultural Research,* **10**, 429–495.

Stakman E.C., Parker J.J. & Piemeisel F.J. (1918) Can biologic forms of stem rust on wheat change rapidly enough to interfere with breeding for rust resistance. *Journal of Agricultural Research,* **14**, 111–123.

Stakman E.C., Piemeisel F.J. & Levine M.N. (1918) Plasticity of biologic forms of Puccinia graminis. *Journal of Agricultural Research,* **15**, 221–250.

Taylor A.E.R. & Muller R. (1976) *Genetic Aspects of Host-Parasite Relationships.* Symposia of the British Society for Parasitology, Volume 14. Blackwell Scientific Publications, Oxford.

Turner J.R.G. (1977) Butterfly mimicry: the genetical evolution of an adaptation. In *Evolutionary Biology* (Ed. by M.K. Hecht, W.C. Steere & B. Wallace), vol. 10, pp. 163–205. Plenum Publishing Corporation, New York.

Turner M. (1982) Antigenic variation in the trypanosome. *Nature*, **298**, 606–607.

Wahl I., Eshed E., Segal N. & Sobel Z. (1978) Significance of wild relatives of small grains in cereal powdery mildews. *The Powdery Mildews* (Ed. by D.M. Spencer), pp. 83–100. Academic Press, London.

Webster R.G. & Laver W.G. (1975) Antigenic variation of influenza viruses. *The Influenza Virus and Influenza* (Ed. by E.D. Kilbourne), pp. 269–314. Academic Press, New York.

Williams R.O., Young J.R. & Majiwa P.A.O. (1979) Genomic re-arrangements correlated with antigenic variation in Trypanosoma brucei. *Nature*, **282**, 847–849.

Wood R.J. (1981) Insecticide resistance: genes and mechanisms. *Genetic Consequences of Man-Made Change* (Ed. by J.A. Bishop & L.M. Cook), pp. 53–96. Academic Press, London.

Wood R.J. & Bishop J.A. (1981) Insecticide resistance: populations and evolution. *Genetic Consequences of Man-Made Change* (Ed. by J.A. Bishop and L.M. Cook), pp. 97–127. Academic Press, London.

14. GENETIC DIVERSITY AND STABILITY IN PARASITE–HOST SYSTEMS

DAVID PIMENTEL
Department of Entomology and Section of Ecology and Systematics, Cornell University, Ithaca, NY 14853, USA

INTRODUCTION

Stability in parasite–host interactions has ecological advantages to both populations, but especially to the parasite. Because parasites depend upon their hosts for energy and other resources, conditions of severe fluctuations in numbers of parasites and hosts can result in the extinction of the parasite or of both parasite and host. If parasitism reduces the net reproductive rate of the host below 1, the host will decline and eventually become extinct. Reduced host numbers most likely will result in a decline in parasite numbers. Of course, for an obligate parasite, loss of the host population means certain extinction for the parasite.

In parasite–host interactions there appear to be advantages associated with stability and for a balanced supply–demand economy (Pimentel 1961). Parasite demand for energy and other resources should equal the host's ability to supply these resources without a major decline in host numbers (Pimentel, Levin & Soans 1975). Evolution in host and/or parasite toward a balanced supply–demand economy and stability are possible by the genetic feedback mechanism (Pimentel 1961; Person 1968; Levin 1972; Levin & Pimentel 1981). Stability of the parasite and host system can be further improved by appropriate genetic diversity in both species (Browning 1974; Harlan 1976; Pimentel & Bellotti 1976; Nelson 1978; Anikster & Wahl 1979).

The role of genetic diversity in stabilizing parasite–host interactions is extremely complex. This paper assesses genetic diversity and stability in parasite–host systems by evaluating the following factors:

1 genetic instability in parasite and host plants in agriculture;
2 comparative genetic structure of parasites and hosts;
3 genetic implications of parasite dispersal and colonization;
4 impact of host genetic diversity on parasite evolution; and
5 genetic diversity and long-term stability in parasite–host population systems.

GENETIC INSTABILITY IN AGRICULTURE

In contrast to the stability that is normally observed in nature, parasite–host systems in agriculture appear to be relatively unstable (Segal *et al.* 1980). Some parasitic pathogens and insects appear to evolve rapidly in response to genetic changes, such as genes for resistance in crop hosts. For example, stem rust (*Puccinia graminis*) and crown rust (*P. coronata*) are known to repeatedly overcome genetic resistance bred into their oat host. As a result, oat varieties have had to be changed in the corn belt region every 4 to 5 years since 1940 (Stevens & Scott 1950; Van der Plank 1968).

Parasitic insects have also evolved and overcome genetic resistance in their host plants. For example, some resistance bred into various wheat cultivars against the Hessian fly (*Mayetiola destructor*) has been overcome by some fly genotypes (Suneson & Noble 1950; Allan *et al.* 1959; Hatchett & Gallun 1968, 1970; Gallun & Reitz 1971; Gallun 1977). Wheat resistance to the Hessian fly is inherited as a series of single dominant characters, while the ability of the fly to exploit different resistant wheats is inherited as several recessive characters (Hatchett & Gallun 1970).

The spotted alfalfa aphid (*Therioaphis maculata*) and the pea aphid (*Acyrthosiphon pisum*) have evolved genotypes that survive on various aphid-resistant varieties of alfalfa (Cartier *et al.* 1965; Nielson Lehman & Marble 1970; Nielson & Don 1974). Neither the genetic basis of this evolution, nor the proximate mechanisms with which the aphid biotypes overcome the host's resistance, has been fully investigated.

The importance of genetic diversity in host resistance was illustrated in 1970 by corn and the southern corn leaf blight (*Helminthosporium maydis*) (Ullstrup 1972; NAS 1972; Thurston 1973). Use of the Texas sources of cytoplasmically inherited male sterility (TMS) had become so widespread that in 1970, about 85% of the corn grown in the USA had TMS cytoplasm (Moore 1970; Roane 1973). In that year, favourable environmental conditions combined with the evolution of the blight's race T, which is virulent on all plants with TMS cytoplasm, produced a severe outbreak of the blight. The epidemic devastated much of the corn in the USA (Nelson *et al.* 1970).

We may conclude that when resistance in an agricultural host arises from genetically simple traits, correspondingly simple genetic changes in the parasite often serve to overcome resistance. Simple genetic responses in the parasites appear to be typical (V.E. Gracen, unpublished), and this pattern gave rise to the 'gene-for-gene' model in parasite–host interactions (Flor 1956, 1971; Person 1968).

The ability of parasites to overcome a single selective factor in their environment has also been demonstrated with pesticides and pest populations. Pesticides, acting like a single genetic factor on insect and mite pest populations, have resulted in these populations evolving rapidly to overcome this major selective factor in their environment (Brown 1959, 1960; Yanagishima 1961a, 1961b, 1964). Evolved resistance in insect populations often occurs in as few as three to ten generations (Georghiou 1972; D. Pimentel, unpublished results). At present, nearly 400 arthropod species populations have evolved resistance to pesticides throughout the world (Metcalf 1980).

These examples also show that parasites (including herbivores and predators) possess the genetic variability needed to evolve and overcome genetically simple resistance in their hosts. Therefore, parasites of agricultural hosts that have genetically simple resistance, often evolve rapidly and overcome this resistance. This evolution and type of epidemic do not apparently occur often in natural ecosystems where genetic diversity of resistance is believed to be much greater than it is in agricultural ecosystems.

GENETIC MAKEUP OF PARASITES AND HOSTS

Many parasites, especially micro-organisms like viruses, bacteria, and fungi, have genetic structures that are simple compared with the genetic structure of their hosts. For example, a viroid contains less than one gene, tobacco mosaic virus has only three genes, the bacterium *Escherichia coli* has about 1000 genes, and a typical fungus about 30 000 genes (O.D. Yoder, personal communication). In sharp contrast, soybeans, corn, and wheat might have as many as 2, 5 and 9 000 000 genes (estimated on base pairs) respectively (A.A. Szalay pers. comm.). Clearly, these higher plants are genetically much more complex than are viral, bacterial and fungal micro-organisms.

Note that in agricultural crops, almost all instances of a parasite overcoming host resistance have been with the genetically complex fungal rusts and parasitic insects and not with the genetically simple viral and bacterial pathogens (Van der Plank 1982; H.D. Thurston pers. comm.). (Note, earlier when examples were presented of parasites evolving and overcoming single trait resistance, these species were either fungal or insect parasitic types.) Thus, this evidence tends to support the proposition that parasites with few genes are more easily resisted than parasites with many genes such as fungi and insects.

In general, the genetically complex hosts appear to have the capacity to evolve a wide array of resistance genes to pathogen parasites. This tends to give the host several genetic advantages in defending itself from parasites. Complexes of resistant characters are evidently controlled by wide arrays of genes in uncultivated hosts (Wahl 1970; Browning 1974; Zimmer & Rehder

1976). In flax, for example, Flor (1956) reported that at least twenty-five different genetic factors control flax reaction to a rust parasite. He postulated that these factors occur as multiple alleles at five different loci.

If flax possesses five loci and twenty-five different alleles to resist a parasite, the number of possible resistant types would be about 700,000. This is significantly more genetic diversity than a parasite might effectively deal with. Indeed, the great number of combinations is a formidable barrier of genetic diversity for a parasite to overcome.

Let us assume that a parasite genotype had the genetic makeup to parasitize one particular host genotype. The probability that three hosts adjacent to the infected host would include one of these susceptible genotypes suitable for the parasite would be very low.

Suppose that the parasite had the genetic capacity to survive and reproduce on a specific 'resistant' flax genotype and that the host population consisted of a culture of several polygenic types. The probability would be very small that the parasite could evolve the correct specific polygenic combination required to overcome the host's complex defenses. Nelson (1972) pointed out that, if the parasite is to overcome this polygenic resistance, it would have to 'acquire all the genes necessary to overcome' the hosts' numerous defences. He emphasized that this acquisition is a 'matter of genetic probabilities ... when large numbers of genes are involved, the probability of a total overcome is the product of the probability of each individual event'. Knott (1972) stated that it is an 'improbable event' for the parasite to 'accumulate two or three genes' for virulence against a host having only two or three loci for resistance. Thus, a host with a complex group of alleles for resistance is probably sufficiently genetically diverse so the parasite will have major difficulty in overcoming the host's defences.

PARASITE DISPERSAL AND COLONIZATION

Parasitic micro-organisms of plants disperse and infect new hosts by showering their non-motile progeny on nearby hosts. Because these parasites have such limited powers of dispersal and colonization, nearly 80% of the new disease infections occur within 5 m of the infected host (Fig. 14.1). Parasitic microorganisms of animals also have poor powers of dispersal (Anderson & May 1981). Thus, uninfected susceptible hosts must live in close proximity if the parasite is to infect these new hosts. A uniform 'carpet' of susceptible hosts, typical of agriculture, provides the ideal environment for parasite dispersal and colonization. When an infected host is surrounded by resistant hosts, the parasite cannot spread the infection; they are in a sense trapped by unsuitable hosts. Trenbath (1975) termed this trapping the 'fly-paper effect',

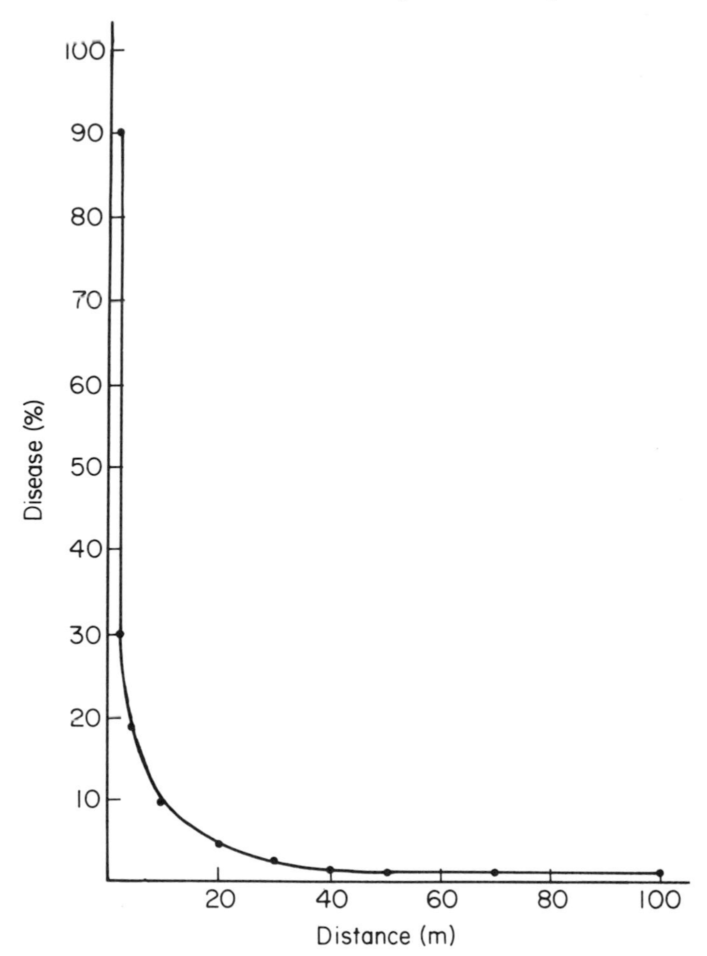

FIG. 14.1. The relationship of distance to incidence of disease or capacity of pathogens to disperse and colonize new hosts (after Nelson & MacKenzie 1973).

and it has been used to control plant pathogens in wheat and oats (Jensen 1965; Browning & Frey 1969; Luthra & Rao 1979). Browning & Frey (1969) reported that a 1 : 1 mixture of two oat lines, each susceptible to one race of oat stem rust, reduced the incidence of rust pustules by 56%. A mixture of three oat isolines, each susceptible to one race of oat stem rust, reduced the incidence of infections by 64%.

Selection of a virulent agricultural parasite takes place best in a resistant crop when the hosts are genetically uniform (Cherewick 1958; Nelson 1972; Browning 1974; Van der Plank 1975). Once a suitable virulent parasite

TABLE 14.1. Percentages of smut-infected plants in different varieties of oats during eight generations of selection of the pathogen parasite (*Ustilago avenae*) on the single resistant oat variety Monarch. (Date from Cherewick 1958).

	Year and percentage infection			
Oat variety	1948*	1949	1950	1951
Monarch	8	12	26	58
Anthony	65	10	19	32

*The field collection was tested in 1948; inoculum in all other years was produced in preceding year only on variety Monarch.

genotype evolves in such a uniform 'carpet' of hosts, it will spread rapidly through the host population.

A strain of parasite that evolves virulence on a resistant host may lose some of its fitness (Table 14.1). Note that the loose smut pathogen strain that was virulent on oat variety Anthony lost some of its fitness when selected for virulence on oat variety Monarch. As the parasite evolved virulence toward Monarch, the parasite regained some of its virulence on Anthony.

The virulent parasite introduced in a genetically diverse host population may have reduced fitness (Frey, Browning & Simons 1977; Trenbath 1977; Browning *et al.* 1979). In such an environment there are few susceptible hosts and parasite movement and colonization is limited. Such a specialized virulent parasite in a genetically diverse host population may have a lower mean fitness than an avirulent parasite genotype has (Watson 1970; Van der Plank 1975; Fenner & Myers 1978; Levin & Pimentel 1981). This occurs if the less virulent parasite, by killing its host more slowly than the virulent parasite does, has a longer time to produce progeny thereby producing a larger total number of offspring.

A classic example of this disadvantage of virulence arose in the myxoma virus that infects rabbits in Australia. Rabbits infected with the avirulent 'Grade IV' survive for 29–50 days, whereas the virulent 'Grade I' kills rabbits in 9–13 days (Table 14.2). The virus parasite in this system requires a mosquito for transport from host to host. The longer the rabbits live, the greater the opportunity for the virus progeny to be carried to other hosts by the mosquitoes. Mosquitoes never feed on dead rabbits. The avirulent virus types have therefore been selectively favoured and have become the dominant types in the system, types 'IIIB' and 'IV' now account for 64% of all infections (Table 14. 2).

Clearly a parasite colony that has an intermediate rate of increase allows

TABLE 14.2. Virulence of field myxoma virus types in rabbits in Australia (Fenner & Myers 1978).

	Virulence type grade					
	I	II	IIIA	IIIB	IV	V
Mean survival times of rabbits (days)	9–13	14–16	17–22	23–28	29–50	—
Case-mortality rate (%)	> 99	95–99	90–95	70–90	50–70	< 50
Australia						
1950–1951	100	—	—	—	—	—
1958–1959	0	25	29	27	14	5
1963–1964	0	0.3	26	33	31	9

its host to live for a longer period and thus enables its host to produce greater quantities of energy and resources than if the host lived for a short period (Anderson & May 1981; Levin & Pimentel 1981). In total this means greater quantities of resources are available to the parasite population, and it can produce greater numbers of progeny over time than the virulent parasite types that quickly destroy their host and resources.

Also an intermediate parasite type, one that produces larger numbers of progeny than the virulent parasite, has a greater chance of having its dispersing progeny colonize new susceptible hosts than the virulent type. With dispersal and colonization being a critical factor in the fitness of parasite genotypes, then host spatial distribution plays a major role in parasite ecology. A genetically diverse host population or a sparsely distributed host population favours an avirulent parasite over a virulent genotype.

IMPACT OF HOST GENETIC DIVERSITY ON PARASITE EVOLUTION

Genetic diversity helps stabilize population fluctuations in parasite–host systems. Fluctuations are not eliminated, however, and evolution continues and tends to reduce the intensity of the fluctuations.

Relative to evolution and the genetic makeup of organisms, it should be emphasized that no organism is truly a 'generalist'. No single organism can make use of all forms of energy and nutrient resources, attack all host genotypes, survive in all temperature and moisture conditions, or itself resist all parasitism and predation, although certain organisms may be able to tolerate a slightly wider array of environmental conditions than others. Typically, an organism that has the specific genetic makeup to be either a virulent parasite or a resistant host lacks the capacity to deal effectively with some other environmental pressures (Pimentel 1961; Simons 1979). As a result, virulence in a parasite or resistance in a host 'costs' the parasite and host in fitness (Levin 1976; McLaughlin & Shriner 1980; Browning 1981). There will be a net cost, of course, only if the parasite and host are in ecosystems where virulence and resistance do not have a large enough selective advantage (Pimentel 1961; Van der Plank 1968, 1975; Watson 1970; Leonard 1977; Marshall 1977).

Another factor that tends to stabilize parasite–host population interactions is the complex structure of the parasite and host genome. The associa-

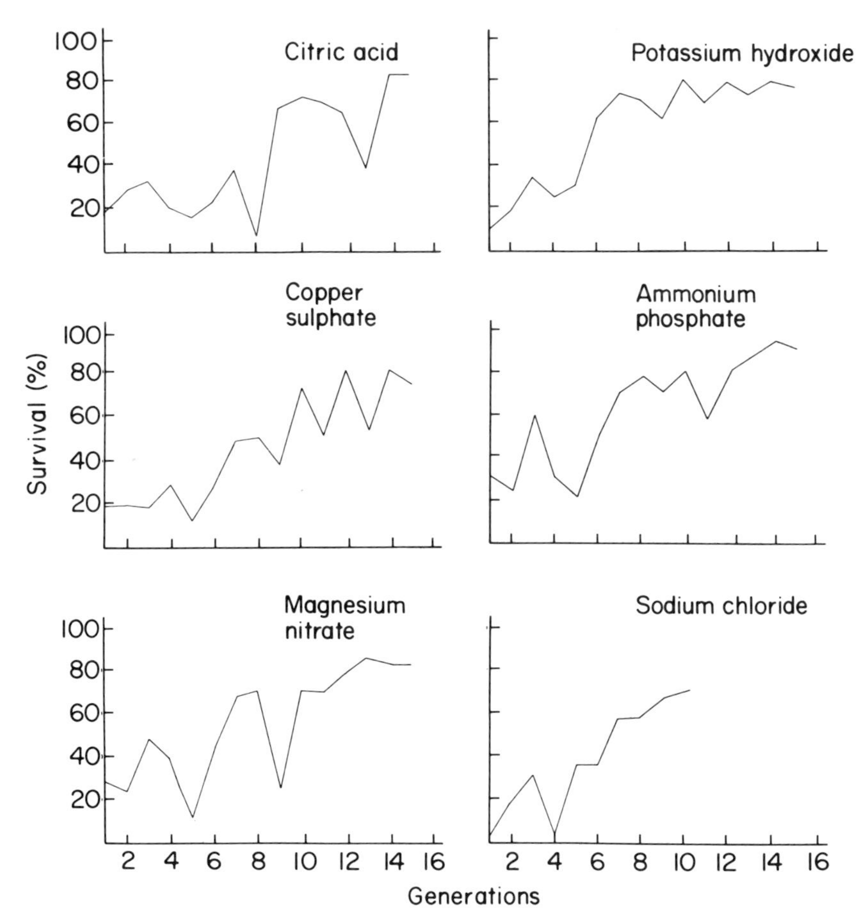

FIG. 14.2. The survival of parasite populations (houseflies) associated with genetically uniform (citric acid, $H_3C_6H_5O_7H_2O$; copper sulphate, $CuSO_4$; magnesium nitrate, $Mg(NO_3)_2$; sodium chloride, NaCl; ammonium phosphate, $NH_4H_2PO_4$; and potassium hydroxide, KOH) characters in host–plant population (Pimentel & Bellotti 1976).

tion of genes in these organisms is highly integrated and has evolved over thousands of generations. Although it may be possible to change a single gene relatively easily, to change ten to twenty genes in an organism at one time and retain its overall fitness is nearly impossible in the short term (Knott 1972). Changing a single gene for an organism to enable it to become virulent or resistant may use some of the organism's resources. Hence, altering the structure and function of an organism to refocus ten to twenty genes would require major re-organization and re-integration of the genome and a significant quantity of the organism's resources.

Although parasites often have shorter generation times than their hosts and therefore may be able to evolve and change their relatively simple genome, this same simple genome limits their capacity to deal with host genetic diversity. An individual host itself can be more genetically diverse than its parasite. This diversity facing the parasite can be greatly increased if the host population itself is genetically diverse (e.g. flax resistance and its parasite).

Genetic diversity for resistance in the host may often retard or prevent the evolution of virulence in the parasite. This effect is illustrated in an experimental host–parasite model system. In this model, house-flies served as parasites; hosts were simulated 'glass plants'; vials were filled with sugar water either pure or tainted with one of six chemicals (citric acid, copper sulfate, magnesium nitrate, sodium chloride, amonium phosphate, and potassium hydroxide) (Pimentel & Bellotti 1976). House-fly populations that were exposed to any one of these chemicals alone evolved substantial resistance within eight to ten generations (Fig. 14.2). Flies from the individual populations were especially resistant to the particular chemical that they had been exposed to. Thus, each chemical affected a unique set of genes.

In contrast to the populations that were exposed to individual chemicals, the population that was exposed to all six chemicals simultaneously failed to evolve resistance to any of them after thirty-two generations (Fig. 14.3). Thus, exposure to six diverse factors individually indicated that the fly population had the capacity to re-combine a small set of genes to overcome each chemical individually, but the fly was incapable of modifying a wide array of its own genetic characters to deal with all six chemicals at one time.

In this laboratory study many of the numerous diverse environmental variables that affect populations in nature were absent. Additional selective variables would probably further inhibit even the adaptation of a parasite to a genetically diverse host. Also, if the parasitic micro-organism had less genetic variability than is typical of the house-fly, the population would probably have greater difficulty in evolving to overcome six resistant factors.

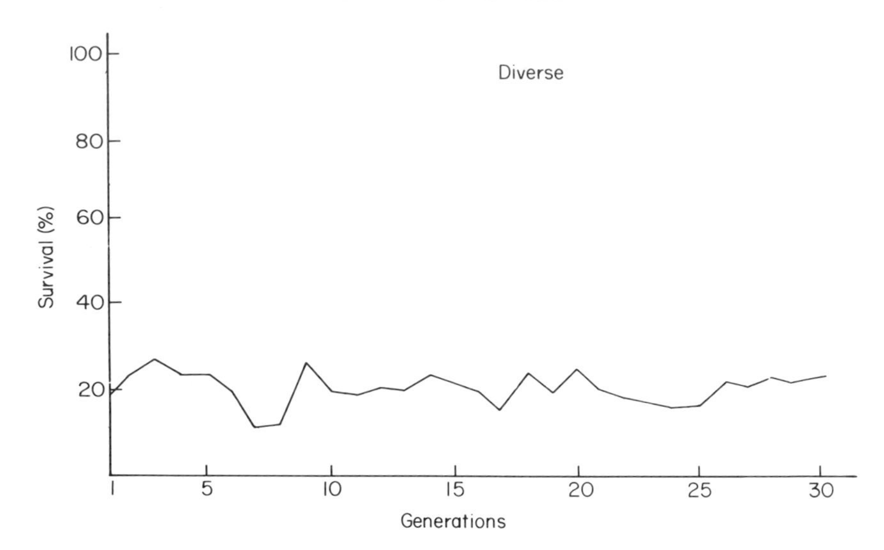

FIG. 14.3. The percentage survival per generation of the genetically diverse population system containing six resistant characters (citric acid, copper sulphate, magnesium nitrate, sodium chloride, ammonium phosphate, and potassium hydroxide) in the simulated host–plant population generations (Pimentel & Bellotti 1976).

GENETIC DIVERSITY IN BOTH PARASITE AND HOST POPULATIONS

In the Fertile Crescent of the Middle East, where progenitors of wheat, barley and oats originated and grow naturally, the boom and bust cycles that typify pathogenic parasites in US cereals do not occur (Browning 1974). The host populations in the Middle East have both susceptible and resistant genotypes, including oligogenic resistant types, and the parasite–host systems are relatively stable. Thus, a wide array of genetic races of parasites, ranging from avirulent to virulent, exist in the Crescent (R.R. Nelson pers. comm.). However, no epidemics of the parasites have been reported. The broad polygenic resistance that exists in the host–grain population apparently stabilizes the parasite–host system (Browning 1974). No single parasite genotype can either increase rapidly enough or evolve sufficiently to overcome the genetically diverse host population (Nelson 1980).

Although polygenic resistance is predominant in the Fertile Crescent ecosystem in which wild oats (*Avena sterilis*) exist, genes for specific resistance are also common (Browning *et al.* 1979). Genes for resistance to crown rust parasite are abundant; most parasitic genotypes are present, including the most virulent race 276 (Browning 1980). On the average, 40% of the infections of wild oats are due to this race (Browning 1980).

Twenty-nine per cent of the wild oat population carries the gene that confers resistance to race 276 (Wahl 1970). Thus, a resistant gene frequency of about 30% provides 'adequate protection against the most virulent and prevalent' race group of the pathogen (Browning *et al.* 1979). Under such conditions the virulent pathogen could not increase rapidly on the susceptible hosts in the oat population. Further, the proportion (70%) of the susceptible hosts did not decline because of parasite population selective pressure.

These studies indicate that the virulent race loses sufficient numbers of its progeny on resistant hosts (the 'fly-paper effect') so that it cannot explosively increase on the susceptible hosts. Other selective factors operating simultaneously presumably prevent the virulent race group from evolving to overcome the resistant genotypes present in the oat population.

Browning & Frey (1969) used pure and multi-line cultivated oats to demonstrate the effect of the host's genetic diversity on a virulent parasite. On a susceptible pure line, the parasite had a high rate of increase, and in time it infected a large proportion of the population. However, in two susceptible pure lines mixed with eight other lines that carried genes resistant to the virulent parasite race, the virulent parasite genotype had a low rate of increase. In this situation with both susceptible and resistant hosts, the parasite lost 80% of its spores on the resistant lines—those that failed to land on susceptible lines.

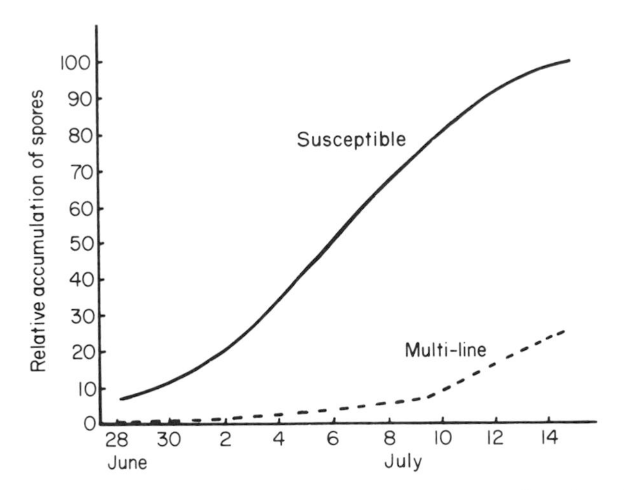

FIG. 14.4. Relative accumulation of spores of crown rust spores collected outside of test plots planted to susceptible isoline and a multiline resistant cultivar of oats that had been inoculated with four rust races (after Frey *et al.* 1977; data from Cournoyer 1967).

The rapid increase of virulent crown rust race 264 on an oat multi-line as documented by Frey *et al.* (1977) also demonstrates the impact of host genetic diversity on a virulent parasite (Fig. 14.4). At the end of an 11-day period, the number of spores on the multi-line were less than half as abundant compared with the pure line.

The combination of reduced host density and difficulty of the parasite colonizing susceptible hosts sometimes results in a selection on the parasite that favours avirulent genotypes. This occurred with the myxomatosis virus parasite in the introduced European rabbit, *Oryctolagus cuniculus*, in Australia. To control the rabbit, the myxoma virus was introduced from the South American tropical forest rabbit, *Sylvilagus brasiliensis* (Ratcliffe *et al.* 1952; Fenner & Marshall 1957), in which it usually produces no generalized disease.

In contrast to its interaction with the forest rabbit, the introduced virus strain produced a 'rapidly lethal disease' in the European rabbit (Fenner & Myers 1978). Rabbit death usually occured within 9–13 days after infection

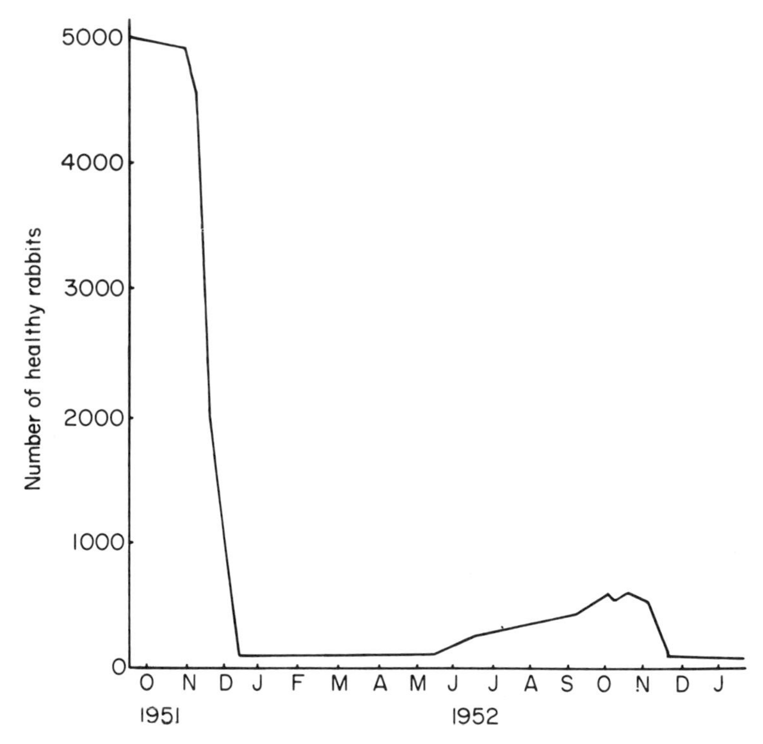

FIG. 14.5. Number of healthy rabbits per standardized transect counts at Lake Urana region immediately after the introduction of the myxoma virus into the host rabbit population (after Myers, Marshall & Fenner 1954).

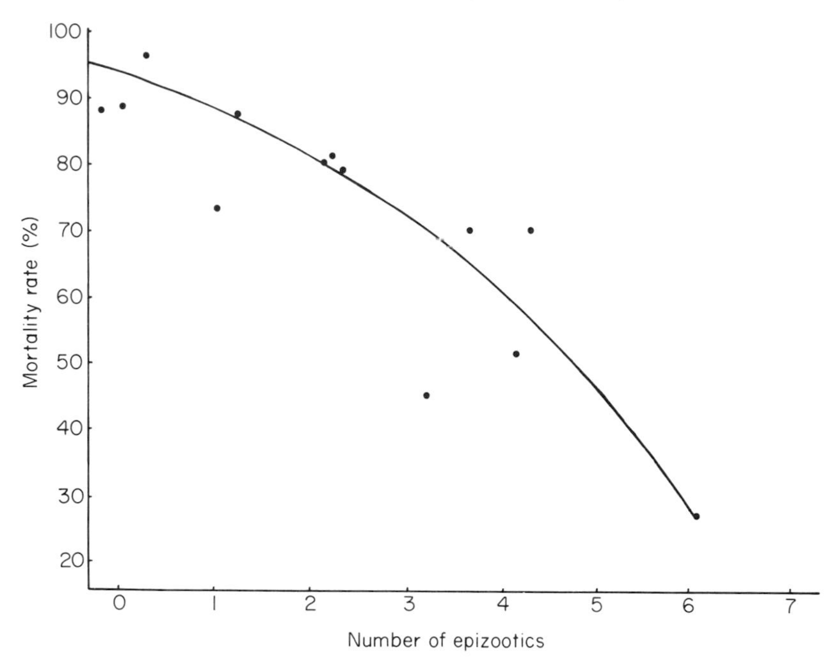

FIG. 14.6. Mortality rates of wild rabbits from Lake Urana region after exposures to several epizootics of myxoma virus, after challenge infection with strain of myxoma virus grade III virulence (after Fenner & Myers 1978).

with even a small dose of virus. At first, in the Lake Urana region the virus caused a catastrophic drop in rabbit numbers (Fig. 14.5). This substantiated the pathogenic nature of this new association of myxoma virus and the European rabbit population.

Subsequently, however, the virus parasite evolved and gradually changed from the virulent type I toward less virulent types (Table 14.2). Because of this change in viral genotypes, there was a drop in rabbit mortality (Fig. 14.6). Investigations confirmed that the virus was evolving from virulent type I toward types II, IIIA, IIIB, IV, and V that were less virulent grade types than type I (Table 14.2). Fenner (1965) said 'The critical factor which favoured the strains of slightly reduced virulence and which accounted for their dominance in Australia ... was the longer survival time of rabbits infected with strains of Grade III and to a lesser extent Grade IV virulence, in a highly infectious condition'.

The evolution that occurred in the viral parasite was probably due to interdemic selection (Lewontin 1970; Levin & Pimentel 1981). Transmission of the virus parasite depended upon *Aedes* and *Anopheles* mosquitoes that fed only on living rabbits (Myers, Marshall & Fenner 1954; Day 1955).

Rabbits infected with the virulent grade type I virus lived for 9–13 days, whereas rabbits infected with avirulent type IIIB lived for about 26 days (Table 14.2). Therefore mosquitoes could feed on rabbits infected with the avirulent virus for at least twice as long as on rabbits infected with the virulent strain. More avirulent progeny were produced and were available for transmission and colonization of other susceptible rabbits (Fenner & Myers 1978). Hence, selection favoured the avirulent genotype colonies over the virulent genotype colonies.

ACKNOWLEDGEMENTS

I thank D. Andow, J. A. Browning, D. Gallahan, E. Garnick, V.E. Gracen, H. Hokkanen, N.F. Jensen, S.A. Levin, G. Mark, R.R. Nelson, S. Risch, M.E. Sorrells, and H.D. Thurston for reviewing an early draft of the manuscript and for their valuable comments and suggestions. This investigation was supported in part by NSF grants DEB 77-05332 and DAR 7922497.

REFERENCES

Allan E.R., Heyne E.G., Jones E.T. & Johnston C.O. (1959) Genetic analysis of ten sources of Hessian fly resistance. Their interrelationships and association with leaf rust reaction in wheat. *Kansas Agricultural Experiment Station Technical Bulletin 104.*

Anderson R.M. & May R.M. (1981) The population dynamics of microparasites and their invertebrate hosts. *Philosophical Transactions of the Royal Society of London B. Biological Sciences*, **291**, 451–524.

Anderson R.M. & May R.M. (1982) Directly transmitted infectious diseases: control by vaccination. *Science*, **215**, 1053–1060.

Anikster Y. & Wahl I. (1979) Coevolution of the rust fungi on Gramineae and Liliaceae and their hosts. *Annual Review of Phytopathology*, **17**, 367–403.

Brown A.W.A. (1959) Inheritance of insecticide resistance and tolerance. *Miscellaneous Publication of the Entomological Society of America*, **1(1)**, 20–26.

Brown A.W.A. (1960) Mechanisms of resistance against insecticides. *Annual Review of Entomology*, **5**, 301–325.

Browning J.A. (1974) Relevance of knowledge about natural ecosystems to development of pest management programs for agro-ecosystems. *American Phytopathological Society Proceedings*, **1**, 191–199.

Browning J.A. (1980) Genetic protective mechanisms of plant-pathogen populations: their coevolution and use in breeding for resistance. *Biology and Breeding for Resistance to Arthropods and Pathogens in Agricultural Plants* (Ed. by M.K. Harris), pp. 52–75. Texas Agricultural Experiment Station, Misc. Publ. 1451.

Browning J.A. (1981) The agroecosystem-natural ecosystem dichotomy and its impact on phytopathological concepts. *Pests, Pathogens and Vegetation* (Ed. by J.M. Thresh), pp. 159–172. Pitman Press, Boston.

Browning J.A. & Frey K.J. (1969) Multiline cultivars as a means of disease control. *Annual Review of Phytopathology*, **7**, 355–382.

Browning J.A., Frey K.J., McDaniel M.E., Simons M.D. & Wahl I. (1979) The bio-logic of using multilines to buffer pathogen populations and prevent disease loss. *Indian Journal of Genetics and Plant Breeding*, **39**, 3–9.

Cartier J.J., Isaak A., Painter R.H. & Sorensen E.L. (1965) Biotypes of pea aphid *Acyrthosiphon pisum* (Harris) in relation to alfalfa clones. *The Canadian Entomologist*, **97**, 754–760.

Cherewick W.J. (1958) Cereal smut races and their variability. *Canadian Journal of Plant Science*, **38**, 481–489.

Cournoyer B.M. (1967) *Crown Rust Intensification within and Disemination from Pure Line and Multiline Varieties of Oats*. Msc Thesis, Iowa State University, Ames.

Day M.F. (1955) Factors influencing the transmissibility of myxoma virus by mosquitos. *Journal of the Australian Institute of Agricultural Sciences*, **21**, 145–151.

Fenner F. (1965) Myxoma virus and *Oryctolagus cuniculus*: two colonizing species. *The Genetics of Colonizing Species* (Ed. by H.G. Baker & G.L. Stebbins), pp. 485–501. Academic Press, New York.

Fenner F. & Marshall I.D. (1957) A comparison of the virulence for European rabbits (*Oryctolagus cuniculus*) of strains of myxoma virus recovered in the field in Australia, Europe and America. *Journal of Hygiene*, **55**, 149–191.

Fenner F. & Myers K. (1978) Myxoma virus and myxomatosis in retrospect: the first quarter century of a new disease. *Viruses and Environment* (Ed. by E. Kurstak & K. Maramorosch), pp. 539–570. Third International Conference on Comparative Virology, Mont Gabriel, Quebec.

Flor H.H. (1956) The complementary genic systems in flax and flax rust. *Advances in Genetics*, **8**, 29–54. Academic Press, New York.

Flor H.H. (1971) Current status of the gene-for-gene concept. *Annual Review of Phytopathology*, **9**, 275–296.

Frey K.J., Browning J.A. & Simons M.D. (1977) Management systems for host genes to control disease loss. *New York Academy of Sciences*, **287**, 255–274.

Gallun R.L. (1977) Genetic basis of Hessian fly epidemics. *New York Academy of Sciences*, **287**, 223–229.

Gallun R.L. & Reitz L.P. (1971) Wheat cultivars resistant to races of Hessian fly. *Production Research Report 134*. U.S. Department of Agriculture, Washington D.C.

Georghiou G.P. (1972) The evolution of resistance to pesticides. *Annual Review of Ecology and Systematics*, **3**, 133–168.

Harlan J.R. (1976) Diseases as a factor in plant evolution. *Annual Review of Phytopathology*, **14**, 31–51.

Hatchett J.H. & Gallun R.L. (1968) Frequency of Hessian fly, *Mayetiola destructor*, races in field populations. *Annals of the Entomological Society of America*, **61**, 1446–1449.

Hatchett J.H. & Gallun R.L. (1970) Genetics of the ability of the Hessian fly, *Mayetiola destructor*, to survive on wheats having different genes for resistance. *Annals of the Entomological Society of America*, **63**, 1400–1407.

Jensen N.F. (1965) Multiline superiority in cereals. *Crop Science*, **5**, 566–568.

Knott D.R. (1972) Using race-specific resistance to manage the evolution of plant pathogens. *Journal of Environmental Quality*, **1**, 227–231.

Leonard K.J. (1977) Selection pressures and plant pathogens. *New York Academy of Sciences*, **287**, 207–222.

Levin D.A. (1976) The chemical defenses of plants to pathogens and herbivores. *Annual Review of Ecology and Systematics*, **7**, 121–159.

Levin S.A. (1972) A mathematical analysis of the genetic feedback mechanism. *American Naturalist*, **106**, 145–164.

Levin S.A. & Pimentel D. (1981) Selection of intermediate rates of increase in parasite–host systems. *American Naturalist*, **117**, 308–315.

Lewontin R.C. (1970) The units of selection. *Annual Review of Ecology and Systematics*, **1**, 1–18.

Luthra J.K. & Rao M.V. (1979) Escape mechanism operating in multi-lines and its significance in relation to leaf rust epidemics. *Indian Journal of Genetics and Plant Breeding*, **39**, 38–49.

Marshall D.R. (1977) The advantages and hazards of genetic homogeneity. *New York Academy of Sciences*, **287**, 1–20.

McLaughlin S.B. & Shriner D.S. (1980) Allocation of resources to defense and repair. *Plant Disease* (Ed. by J.G. Horsfall & E.B. Cowling), vol. 5, pp. 407–431. Academic Press, New York.

Metcalf R.L. (1980) Changing role of insecticides in crop production. *Annual Review of Entomology*, **25**, 219–286.

Moore W.F. (1970) Origin and spread of southern corn leaf blight in 1970. *Plant Disease Reporter*, **54**, 1104–1108.

Myers K., Marshall I.D. & Fenner F. (1954) Studies in epidemiology of infectious myxomatosis of rabbits. III. Observations on two succeeding epizootics in Australian wild rabbits on the Riverine Plain of south-eastern Australia 1951–1953. *Journal of Hygiene* **52(3)**, 337–360.

NAS (1972) *Genetic Vulnerability of Major Crops*. National Academy of Sciences, Washington, D.C.

Nelson R.R. (1972) Stabilizing racial populations of plant pathogens by use of resistance genes. *Journal of Environmental Quality*, **1**, 220–227.

Nelson R.R. (1978) Genetics of horizontal resistance to plant diseases. *Annual Review of Phytopathology*, **16**, 359–378.

Nelson R.R. (1980) The evolution of parasitic fitness. *Plant Disease* (Ed. by J.G. Horsfall & E.B. Cowling), vol. 5, pp. 23–46. Academic Press, New York.

Nelson R.R. & MacKenzie D.R. (1973) The detection and stability of disease resistance. *Breeding Plants for Disease Resistance* (Ed. by R.R. Nelson), pp. 26–39. Pennsylvania State University Press, University Park.

Nelson R.R., Ayers J.E., Cole H. & Petersen D.H. (1970) Studies and observations on the past occurrence and geographical distribution of isolates of race T of *Helminthosporium maydis*. *Plant Disease Reporter*, **54**, 1123–1126.

Nielson M.W. & Don H. (1974) A new virulent biotype of the spotted alfalfa aphid in Arizona. *Journal of Economic Entomology*, **67(1)**, 64–66.

Nielson M.W., Lehman W.F. & Marble V.L. (1970) A new severe strain of the spotted alfalfa aphid in Calfornia. *Journal of Economic Entomology*, **63(5)**, 1489–1491.

Person C. (1968) Genetical adjustment of fungi to their environment. *The Fungi—An Advanced Treatise* (Ed. by G.C. Ainsworth & A.S. Sussman), pp. 395–415. Academic Press, New York.

Pimentel D. (1961) Animal population regulation by the genetic feedback mechanism. *American Naturalist*, **95**, 65–79.

Pimentel D. & Bellotti A.C. (1976) Parasite–host population systems and genetic stability. *American Naturalist*, **110**, 877–888.

Pimentel D., Levin S.A. & Soans A.B. (1975) On the evolution of energy balance in exploiter–victim systems. *Ecology*, **56**, 381–390.

Ratcliffe F.N., Myers K., Fennessy K. & Calaby J.H. (1952) Myxomatosis in Australia. A step towards the biological control of the rabbit. *Nature*, **170**, 7–11.

Roane C.W. (1973) Trends in breeding for disease resistance in crops. *Annual Review of Phytopathology*, **11**, 463–486.

Segal A., Manisterski J., Fischbeck G. & Wahl I. (1980) How plant populations defend themselves in natural ecosystems. *Plant Disease* (Ed. by J.G. Horsfall and E.B. Cowling), vol. 5, pp. 75–102. Academic Press, New York.

Simons M.D. (1979) Influence of genes for resistance to *Puccinia coronata* from *Avena sterilis* on yield and rust reaction of cultivated oats. *Phytopathology*, **69**, 450–452.

Stevens N.E. & Scott W.O. (1950) How long will present spring oat varieties last in the central corn belt? *Agronomy Journal*, **42**, 307–309.

Suneson C.A. & Noble W.B. (1950) Further differentiation of genetic factors in wheat for resistance to the Hessian fly. *U.S. Department of Agriculture Technical Bulletin 1004*. Department of Agriculture, Washington D.C.

Thurston H.D. (1973) Threatening plant disease. *Annual Review of Phytopathology*, **11**, 27–52.

Trenbath B.R. (1975) Diversity or be damned? *Ecologist*, **5**, 76–83.

Trenbath B.R. (1977) Interactions among diverse hosts and diverse parasites. *New York Academy of Sciences*, **287**, 124–150.

Ullstrup A.J. (1972) The impacts of the southern corn leaf blight epidemics of 1970–1971. *Annual Review of Phytopathology*, **10**, 37–50.

Van der Plank J.E. (1968) *Disease Resistance in Plants*. Academic Press, New York.

Van der Plank J.E. (1975) *Principles of Plant Infection*. Academic Press, New York.

Van der Plank J.E. (1982) *Host–Pathogen Interactions in Plant Diseases*. Academic Press, New York.

Wahl I. (1970) Prevalence and geographic distribution of resistance to crown rust in *Avena sterilis*. *Phytopathology*, **60**, 746–749.

Watson I.A. (1970) Changes in virulence and population shifts in plant pathogens. *Annual Review of Phytopathology*, **8**, 209–230.

Yanagishima S. (1961a) $CuSO_4$ resistance in *Drosophila melanogaster*. III. Various changes of characters accompanied with acquisition of resistance to copper. *Memoirs of the College of Science, University of Kyoto, Series B*, **28(1)**, 9–31.

Yanagishima S. (1961b) $CuSO_4$ resistance in *Drosophila melanogaster*. IV. Are there any cross resistance phenomena among various chemical agents? *Memoirs of the College of Science, University of Kyoto, Series B*, **28(1)**, 33–52.

Yanagishima S. (1964) $CuSO_4$ resistance in *Drosophila melanogaster*. VI. Comparative studies in resistant variants induced by various kinds of bivalent metallic salts. *Biology Laboratory Kyoto University Contribution 17*. Kyoto University, Kyoto.

Zimmer D.E. & Rehder D. (1976) Rust resistance of wild *Helianthus* species of the North Central United States. *Phytopathology*, **66**, 208–211.

15. DARWIN'S COFFIN AND DOCTOR PANGLOSS—DO ADAPTATIONIST MODELS EXPLAIN MIMICRY?

JOHN R.G. TURNER
Department of Genetics, University of Leeds, Leeds LS2 9JT

NEW COFFINS FOR OLD THEORIES

Once marked for burial in Westminster Abbey, Darwin's body was taken from the plain oak coffin he had ordered from the village carpenter at Downe and placed in something grander. Ever since, we biologists have been trying to fit him into boxes of our own devising. The neo-Darwinian school likes to see, as the central insight among its multifarious contributions to biology, the linking together of two widely recognized but hitherto unconnected phenomena: adaptation and organic diversity.

The newest alternative resting place for Darwin is a theory (Stanley 1979; Gould 1980) which denies that organic diversity and adaptation are connected. According to this view, adaptation is a concept which is not operationally definable, depending usually on unsupportable, imaginative scenarios to explain why the organism has a particular feature. These scenarios have been lampooned as 'just-so stories' (after Kipling), or as the naive optimism of Dr Pangloss (after Voltaire) (Gould & Lewontin 1979). Most features of most species can be ascribed neither to adaptation, nor to the more rigorous concept of natural selection between individuals, but must be seen as the outcome of a separate evolutionary process unknown to neo-Darwinists. The process of natural selection is held to produce only minor adjustments *within* species. The major part of evolutionary diversity arises at the time of speciation by a not clearly understood process, and its distribution is then sorted, not by Darwinian selection but by a process of selection at the species level. The theory implies first that there is an unrecognized mechanism for regulating phenotypic change, over and above genetic mutation—proponents of the theory invoke Lamarck's *milieu intérieur* and Goldschmidt's macro-mutations or hopeful monsters (Gould, in introduction to reprint of Goldschmidt 1940)—and second that in some way populations or species can be 'adapted'.

I shall here put this theory to the test, by taking a real example of an adaptation, and asking first whether we do have a convincing scenario for its evolution, and second whether it can give rise to real evolutionary diversity. Mimicry is probably the best understood of all adaptations: in its various forms it is widespread among animals and plants (Wickler 1968; Wiens 1978; Pasteur 1982), and the experimental evidence for its effectiveness as an adaptation is extensive (reviews by Rettenmeyer 1970; Turner 1977). Accounting for it has presented such great difficulties for Darwinists that it has been used by two previous authors (Punnett 1915; Goldschmidt 1945) to argue in favour of a macro-mutational theory of evolution analagous to the current one. On the other hand, arguments based on its 'adaptiveness', both for the individual and the species, have frequently been used rather uncritically in place of fully worked schemes of natural selection. If the 'modern synthesis' of neo-Darwinism is valid, this adaptation should be explicable by natural selection acting on genetic variation, and selection should be able to account for the diversity of mimetic species in a 'gradualistic' way.

The first step is to understand the form of natural selection which acts on mimics. I shall restrict the argument to mimicry of distasteful, protected butterflies by other unpalatable species—*Muellerian mimicry*—and by unprotected palatable butterflies—*Batesian mimicry*. As I have recently reviewed experimental evidence on the evolution of mimicry (Turner 1977, 1981, 1983a, b), this chapter concentrates on some questions of theory.

THE NATURE OF SELECTION—MODELS OF PREDATOR BEHAVIOUR

Inverted Russian roulette

We know from extensive work on learning that vertebrates avoid stimuli associated with an unpleasant experience, and seek out those associated with gratification. Without the jargon, they learn what is nasty and what is nice, and act accordingly. Figure 15.1 shows a flock of garden birds feeding on a mixture of artificial prey, red and unpalatable or green and palatable. As can be seen, at intervals 'mistakes' happen, and palatable prey are ignored, or, more significantly unpalatable prey are attacked. Although this could be due in this case to the arrival of a naive bird, experienced individuals do in fact make just such 'mistakes', which are no doubt an adaptive way in which the predator probes a changing and uncertain world. We could say either that the predator has changed its mind about avoiding that species, or has forgotten, temporarily that it is nasty.

With the help of honours finalists at Leeds I have been working with a computer model which simulates this behaviour (J.R.G. Turner, L.S. Exton &

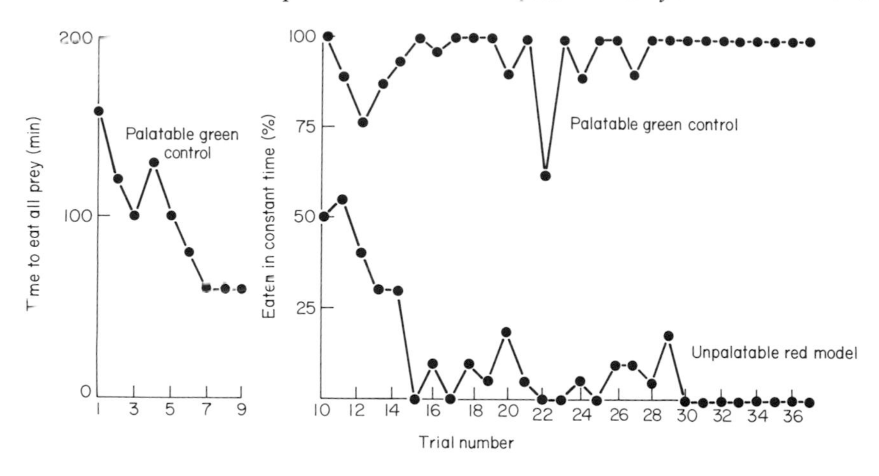

FIG. 15.1. Garden birds learning what is nice and what is nasty: the rate of consumption of palatable pastry increases, while the quinine-flavoured pastry is left increasingly alone (after Morrell & Turner 1970).

E.P. Kearney, unpublished.) It assumes that a predator has a fluctuating probability of attacking any particular species of prey. A pleasant experience increases, and a nasty experience decreases the probability of attacking that species on a later occasion. Further, the predator slowly forgets the experience, so that in any one time unit the probability of attack declines (or increases) asymptotically towards the value for a naive predator. The decision to attack or not to attack is made by testing the current probability against a pseudo-random number. I call the model 'inverted Russian roulette', as it uses a random sequence to gamble with the life, not of the gambler, but of his food-supply. It generates learning, forgetting, and intermittent 'mistakes' which reinforce the learning, as well as an approximation to searching image behaviour for palatable, abundant prey.

The results of a series of runs of this program are shown in Fig. 15.2. The predator is confronted by four species of insect: the 'nasty' prey always decreases the probability of subsequent attack when eaten; confining ourselves to the right half of the graph, the 'nice' prey increases the attack probability, although to varying extents as indicated on the x axis. (The nastiness of the 'nasty' is constant at all points in the graph.) As expected, the average probability of attack for the 'nasty' is rather low, whereas most of the 'nice' are consumed, except when they are rather close to neutral palatability. The predator is also confronted by a model and a mimic, which it is unable to distinguish; the model is as unpleasant as the 'nasty', and the mimic as pleasant as the 'nice', and the model-'nasty' and mimic-'nice'

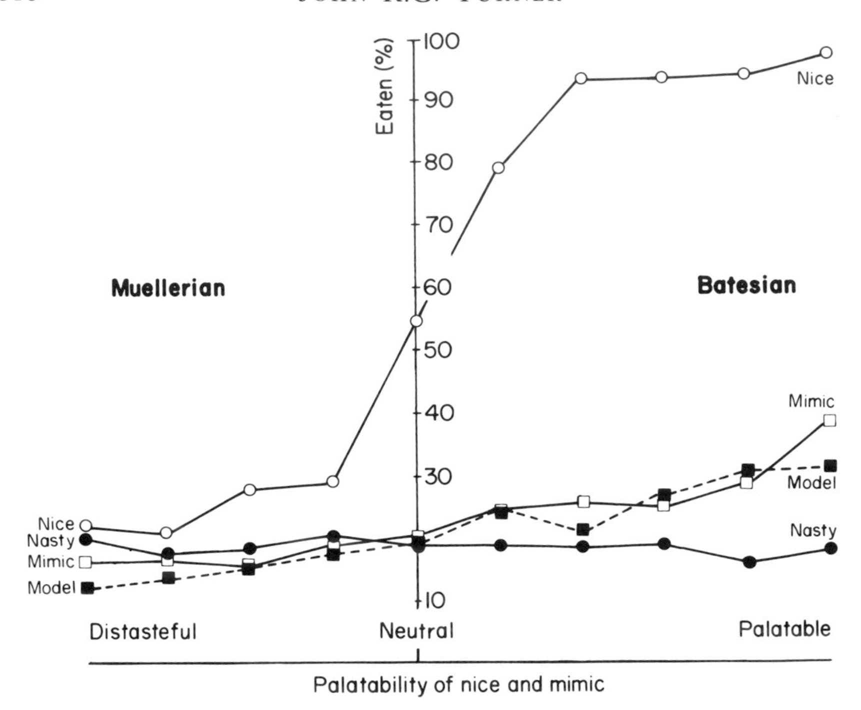

FIG. 15.2. Mimicry and the palatability spectrum: rate of predation on four types of prey, simulated by a Monte Carlo model of predator learning and forgetting. The 'nasty' and 'model' have a constant high level of distastefulness; the 'nice' and 'mimic' vary from being as distasteful as the 'model' and 'nasty' at the extreme left, to being very palatable at the extreme right. The 'nasty' and 'nice' have distinctive patterns; the 'mimic' has a perfect resemblance to the 'model' (from Turner 1983b).

densities are also the same. Again as expected, the mimic is protected in comparison with the 'nice', but the model suffers from the mimicry, although only slightly, by being predated more than its 'nasty' control. In short, the simulation produces the conventional effects of Batesian mimicry.

On the left side of the graph all four species are unpalatable, the 'nasty' and the model in the same degree as on the right, but the 'nice' and the mimic varying from neutrality in the middle to being as nasty as the other two species on the extreme left. The mimicry now protects *both* species, for the model-mimic pair are always the least predated. This is the conventional wisdom about Muellerian mimicry.

Therefore although palatability is indeed a continuous phenomenon (Huheey 1976; Rothschild 1980), this spectrum does divide into two kinds of mimicry, which are qualitatively different: Batesian mimicry, in which the model loses and the mimic gains; and Muellerian mimicry, in which both of the co-mimics gain from the resemblance (Benson 1977; Sheppard & Turner

1977). In theory, when the mimic is of neutral palatability, and in practice under a wider range of conditions, such as extreme rarity of the mimic, or a mimic which has an annual flying season entirely after that of the model, there is a third kind of mimicry in which the mimic gains protection without any effect on the model (Sbordoni *et al.* 1979, Turner 1983c).

Do birds count?

There are other ways of modelling predator behaviour. Suppose that the predators 'decide' to ignore the next n members of a prey species which they encounter, after eating one and finding it unpleasant; the value of n will vary with the unpleasantness of the prey. As Huheey (1976) has shown, this model gives some unconventional results. To take a simple example, if there are two equally abundant, distinct species, for one of which $n = 4$ and for the other $n = 9$, then 20% of the first and 10% of the second will be attacked. If they are perfect Muellerian mimics, then an average 6.5 individuals are avoided after one is consumed, so that 13% of each species are attacked. The first species gains from the mimicry, but, in complete contrast to conventional beliefs about Muellerian mimicry, the nastier of the species suffers from the resemblance, just as if it were the model of a Batesian mimic (also Huheey 1980b).

Huheey's model will also produce a startling conclusion about mimicry between two palatable species. Suppose that the predator decides to eat the next n of a palatable prey it encounters. It is not perhaps unlikely, if we remember that children will fill themselves nearer to capacity with nice cream cakes than boring biscuits, that n increases with the desirability of the prey. By an argument analogous to the one just rehearsed, the more desirable of a pair of palatable species will benefit if it mimics the less desirable, even though that is itself highly palatable and, once tasted is always sought again. (Because the average percentage attacked in a mixture will always exceed the percentage for the less desirable species alone, but fall short of the percentage of the more desirable that would be attacked if the pair were not mimics.)

Choosing one or the other of these models will make some profound differences to our conclusions about the evolution of mimicry. What is the critical difference in their assumptions? It is not, as appears at first thought, the matter of decision-making versus the inverted Russian roulette wheel. A model in which the predator makes definite decisions to attack or not to attack a prey species will produce results qualitatively similar to Fig. 15.2, provided that the decision is reversed after a pre-determined *time period*, rather than after the sighting of a predetermined *number of prey*. The roulette model can in fact be regarded as a kind of decision model in which the length of the avoidance (or acceptance) decision varies about its mean value.

It is the counting of the prey, as distinct from forgetting (or changing the mind) after a period of time which crucially distinguishes the two models. It should be possible to get empirical evidence on the relative importance of time and counting in the 'forgetting' of predators. At the moment I prefer to believe that the process usually depends on time, as it is difficult to imagine that predators carry a running tally of the numbers of several dozen prey species that they observe, particularly above a maximum of seven. It is certainly very unlikely, if the process is one of forgetting rather than mind-changing, that it can depend on numbers, as this leads to the evident self-contradiction that simply seeing an unpalatable prey helps the predator to *forget* how nasty it was! Commonsense tells us that, if anything, the predator should be *reminded*.

Therefore, although real predators may be have in a way ascribable to a mixture of the two models, I believe that a purely time-dependent system is nearer to the truth than a purely number-dependent one, and all further discussion is based on this first model.

Generalization

As all experiences are, by definition, unique, learning would be impossible without generalization. There is excellent experimental evidence that a predator encountering a nasty prey generalizes by avoiding prey that are

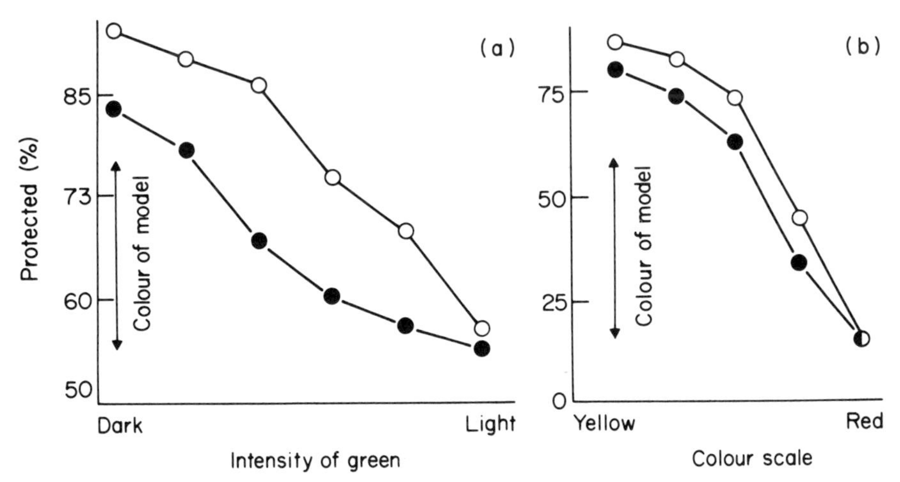

FIG. 15.3. Generalization: after learning to associate a particular colour with an unpleasant stimulus, electric shock in (a) and unpleasant taste in (b), predators also avoid colours which depart from the learned colour, but avoid them less the further they depart. (●) Experiments with weaker shock or less unpleasant model ([a] after Duncan & Sheppard 1965: [b] after Goodale & Sneddon 1977.)

similar in colour, the probability of attack increasing as the colour departs from the colour of the original unpleasant prey (Fig. 15.3). This is true both when the unpleasantness is a mild electric shock, and when it is a nasty taste (Duncan & Sheppard 1965; Goodale & Sneddon 1977).

We can therefore conceive of any particular model species as generating round itself an envelope of protection, represented by the heavy curve in Fig. 15.4, which gives maximum fitness to the pattern of the model itself, but will protect mimics which resemble the model closely enough to deceive some of the predators some of the time. It is known that the extent of generalization depends on a subtle mixture of perception and calculation on the part of birds: if highly attractive alternative food is plentiful, they will avoid a mimic that is attacked when the alternative food is less attractive (Schuler 1974), thus showing that they *can* distinguish between the model and the mimic, but do not always choose to do so.

PATHWAYS TO MIMICRY

Bridging the unbridgable gap

The first problem encountered in accounting for mimicry by an acceptable selectionist argument is the one beloved of creationists and all sceptics of evolution since Mivart: the non-adaptive or maladaptive intermediate. This sometimes takes assinine forms, such as 'What were eye sockets for before there were eyes?' (N.V. Halliday, letter to *The Times*, 1 May 1982) but in other cases, the strong electric shock of electric eels, the wings of insects, and Batesian mimicry, the question is entirely proper and must be answered by evolutionists. Since the time of Darwin it has been recognized that Batesian mimics often differ greatly from the presumed ancestral pattern shown by their close relatives or their own non-mimetic forms. Yet it is difficult to believe that predators generalize so widely that, for instance, a small increase in the amount of black on a predominantly yellow swallowtail butterfly would lead them to avoid it as a mimic of the black pipe-vine swallowtail. This difficulty led Darwin to propose the ingenious theory that the resemblance commenced at some remote period when model and potential mimic were still rather similar and lacked their warning colours; Punnett (1915) and Goldschmidt (1945) took the view that mimicry must arise by a large mutational step, by which the phenotype would be taken clear across the maladaptive gap between the two patterns. Punnett interpreted the existing evidence on the inheritance of mimicry as showing that the resemblance was achieved solely by this single mutation, and explained this as the result of homologous genes acting in both model and mimic.

If Punnett's interpretation was an intelligent synthesis in the light of current knowledge, Goldschmidt's version, in which the unlikely achievement of high quality mimicry by a single mutation was attributed to homologous developmental pathways in the model and mimic, showed much logical confusion (cf. Turner 1983d). At the simplest level this hypothesis fails to explain instances where mimicry occurs purely as an optimal trick which produces a resemblance between non-homologous structures. Even Goldschmidt did not attribute the resemblance of leaf butterflies and leaves to homology, but the problem occurs at much closer taxonomic quarters than that, with bright patches on the *wings* of the mimic often being used to give the impression of a mark of the same colour on the *body* of the model (Poulton 1931; Carpenter 1946; Rothschild 1971). A much more satisfactory hypothesis had already been proposed by Nicholson (1927) and Poulton (1912), and Goldshmidt's refusal to accept it arose from the need to find a demonstrable example of the saltational mutations which he (Goldschmidt 1940) held were the basis of macro-evolution.

This other model, which I like to call the Nicholson–Poulton two-phase theory, can be illustrated for a pair of Muellerian mimics. Suppose there are two distasteful, warningly coloured species, with rather different patterns (Fig. 15.4). Although generalization by the predators potentially protects a wider range of patterns than actually exist, they never mistake one species for the other; selection on both species is normalizing, and they will not converge in their patterns. But if species *B* is better protected that species *A*, because it is nastier or commoner, a mutation of *A* which produces a pattern

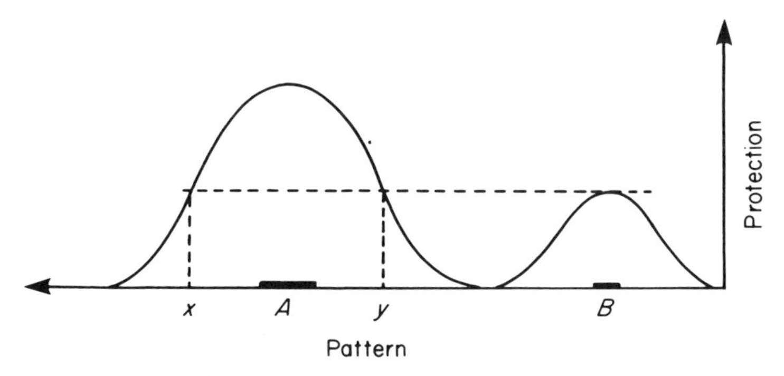

FIG. 15.4. Evolution of mimicry by a major mutation. The horizontal scale represents a range of potential patterns, two of which, *A* and *B*, exist. If both species are distasteful, then generalization by predators will confer protection on any patterns sufficiently resembling *A* or *B*, as shown by the heavy curved lines. Species *B* can become a Muellerian mimic of species *A* by producing a mutation with a pattern anywhere between points *x* and *y*, but species *A* cannot evolve mimicry of species *B*, as no such mutant of *A* will be better protected then *A* is already (after Sheppard *et al.* 1983).

not necessarily resembling *B* very closely, but coming within the construction *x y* in Fig. 15.4, is fitter than the original pattern of *A*, and will spread through the population.

There are three important things to notice about this model. The non-adaptive gap is indeed bridged by a single mutation, but not one which, as Punnett and Goldschmidt proposed, produces perfect mimicry; generalization by the predators will cause the spread of a mutant giving only a tolerable resemblance. Second, the direction of evolution must always be of the less protected towards the better protected species; just as in the evolution of Batesian mimicry the palatable species will converge towards the model. Third, and contrary to what most workers have believed from a misleading passage in Fisher (1930), the development of Muellerian mimicry, just as of Batesian mimicry, is often likely to involve a major mutation.

Convergence

The second phase in the evolution of a Muellerian resemblance will occur once the initial mutation has become fairly common in the population of *A*. It will, by definition, resemble *B* so closely that predators will rather frequently mistake one for the other. Under these conditions, even small variations of *A* in the direction of *B*'s pattern will be favoured, as will small variations of

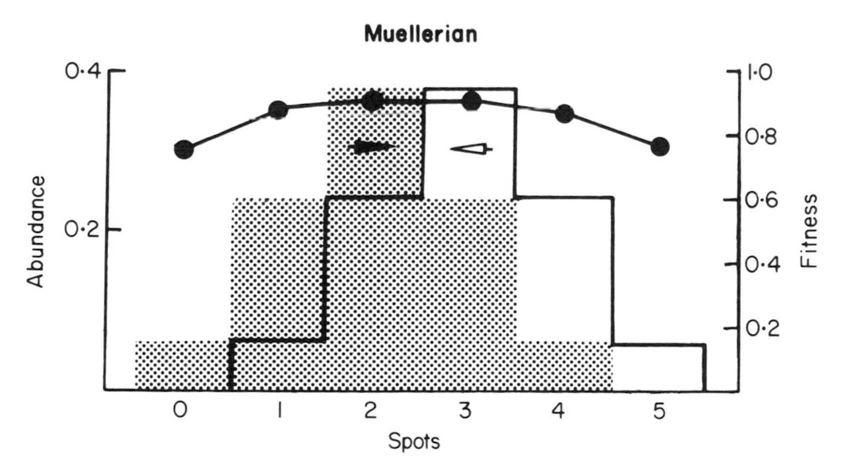

FIG. 15.5. Gradual convergence between two Muellerian mimics (second phase in the evolution of mimicry). The species *A* (▩) and *B* (□) are very similar, but differ slightly in the number of spots on the wing. A predator, operating on the Monte Carlo system of Fig. 15.2, cannot distinguish the two species, but does recognize differences in spot number, only generalizing its learning to the next higher or next lower spot class after encountering a prey. When both species are distasteful the maximum fitness (heavy line) lies between the present patterns, and they will converge.

of *B* that are more readily mistaken for *A*. Gradual mutual convergence of the two species will occur, and provided one or both has, or can produce, the appropriate genetic variations, the mimicry will be improved with time, until a quite close resemblance is achieved.

To illustrate this process, a simple simulation of a later phase in it has been set up, using the inverted Russian roulette model, with the additional feature that the predator generalizes an experience with an insect in any one phenotype class to the two adjacent classes. It is supposed for simplicity that *A* and *B* are equally unpalatable and have become equally abundant; they are good mimics, but differ quantitatively in that *A* has on average one white spot less on the wings than *B*. Figure 15.5 shows the fitness of each of the spot-classes, which is naturally the same for both species. Selection on both species is a mixture of directional and normalizing selection, with *A* being selected for increased, and *B* for decreased spot numbers. Given the appro-

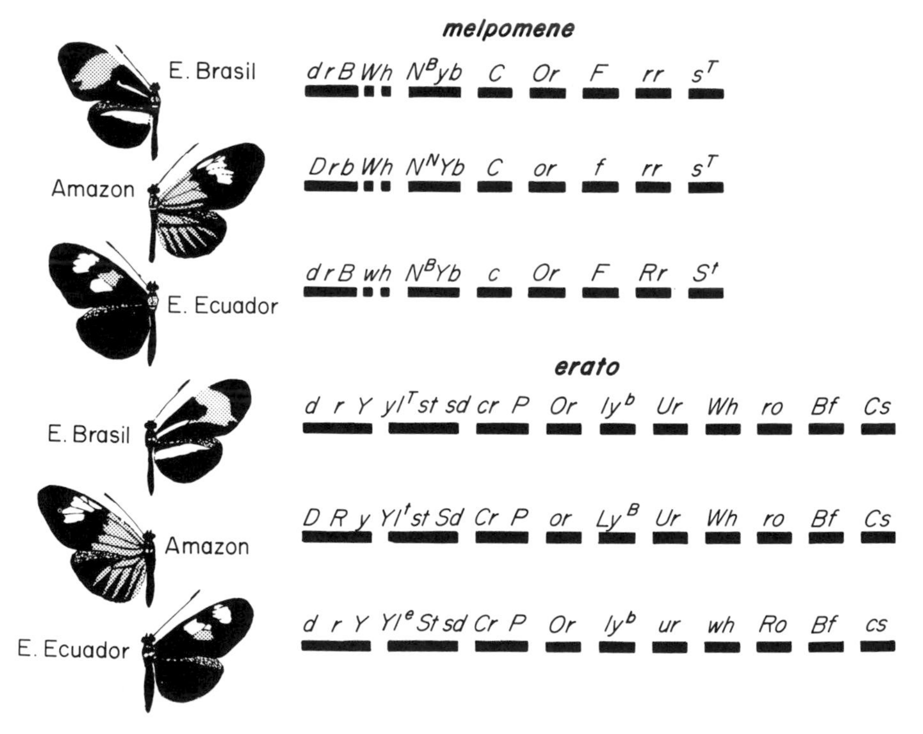

FIG. 15.6. Genetic make-up of three of the many races each of the Muellerian mimics *Heliconius melpomene* and *Heliconius erato*. Heavy bars are chromosomes. Note the comparative lack of linkage between the genes (cf. Fig. 15.8) (data from Sheppard *et al.* 1983; after Turner 1983c).

priate inherited variation in spot numbers, the distributions will converge until eventually they are the same in both species.

The clearest evidence for the two-phase model of evolution comes from genetic studies on Muellerian mimics, the moth *Zygaena ephialtes* (Bullini, Sbordoni & Ragazzini 1969) and butterflies of the genus *Heliconius* (Sheppard *et al.* 1983, reviews by Turner 1981, 1983a, b, c). Unlike the Batesian mimics previously studied (e.g. Sheppard 1961) these are not usually polymorphic: instead geographical races tend to mimic different models, and many of the differences in colour pattern are produced by major genetic mutations which alter quite large parts of the pattern. Figure 15.6 shows the genetic composition of various races of the mutual Muellerian mimics *Heliconius melpomene* and *Heliconius erato*. The smaller alternations brought about in the second phase are seen most obviously in *Z. ephialtes*, where after the alteration of the major parts of the pattern the final *rapprochement* to the model, *Amata phegea*, has been achieved by changing a few remaining red spots to yellow. Similarly, backcrossing in *Heliconius* reveals that the patterns produced by the major genes have been subjected to extensive modification from the polygenic genetic background of the race which contains them.

Hopeful monsters or linked modifiers?

In the evolution of Batesian mimicry we can expect the same two phases: the establishment of an adequate but approximate resemblance by means of a major mutation in the mimic, followed by the gradual modification of the pattern toward better and better mimicry. But there are some important differences. As is widely recognized, a warning pattern or Muellerian mimic becomes fitter as it becomes commoner, as the predators' experience is more frequently reinforced; it is this effect which makes Muellerian mimicry advantageous, and also produces normalizing selection on warning patterns, by selectively removing rare, unfamiliar variants (see the experiments of Benson (1972) with re-painted *Heliconius*). Selection on Batesian mimics is by contrast negatively frequency-dependent: as they become commoner they are attacked more frequently, as the predators either come to believe that the warning pattern indicates palatability, or learn to discriminate model and mimic. The inverted Russian roulette routine clearly shows these two kinds of frequency dependence (Fig. 15.7).

The consequence for a newly evolving Batesian mimic is that instead of its romping through the population like a new Muellerian pattern, the advantage of the new pattern is reduced as it spreads, until it becomes equal to the fitness of the original non-mimetic pattern. At this point, as Fisher first noted, the forms are at a stable equilibrium (Barrett 1976; Turner 1980),

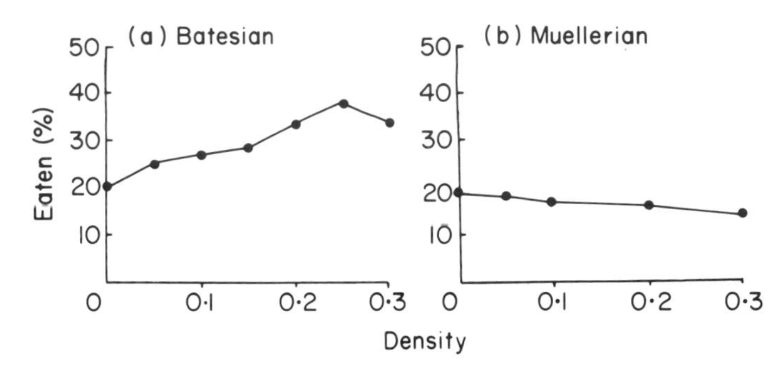

FIG. 15.7. Dependence of the rate of predation on the *numbers* of a mimic, with a constant number of models: (a) a Batesian mimic, showing an increase in predation with density and (b) a Muellerian mimic, showing a decrease.

and the species has a mimic-non-mimic polymorphism. A familiar example is the North American tiger swallowtail, *Papilio glaucus*, in which the female occurs in a yellow non-mimetic, and a black mimetic form.

This polymorphism has some profound and rather complicated effects on the second 'modification' phase. As Charlesworth & Charlesworth (1976) were first to make clear, we must distinguish between those genes which improve the mimicry but reduce the adaptiveness of the non-mimetic pattern, called *non-specific modifiers*, and those which alter *only* the mimetic pattern, called *specific* modifiers. The first type are favoured in company with the mimic, but selected against in the company of the other pattern, and this gives them some curious dynamics.

If unlinked to the mimicry gene, non-specific modifiers either do not spread, or spread to fixation, at the same time 'flipping' the selection regime on the major mimicry gene so that it too goes to fixation, producing a purely mimetic species. If the modifier is tightly linked to the major locus, then it spreads to an equilibrium in which both it and the major gene remain polymorphic and in strong gametic disequilibrium. As this same rule about tight linkage applies to genes which convert the pattern to mimicry of a second model, it follows that in polymorphic Batesian mimics, the different patterns are found to be controlled by tightly linked clusters of genes, all apparently alleles of a single locus. As these 'supergenes' switch the butterfly from one complex pattern to another, they give the impression, which readily misled both Punnett and Goldschmidt, that the whole of an elaborate mimetic pattern has been produced by a single 'saltatory' mutation.

Again, there is ample experimental evidence in support of this theory. Birds are capable of discriminating mimics which have parts of the warning patterns of two models recombined (say the black marks of one model with the yellow colour of another), as is required for the theory to operate (Ikin &

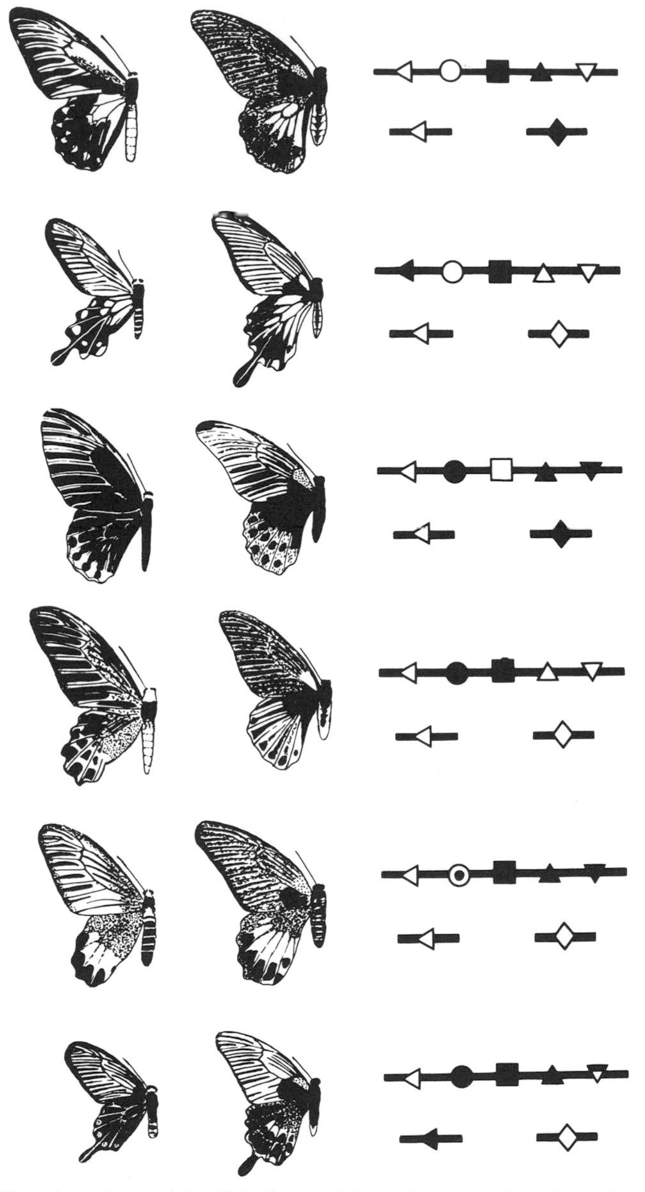

FIG. 15.8. Genetic makeup of six of the forms of the polymorphic Batesian mimic *Papilio memnon*. The models are on the extreme left, the forms of *memnon* in the centre, and a diagrammatic version of the genetics on the right. The five tightly linked genes control respectively tails on the hindwing, white on the hindwing, white on the forewing, colour of 'shoulder patch', and colour of abdomen; the two independent genes control tails on the hindwing and yellow suffusion of the hindwing (data of C.A. Clarke, P.M. Sheppard & I.W.B. Thornton; after Turner 1983b).

Turner 1972; Sargent & Turner 1981). The classic work of Clarke & Sheppard on three of the famous polymorphic swallowtails, *Papilio dardanus*, *P. polytes* and *P. memnon* has demonstrated the clusters of tightly linked genes, particularly in *P. memnon* (Fig. 15.8) (reviews by Sheppard 1961, 1975; Ford 1971; Bertram 1966; Turner 1977, 1983b). Finally, modifying genes controlling recombination, which will allow these clusters to become even more tightly linked once they are formed, have been demonstrated in the silkworm (Ebinuma & Yoshitake 1982; Turner 1979).

Specific modifiers, which improve one mimetic pattern only, can be selected without these complications, and obviously once the species is a monomorphic mimic mimicry can be improved, without hindrance, by modifiers of either type. Extensive evidence for this kind of modification has been obtained by the simple expedient of out-crossing mimetic forms into races or species where they do not occur. Usually the resemblance to the model deteriorates considerably, showing that, as predicted, populations which lack the major gene lack the modifiers which improve the mimicry. Specific modifiers are however subject to one restriction: they are positively selected only when they are in a mimetic individual; in the rest of the population they are neutral. Hence the rate of improvement of a mimetic form is, other things being equal, proportional to its frequency in the population: a form at a frequency of 50% will evolve at half the speed of a monomorphic mimic. Clarke & Sheppard (1960) showed that the form of *Papilio dardanus* which mimics *Amaris echeria* was indeed a better mimic in those parts of Africa where it was common than in those areas where it was rare. We shall see that this complication is of crucial significance when we combine it with the next problem: the evolution of the model.

Co-evolution of model and mimic: the life–dinner arms race

The second major problem in accounting for mimicry, although one which has received much less attention than the maladaptive gap, is the problem of the 'arms race'. The model of a Batesian mimic suffers increased predation as a result of the mimicry: why does it tolerate it? Can the model not simply change its pattern and 'lose' the mimic? Similarly, why do gazelles not evolve so that they can run faster than cheetahs? Dawkins & Krebs (1979) and Dawkins (1982) suggest that the explanation lies in an 'asymmetry' between the contestants in the race (but cf. Greenwood in this volume). Similarly, it is tempting to solve the model–mimic arms race by observing that the mimic gains more from the mimicry than the model loses (compare the discrepancy in predation between the mimic and nice with the discrepancy between the model and nasty in Fig. 15.2). The mimic must be under greater

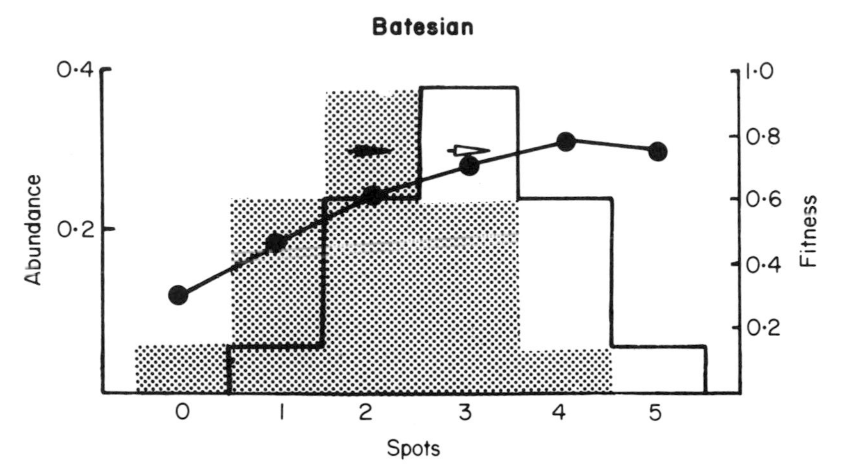

FIG. 15.9. 'Advergence' or 'arms race' between a Batesian mimic and its model. As Fig. 15.5, except species *A* (▩) is now highly palatable so that the optimum fitness for *both* species is achieved at a higher spot number.

pressure to maintain the mimicry than the model is to lose it (Fisher 1930). But this answer at the least fails to go right to the heart of the matter.

If we simulate again the final stages in the convergence of two mimetic species with varying spot numbers, but this time make the one with the lower mean number into a palatable, Batesian mimic, we find a much more satisfying explanation. The peak of fitness (Fig. 15.9) is now skewed to a high spot number so that both species are selected for increased spots. But the model, and this result is obviously general and independent of the particular parameter values, is inevitably closer to the optimum than the mimic. Selection on the mimic is therefore stronger, and other things being equal, it will increase its spot number faster than the model. We will therefore see a process for which Brower & Brower (1972) coined the term *advergence*: both species evolve in the same direction (as distinct from *con*vergence, where they approach from opposite directions), but it is inescapably built into the dynamics of the system that the model moves more slowly than the mimic. The outcome of this purely mechanical process (first pointed out in a verbal argument by Nur 1970) is that the mimic is fastened firmly onto the pattern of the model: *c'est Vénus toute entière à sa proie attachée*.

We can thus explain the paradoxical outcome of the model–mimic arms race purely in terms of evolutionary dynamics. 'Adaptation' has been dispensed with. Furthermore, we can predict the circumstances in which the in which the mimic will *not* win the race. The mimic would for example fail to catch up with its model, if it had a substantially lower heritability in the variation of its pattern than the model did. We will find a very significant wrinkle on this principle later on.

It might appear that the model could escape from the mimic by the same method that the mimic used to initiate the mimicry: a major mutation taking the pattern, this time, *out* of the predators' range of generalization. Major mutations of this kind are used both by hosts and pathogens in their continuing arms races (cf. Barrett in this volume). But normally in butterflies, the strong frequency-dependent selection against rare warning patterns will prevent the establishment of such mutants, whose initial rarity will make them less fit even than the much-mimicked wild-type. The mimic can jump toward the model, and then creep up to it, the model can only creep away, and that more slowly (Nur 1970, Sheppard, 1975). Only if the mimic is very abundant indeed, or if the model can produce a mutation which takes it into another better protected Muellerian group, will the model escape (Turner 1977, Sheppard *et al.*, 1983).

It should be noted that these restrictions on the escape of the model apply to its total phenotype, not simply to its pattern. Escape into another habitat, either in space or time, is equally difficult, for the new mutant of the model will again be initially rare, and will be sampled as an unfamiliar prey item by the predators in that habitat (cf. the suggestion by Gilbert 1983).

It has sometimes been suggested that once a model has many mimics, each not quite resembling it, but in different ways, then escape becomes even harder. Suppose that in the dynamic model (Fig. 15.9) there were a *second* mimic, with a greater spot number than the model. Depending on palatability, abundance, and the means and variances of the phenotypes, the model would evolve *away* from one mimic and *toward* the other. Thus a Batesian mimic with a low heritability that would otherwise find advergence to its model impossible, might find through no virtue of its own that the model was converging with it. In this way much-mimicked models may become more mimicked.

Modification versus pre-adaptation

The concept of the modification of the effects of a major gene therefore allows us to produce a satisfactory selectionist account of the evolution of mimicry; as major mutations are involved, the theory is not purely 'gradualist', but it does provide a more satisfactory account than the purely 'saltationist' views of Punnett and Goldschmidt (cf. Kimler 1983, Turner 1983d); and, what is more, an account which is confirmed extensively by detailed work on the genetics of mimetic butterflies.

But the power of Fisher's theory of modification, and its importance to the Oxford–Cambridge school of population genetics, did lead to important alternatives being not so much ignored as dismissed as improbable. It was for instance generally supposed that the supergenes of mimetic butterflies arose from the tightening of linkage between widely spaced loci, and even the transfer of genes from one chromosome to another by structural interchange (Ford 1971). The alternative to this modification model, which we can call a pre-adaptation or sieve model, was that the loci were already tightly linked, and that far from polymorphic mimicry generating the linkage it was the linkage that permitted the evolution of the polymorphism. The theoretical work of Charlesworth & Charlesworth (1976) has shown that this apparently unlikely alternative is the correct one, and Rothschild (1980) has recently written eloquently in support of their position. The answer to the intellectual difficulty faced by everyone who first encounters this theory, 'how do the butterflies 'know' that their genes must be linked in this way?' is simple: they don't! That is why spectacular mimetic polymorphism is found in fewer than a dozen butterflies: most species do not have the appropriate pre-adaptation of their genetic architecture.

An element of the same pre-adaptation can be seen in the modification of some of the mimetic patterns themselves. The form "cenea", which mimics *Amauris echeria echeria*, and "ochracea", which mimics a local sub-species (*septentrionalis*) of the same model confined to an isolated area in northern Kenya, are produced by different alleles, H^c and H^o, of the supergene which controls mimicry in *Papilio dardanus*. Yet on the genetic background of the main African populations of *P. dardanus* these alleles produce indistinguishable phenotypes. Only on the genetic background of the north Kenya isolate is the difference apparent (Clarke & Sheppard 1962). Thus it is unlikely that the north Kenyan 'modifiers' were selected *after* the new form arose, and much more likely that, in order that H^o should become established in the population, some of the 'modifiers' were already present. Similarly, there is no reliable way of distinguishing the alleles H^c and h on a Malagasy genetic background, although they produce different mimetic forms ("cenea" and "hippocoon" mimicking *A. echeria* and *A. niavius*) on a South African genetic background, which likewise must been pre-adjusted to express at least some of this difference between the alleles. Again, there is nothing mysterious about this: selection is purely opportunistic, and mutations must occur on *some* genetic background. Those which happen to interact with their local genetic background in such a way as to produce an adequate mimetic resembalance are the ones which establish themselves.

When two feature of an adaptation are found together, therefore, we have the choice of believing that, say, the adaptation altered the genetic architecture, or that the genetic architecture pre-adapted the species to develop that adaptation. An intriguing case in which, I think, we have no evidence as to which, if not both, of the effects caused the other is the correlation of mimicry and habitat. It has been shown in an important unpublished study that in the intricately structured environment of the rain forest, mutually mimetic *Heliconius* really do fly together in the same microhabitats (Smiley 1978). Has the possession of the same Muellerian pattern caused the butterflies to adjust their flight habits, or have butterflies which flew in the same places tended to become mimics? At the moment, we do not know.

CO-EVOLUTION OF MODEL AND MIMIC

Abundance and apparency

Hence for any particular co-adaptation between model and mimic, we must always consider whether it is plausible that mimicry has generated the co-adaptation, or that the pre-existing relationship has itself caused the mimicry to evolve. Rothschild (1971; 1980) has pointed out some relations between model and mimic, either observed directly or predicted from mimicry theory, and has suggested that these co-adaptations are connected with the problem of population size in mimics. Roughly the idea is that both Batesian mimicry and large population size are adaptive, the one to individuals and the other, by reducing the rate of stochastic extinction, to the population. But as is well known, mimicry fails if the mimic is too common relative to its model, as predators either increase their attacks on both species, or improve their discrimination of the mimics from the models (i.e. negative frequency dependence as in Fig. 15.7); in a sense the two adaptations are antagonistic. Rothschild (1980) therefore proposes that there are a number of 'dodges' which permit both large population size and Batesian mimicry; these are summarized in Table 15.1, in the form of a series of differences between the model and the mimic.

Mimicry according to this theory tends to arise in rare species, which then avoid extinction, if they can, by adopting or already possessing, some of the 'dodges' such as sex-limitation or emerging later than the model. In modern terminology, it is predicted that the mimics will become less 'apparent' to predators than the models.

Dr Rothschild (1980, 1981) feels that an earlier attempt of mine to criticize her theory has seriously misrepresented her position. I therefore quote here both of Dr Rothschild's presentations of her theory; the reader may judge how far the present paper is a fair commentary on her position.

TABLE 15.1. Observed, or postulated, differences between Batesian mimics and their models

Model	Mimic
Unpalatable	Palatable
Common	Rare
Monomorphic	Polymorphic
Both sexes alike	Mimicry often only in female
Flies early	Flies late
Flies slowly	*Flies rapidly
Aggregated	*Dispersed
Larger	*Smaller
Full warning colour	*Cryptic when resting
Bright marks on body	*Bright marks on wings
Flaunts itself	*Secretive
More 'apparent'	Less 'apparent'
Share same habitat	

*Not discussed in detail in this chapter

'It was proposed by Bates and generally accepted by subsequent writers ... that in order for Batesian mimicry to succeed, the numbers of the model must considerably exceed those of the mimic. At the same time a species so situated that scarcity of numbers *is selected for*, must find itself ... under the constant threat of extinction. All Batesian mimics are therefore under pressure to evolve specializations whereby they can increase their numbers *vis-à-vis* the model, without interfering with, or seriously altering, the learning pattern of the predators' (Rothschild 1971).

'A number of features characteristic of Batesian mimicry have been described, ... and it is often assumed that they are the *result* of Batesian mimicry. In many cases they were prerequisites or, in some cases, preadaptations ... without which the mimicry situation could not have originated in the first place. One of the most obvious of these conditions concerns the relative numbers of model and mimic ... The learning capacity of the predator demands that the unpalatable model outnumbers the palatable mimic. It is also obvious that, in this situation, there can be no selection "for rarity" (... since this would eventually lead to the extinction of the mimic. It must be "pre-adaptedly" rare for some other reasons, or else resort to other "dodges". One such is female sex-limited mimicry which enables the species to double its numbers without impairing the learning mechanism of the predator. Turner (1978) completely misunderstood this argument and claims that I advocated "the fallacious assumption that in a mimetic species *scarcity of numbers is selected for*"! [italics added.] I fancy we are getting just a little weary of John Turner's perennial misquotations and misrepresentation of our papers . . .) We must therefore assume that one of the prerequisites or felicitous preadaptations for Batesian mimicry is a rare species flying in the company of a common aposematic species or group of species.' (Rothschild 1980).

This argument seems, at first glance, to require group selection, or even wilful adaptation. But Rothschild (1980) has quite rightly rejected this notion, suggesting instead that most of the characteristic features of Batesian mimics are *pre-adaptations*, already present in the population for other reasons, and leading to the development of mimicry. We have here a very appealing adaptationist argument which requires much detailed consideration of the mechanisms and evolutionary pathways which might lead to it. The problem is a particularly fascinating one, as it is necessary to consider both individual and kin selection, with the further possibility of group selection, and also the action of a double-ended sieve, which causes not only the gain of mimicry in species which already possess certain favourable characteristics, but the possible loss of mimicry in species which cease to show such pre-adaptations.

In attempting to model this system, I shall consider only the questions of emergence date, abundance, polymorphism and sex-limitation; I shall again assume that predators forget their experiences with time, rather than by counting.

Are mimics rare?

Because of the frequency-dependent selection generated by Batesian mimicry (Fig. 15.6), mimics become disproportionately predated, or if one likes, apparent, when they are common. An almost universally accepted tenet of mimicry theory is therefore that the mimic is scarcer than the model.

This piece of scientific folklore has seldom been critically explored. There are indeed documented cases of mimic species being scarcer than their models: most forms of *Pseudacraea eurytus* are less abundant than the *Bematistes* species which they copy (Owen & Chanter 1972), and dealers' prices certainly suggest that many of the most spectacular mimics are in fact quite rare, the mimic, according to Punnett (1915) often costing as many pounds as the model costs shillings. On the other hand my impression is that the flies which mimic bees and wasps are, collectively, rather common. The problem with the empirical evidence is that, as was argued by Fisher (1930): 'Batesian mimicry by a more numerous of a less numerous form, cannot be excluded ... ; for if the model were extremely noxious or the mimic a not particularly valuable source of food, the motive for avoidance may be but little diminished by the increases of the mimic.' The effect was later confirmed experimentally by Brower (1960). Thus the commonness of bee and wasp mimics is not unexpected, and other cases in which mimics were not rare could be 'explained away', at least until there was extensive evidence about the relative unpleasantness not only of the model, but of the mimic itself, as if the mimic is in fact an unpalatable Muellerian mimic a greater abundance would again be in no way surprising. What then of theory? Do we in fact *expect* mimics to be rare?

If the predators which select the mimicry also control the population size of the mimic, then the abundance of the mimic will be regulated to some fixed proportion of the abundance of the model (assuming a fixed density of predators), for as the mimic becomes common, the increased visual predation will reduce its population growth to zero. In this case, folk wisdom will be entirely correct. On the other hand it is probably quite rare for butterfly populations to be strongly regulated by their visual predators. Although this is not, as Vane-Wright (1980) points out, necessarily true, and may not apply to some of the long-lived tropical species, or to mimicry among caterpillars in other species, it is surely the case that in many butterflies the predators which select for mimicry affect population density only in a very minor way (Nicholson 1927; Charlesworth & Charlesworth 1976). In this case the population size of the mimic will be hardly regulated at all by the abundance of the model. How then can the mimic be 'rare'?

Being a Batesian mimic cannot in this case make a species rare, notwithstanding statements that the mimic is 'in a precarious position because of the selection for rarity (Huheey 1976). If mimics are indeed rare, this leaves only pre-adaptation or group selection as explanations. Group selection seems to be ruled out, as it would require common species to have, as a result of being mimics, a higher rate of extinction than rare species. Pre-adaptation could take any of four forms: (a) rare species tend to become Batesian mimics, (b) common species tend to be used as models, (c) common Batesian mimics tend to lose their mimicry, or (d) common Batesian mimics tend to lose their models. The explanation usually found in the literature is (c), the lose of mimicry by a common mimic.

Nicholson, on the other hand, proposed the eminently sensible solution that it is not that *mimics* are rare but that *models* are common: because an abundant warningly coloured species will gain more mimics than a rare one, there will be an overall tendency for mimics to be of merely average abundance, but for their models to be predominantly among the commonest species (Nicholson 1927). Rothschild's suggestion that in addition, rare species become mimics (which would imply that mimics were in fact rarer than related cryptic species) is, I believe, correct, and represents a considerable insight. To discover a mechanism leading to this outcome, we must consider (i) the selective advantage of a new mimetic form, (ii) the speed with which its mimicry can be improved and (iii) the speed with which the model can 'escape'.

Consider first the establishment of a mutation which produces adequate, although far from perfect mimicry in a non-mimetic species. The selective advantage of this gene, when it is newly arisen and represented by only a few individuals, is the same whether the species is rare or common; its initial success will depend only on the accuracy of its mimicry and the abundance and nastiness of its model. In this sense it cannot be the case that rare species tend to become mimics. On the contrary, the probability that an advantageous new mutation survives the risk of stochastic loss in the first few generations after it appears will favour mimicry in large populations, but the effect will be slight as the mutant's survival depends chiefly on its selective advantage, and on the ratio of effective to actual population size (probability of fixation $\simeq 2sN_e/N$—Kimura & Ohta 1971). The next event to consider is the fate of the new gene for mimicry once it has escaped the danger zone of stochastic loss: unless the population is very small, the frequency-dependent loss of fitness which it suffers as it becomes commoner will slow its advance through the population, until it comes to a stable equilibrium at which its fitness is equal to that of the non-mimetic form. An interesting difference now becomes apparent between common and rare species: as was first pointed out by Charlesworth & Charlesworth (1976) this equilibrium has the property that the *number* of mimics is constant irrespective of population size. Thus other things being equal the mimic is a lower *percentage* of the population in the commoner species. Now because the form at the lower percentage frequency exerts less selective pressure on any unlinked specific modifying genes which might improve the mimicry, in the scarcer species the mimicry first equilibrates at a higher percentage frequency, and then improves more rapidly. As the mimicry improves, its selective advantage increases, pushing the equilibrium point even higher, with the mimic in the end perhaps reaching fixation at 100% of the population. Fixation in this way will occur comparatively rapidly if an unlinked *non-specific modifier* (as defined above) that improves the mimicry can spread in the population, for the stable equilibrium point for the main mimicry gene and for the modifier then becomes 100% (Charlesworth & Charlesworth 1976). Again, the advantage is with the rare species, for such modifiers cause fixation of the mimicry in this way only when the main mimicry gene has already reached a high frequency. In an abundant species, with the mimicry at low frequency, the modifier does not spread through the population, and the mimetic form remains both unimproved and rare, whereas in the rare species it is catapulted to success. Thus we expect rare species to have accurate mimicry at high frequencies in the population or even at fixation, and common species to have poor mimetic forms at low frequency.

But this prediction is not borne out by the evidence. There are indeed species with polymorphic mimetic and non-mimetic forms (*Papilio glaucus*, *P. dardanus*, *P. polytes*), but in all of them the mimics appear to be moderately common. We lack examples of abundant, palatable insects with a rather poor mimetic form at very low frequency in the population.

The problem is rather easily solved if we take into account the evolution of the model. A mimetic form at a low frequency in a population is in the same position as one whose advergence towards the model depends on a character with low heritability: selection will push the pattern of the mimic in the right direction, but the mimic will have difficulty in responding. For example, suppose that there are two palatable species, identical in all respects save that *A* is abundant and *R* is rare. Both have mimetic forms which are under pressure to evolve at, say twice the rate at which the model can escape from them. Species *R*, with a mimetic form occupying say 90% of its population, can evolve almost at this rate; but if *A's* mimetic form, which will be represented by the same *number* of individuals as *R*'s mimic, occupies only 10% of the population, it can evolve at only $2 \times 0.1 = 20\%$ the speed of the model. Instead of model and mimic adverging, the model escapes from the mimic, which will gradually lose the selective advantage it started with, and will disappear from the population. The crucial point is that the rate at which the model escapes is governed by the *number* of the mimetic form, which is the same for *A* and *R*, whereas the speed with which the mimic pursues, being governed by its *percentage frequency*, is greater in the rarer species.

There is therefore a tide, or at any rate a watershed, in the affairs of mimics. Those which arise in rare species increase to high equilibrium frequencies and start an arms race with their model, which they will win. As they do so, their selective advantage increases because of their improved resemblance, and their frequency, and hence number rises. Exactly what happens to those which arise in a species which is so common as to put the mimic on the wrong side of the watershed, is not quite so clear, as it is not fully predicted by the argument so far, which has considered only the equilibrium frequency, not the initial conditions. Much no doubt depends on values of parameters, such as the rate of change of frequency of the main mimicry allele compared with the rate of divergence of the model's pattern from the mimic's. However, although limit cycles might occur in some cases, the most likely scenario is that the mimetic form being initially, by definition, at a selective advantage starts to rise in frequency. As it becomes more numerous, it touches off an arms race which the model proceeds to win, and as the model's pattern diverges from the mimic's, the spread of the mimetic allele will halt, then reverse. As the mimic again becomes rarer, its ability to evolve becomes even more limited, and although the rate of divergence of the model will itself slow down, it is likely that the process will continue until the mimic has been eliminated from the population.

An obvious exception to this behaviour may occur (as I argued above in the case of limited heritability) when an otherwise unpromising mimic is lucky enough to have a phenotype 'opposite' to that of a successful mimic, which is pushing the model's phenotype in the direction of the unpromising mimic. Another exception will occur if the modifier is linked to the major gene (even when it is a *specific modifier* which alters the mimetic pattern only). Such a modifier can, if the linkage is tight enough, spread even when the major gene is at low frequency, and will pull the major gene up with itself. This effect will further enhance the tendency of the mimicry to be controlled by a 'supergene'.

In general, therefore, there is probably quite a strong tendency for mimicry to establish itself in rarer species. But, as (Gilbert 1983) says, species do not necessarily remain common or rare for ever. What will happen to a good Batesian mimic if it becomes common, or its model rare? If the species is still polymorphic the answer is obvious. The frequency of the mimetic form declines in the first case (but its numbers remain the same); in the second case both its numbers and frequency decline. In the extreme, when the model is absent, the mimetic form is simply lost from the population, as in *Papilio glaucus*, which is polymorphic for the non-mimetic yellow and mimetic black form within the range of the model *Battus philenor*, but becomes entirely non-mimetic and yellow in the northern part of its range where the model is absent (Platt & Brower 1968). Similarly Charlesworth & Charlesworth (1976) suggest that the presence of yellow forms of *Papilio dardanus* in Kenya, where again the models are rare, results from the persistence of an ancestral allele which produces the non-mimetic yellow colour.

However, if a species is fixed for a mimetic allele mimicry cannot be lost in this way, and we might expect the pattern to remain little altered for a considerable time, even when the mimic is much more abundant than the model. Taking, again for simplicity, the extreme case in which the model is absent, the mimetic pattern will deteriorate slowly for two reasons: (a) mutations which damage the mimetic pattern will not be selected against, and will accumulate, but only very slowly, in the population; (b) mutations which improve the camouflage, thermo-regulation or intraspecific signalling of the pattern will be selected. In this way species which remain abundant for long periods will gradually lose their mimicry. They may also of course lose their models. The persence of an abundant and accurate Batesian mimic will place an extra selective premium on mutations in the model species which allow it to join a new Muellerian mimicry ring. In this case the answer to whether Batesian mimics should be rare is therefore a qualified yes. There is one final possibility: we are inevitably subjective when we recognize a species as a mimic, and naturalists are understandably reluctant to proclaim a merely vague resemblance as Batesian mimicry. Could it be that common fully mimetic species simply do not improve their mimicry fast enough for us to recognize them as mimics, or even lose the race with their model? The answer is not obvious.

To check whether the simplest expectation, that as the mimic becomes commoner its rate of evolution will decrease while the rate of change in the model increases, the process of advergence between a model and mimic has again been simulated, using the same values as before, but varying the density of the mimic relative to the model. As Table 15.2 shows, the commoner the mimic, the greater the change both in model and mimic. But the net *rapprochement* between them, at 0.088 of a spot, remains constant! At least in the range investigated, a common mimic finds it no harder to improve its mimicry than a rare one. However, this calculation depends critically on one simplifying assumption: that in any one spot class, the fitness of the model and mimic is the same, as it is in Fig. 15.9. In fact, they will differ, to the advantage of the model if this tends to be aggregated, and to the advantage of the mimic if it is intimately mixed with a highly dispersed model. If both model and mimic are randomly distributed in space and/or time, then there is a *slight*

TABLE 15.2. Rates of evolution of model and mimic varying with abundance of mimic

Increase (spots)	Abundance of mimic (model = 1)								
	0	$\frac{1}{4}$	$\frac{1}{2}$	$\frac{3}{4}$	1	$1\frac{1}{2}$	2	$2\frac{1}{2}$	3
Model	0.005	0.021	0.037	0.080	0.104	0.154	0.159	0.178	0.232
Mimic	0.082	0.110	0.135	0.161	0.194	0.240	0.250	0.263	0.317
Mutual	0.077	0.089	0.098	0.081	0.090	0.087	0.091	0.085	0.085

Monte Carlo system the same as in Fig. 15.9, except abundance of mimic is varied. Figures are increase in average spot number within the course of one generation; the realized evolutionary change would depend also on the heritabilities in the two species. At the beginning of a generation there is a difference of one spot between the means of the two species; the bottom line shows 1 minus the difference between the means at the end of the generation (i.e. amount by which the means have approached each other). Calculated from two runs of 5000 cycles each.

Regression analysis gives for the model $b = 0.074$, $a = 0.014$, F (regression) = 160, $P \simeq 10^{-5}$, for the mimic $b = 0.074$, $a = 0.100$, F (regression) = 150, $P \simeq 10^{-5}$, for rate of mutual approach $b = 0.0002$, $a = 0.087$, F (regression) = 0.005 (no test), degrees of freedom (regression)1/7.

advantage to the mimic. The net result is that the mimic is less strongly and the model more strongly selected than is shown in Table 15.2. This discrepancy becomes greater as the mimic increases in density, between the two obvious limiting cases of model common, mimic very rare (mimic evolves, model unaltered), and model very rare, mimic common (mimic unaltered, model evolves), so that, subject to critical assumptions about the dispersal patterns of model and mimic, common mimics will indeed be less accurate than rare mimics, and extremely abundant ones are at risk of losing their models altogether.

So on the whole, Batesian mimetic species will be less numerous than their models and less numerous than their cryptic relatives. A similar, simpler conclusion applies to Muellerian mimicry: when the initial mutation is established, it will normally be in the rarer species, although that principle can be violated by differences in palatability; there is no subsequent restriction on changes in abundance.

Mimetic polymorphism

Polymorphism in Batesian mimics is readily conceived in terms of a mechanistic pathway: the advantage of mimetic forms when rare, declining as they become common (Fig. 15.7) is easily seen to give rise to a stable equilibrium with two or more mimetic forms persisting in the population. The stability of the polymorphism depends also on tight linkage between the genes which produce the different colour patterns (Charlesworth & Charlesworth 1976), thus accounting for the clustering of the genes into tightly linked groups or supergenes, which is so characteristic of the polymorphic *Papilio* species (Fig. 15.8) (Sheppard 1975).

Can we also see polymorphism as a 'dodge' by which the population size of the mimetic species is increased? Insofar as predation on adult butterflies decreases population size, a species will certainly have a higher equilibrium density when polymorphic than when monomorphic. Should we then think of polymorphism as an adaptation at the population level? Most population biologists would answer 'no', unless perhaps group selection was effective in preserving the dense polymorphic species at the expense of the rarer, and more easily exterminated, monomorphic ones. The 'problem', which boils down to asking whether species, as well as individuals, can be said to have adaptations, simply emphasizes the difficulty of the concept of adaptation. There are indeed things which we recognize as *adaptations* which have not been selected: when Muellerian mimicry is initiated, both species gain in protection, even though initially only the species which produced the major mutation is subject to selection. The other species gets 'something for nothing' and can be said to have gained an adaptation without selection. There is therefore no logical reason why, even if there is no selection at the population level, we should not think of populations as being adapted.

But most of us would prefer the simple view that mimetic polymorphism presents us with a special case of the extension of Fisher's fundamental theorem to a density-regulated population (cf. Christiansen in this volume): under the most common side-conditions, the spread of a successful mutuation will increase the standing population density.

Sex limitation

The limitation of mimicry to the female, found in many but by no means all butterfly species (five of the sixteen mimetic species of *Hypolimnas* for instance [Vane-Wright, Ackery & Smiles 1977]) presents an interesting twist to this problem. If the density of the population is affected more by predation on females than on males, then halving the number of mimics by rendering the males non-mimetic will in decreasing the rate of predation on the females, increase the population density. But selection on the mimicry alone will not lead to this apparently adaptive end point: from the point of view of a 'selfish' male it is better for him to be mimetic, unless the population is highly structured and the increased risk of predation is more than offset by the protection gained by closely related females. It is more likely that any gain in population density is a fortuitous outcome of some kind of sexual selection (either by male–female signalling [cf. Turner 1978] or male–male signalling [Silberglied, 1983]) which inhibits the expression of the mimetic pattern in males. If we are bent on applying the concept of adaptation, we might then stretch it to the point of saying that sexual signalling when allied to mimicry is a long-term pre-adaptation for increasing population density!

What is more interesting about sex-limitation is the genetic pathway which leads to it. We might expect the initial mimetic allele to be expressed in both sexes, and that the combined effects of natural and sexual selection would enhance its expression in females and suppress it in males. This would lead to the conditions actually observed (males non-mimetic, females mimetic, with or without a non-mimetic male-like form), but such a model predicts also the existence of species with mimetic females and males which are polymorphic for the mimetic and non-mimetic form, in additon to the persistence for some time of the early condition, in which both sexes are polymorphic for mimics and non-mimics (Fig. 15.10). As neither of these conditions is known, it seems more likely that the mimicry is limited to the female even when the mutant first appears, and that in many species such mutants are selected in preference to those expressed in both sexes (e.g. Turner 1978). This limitation is probably achieved more readily in butterflies than in other insects on account of the lack of dosage compensation of the sex chromosomes: any gene whose product interacts with a product from an X-linked gene may be able to 'know' whether it is in a male or a female (Johnson & Turner 1979). We have yet another case where pre-adaptation is a better explanation than modification.

However, the idea that most of the genes for mimicry just happen to be female-limited in expression makes little sense when we look at Muellerian mimics, where sex-limitation is a rare exception. What is happening to all the unimodal (expressed in both sexes) mimicry genes in Batesian mimics, if as is suggested by the absence of the 'polymorphic male, mimetic female' class, they are *not* being made sex-limited by modification? The answer may lie in an even more elaborate version of the watershed effect already described. In the absence of sexual selection, at equilibrium, the *number* of mimics (male plus female) in a unimodal mimic equals the *number* (females only) in a sex-limited one (provided mimicry has not reached fixation). The model–mimic arms race will have the same outcome in both cases. But if sexual selection acts more strongly against the mimetic pattern in males, then its interaction with frequency-dependent selection produces a rather unexpected effect. The number of mimics at equilibrium for a female-limited mimic is *several times* the number of mimics (both sexes) for a unimodal mimic (Turner 1978). It must therefore often come about that unimodal mimics are unable to catch up with their model because they cannot put enough pressure on their modifiers, whereas the female-limited mimic has little difficulty in homing in on its slowly moving target.

This leaves one outstanding problem in the theory of sex-limitation. On the way to female-limited mimicry we can find the expected intermediate condition, with males non-mimetic, females polymorphic for a mimetic and a male-like form (*Papilio dardanus* has populations exhibiting both conditions); but while there are many unimodal (both-sex) mimics, the intermediate stage (polymorphic for mimics and non-mimics in *both* sexes) is *not known* (Fig. 15.10). What I suggest is that unimodal mimics are predominantly species in which the colour pattern served only a limited function as a signal in both sexes. In such a species a unimodal mutant might readily become established. Then the resulting polymorphism would be comparatively short lived, as the rather small selective difference between the mimic and non-mimic at high frequencies of the mimic would result in the equilibrium point rising readily to 100% mimics when either the model or the predator became common, or the mimetic species rare. Alternatively, or also, they may include species which commenced with female-limited mimicry in which the signalling behaviour became modified, and species which were so rare that even a unimodal mimic was able to catch up with its model.

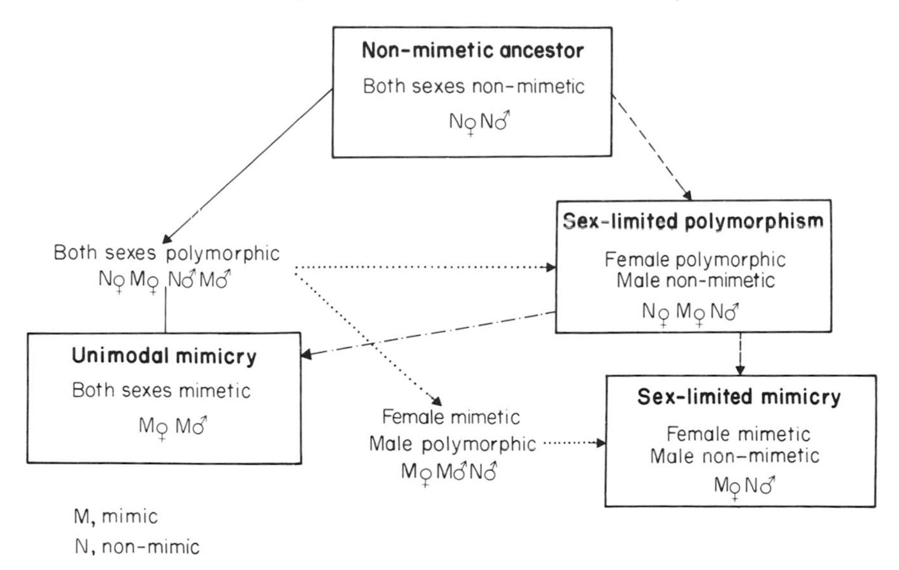

FIG. 15.10. Pathways leading from a non-mimetic species to a species with sex-limited mimicry. Only the conditions shown boxed have been discovered. ⟶, Gene substitution by natural selection; — — ⟶, with sexual selection also ··········→, modification of expression by sexual selection; ·—·—·→, Modification of courtship behaviour (via M♀ M♂ N♂ or N♀ M♀ N♂ M♂).

A further very interesting possibility arises from the observations of Vane-Wright (1980) who suggests that it is the male pattern that controls the evolution of signalling in the species: as males are much more prone to approach other males in territorial combat, there is, according to this theory, a sexual advantage to females in resembling the males. Then if the 'male mimetic, female polymorphic' state arose, it would rather soon become modified into both sexes being entirely mimetic.

Time of emergence

Rothschild (1971) predicted that 'Selection in a seasonal climate will tend to push the emergence dates of the most distasteful species forward, and hold back or delay the less distasteful.' Huheey (1980a) confirmed this prediction with a model of time-independent predator learning, concluding that for the mimic to emerge after the model was not simply an adaptation of the mimic, but a co-adaptation which benefited the model as well: '*If a Batesian mimetic complex can avoid the noise of mixed models/mimics by 'programming' predators through prior exposure to a population composed solely of models, both the Batesian models and mimics benefit.*'

Early emergence will indeed benefit the model, but our findings about advergence (above) might lead us to suspect that the model cannot achieve it. That late emergence benefits the mimic is an appealing idea, but goes against the equally appealing idea that the mimic should come to resemble the model more and more closely!

Suppose that a distasteful, warningly coloured species emerges in late

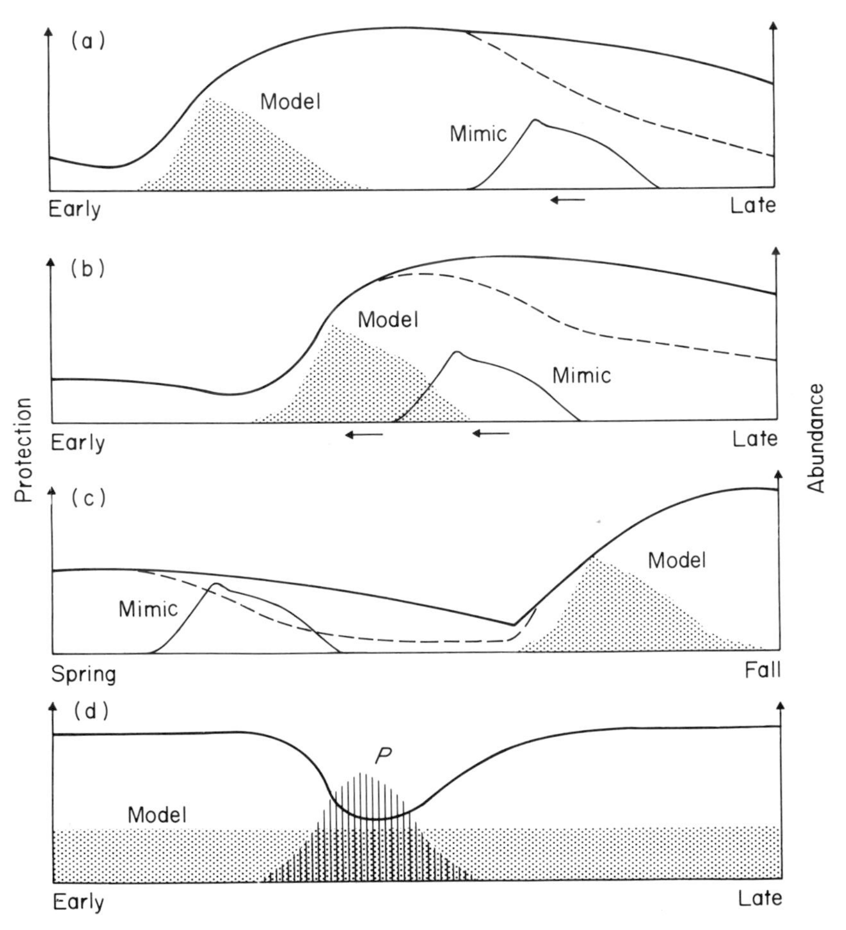

FIG. 15.11. Advergence of a mimic toward the flying season of its model. The heavy line denotes the level of protection afforded to the pattern of the model, the dotted line the effect of the mimic in reducing it. In (a) the mimic is selected to emerge earlier, in (b) both model and mimic are selected to 'adverge' on an earlier emergence date; in (c) a spring mimic which depends on the predators' memory of the model from the previous fall, is selected to emerge earlier, but may be unable at least temporarily to 'skip' through the winter or summer, and thus, paradoxically, appears to be 'avoiding' the model. In (d) a peak of predator activity or fledging (*P*) will select the mimics of a perennial model to emerge at a different period.

spring. The protection it receives will build up during its emergence period and then, a little before its flying season is ended, will start to decline as predators gradually forget their experiences with it, and as new predators are recruited to the population. This protection is shown by the heavy line in Fig. 15.11, where it is assumed also that some of the predators surviving from last year remember that the insect is nasty to eat. It is easy to see that the advantage of mimicry for a species which emerges before the model is only slight, but is much greater for those that happen to emerge after it. Later-emerging species are therefore pre-adapted to become mimics. This applies obviously to a Batesian mimic, but equally to a distasteful species. Thus imagine that *B*(efore) emerges before the equally distasteful *A*(fter). There is a selective advantage to *A* in converging to *B*, but little advantage to *B* in adopting *A*'s pattern. Bullini *et al.* (1969; also Sbordini *et al.* 1979) have shown most convincingly that *Zygaena ephialtes*, although it is the more distasteful, has converged to *Amata phegea*, and have suggested that this has in part been brought about by *ephialtes* being the later to emerge.

We have therefore predicted that sieving will produce a difference in the emergence times of models and mimics, with Batesian mimics, and the member of a Muellerian pair which has undergone the larger amount of genetic change, emerging second. But, using the same model of predator behaviour, we can see that this is not the stable condition. Provided predators continue to sample models and mimics from time to time, the mimic depresses the fitness of both itself and the model, as indicated by the dotted line in Fig. 15.11: individual selection favours mimics which emerge earlier than the norm, and other things being equal, the mimic will evolve an earlier and earlier emergence date. Even when it emerges so early as to overlap with the flying season of the model, and hence causes the model to be selected for earlier emergence, selection on the mimic continues to be greater than selection on the model (Fig. 15.11). In short, model and mimic will adverge for emergence date just as they do for colour pattern, and for the same reason. The stable end-point will be that the species have almost identical flying seasons, although if predators can sometimes distinguish between them we might expect a Batesian mimic to have a lower variance of emergence date than its model.

Thus the adaptationist argument predicts one outcome (models emerge before mimics, to the benefit of both), and modelling predicts another: in so far as mimics are found to emerge after their models, this will represent a disequilibrium generated by the tendency of mimics to evolve among species which emerge after rather than before the model, and which either have not had sufficient time to converge on the model's emergence time or have been restrained from doing so by other requirements of the life cycle.

The size of the mimic's population is likewise related to the time of emergence but in the reverse relation to that predicted by the adaptationist argument, which says that mimics emerging behind their models will have reduced apparency and increased population size. On the contrary, the mimics will be predated least when they emerge with the models, and if predation has any effect on population size it is then that the mimic's population will be largest.

It is indeed true that if a mimic can reduce its apparency to predators then (insofar as predation regulates the population) it will become commoner. But it must be a general principle that the mimic is least apparent to predators when it most closely resembles the model, not just in pattern but in behaviour and phenology. And when it comes to emergence time, hiding *behind* the model does not mean emerging *after* the model. The mimic should no more fly apart from the model in time than it should in space.

Selection and adaptation

The mechanisms required to produce at least some of the peculiarities of Batesian mimics listed in Table 15.1 can therefore be described most readily in terms of individual selection, especially when we consider the complex but elegant effect that population size and sex-limitation can have on the model–mimic arms race. However, such a simple description does miss some interesting additional possibilities suggested by Rothschild's adaptationist argument. The mechanisms required by the adaptationist scenario are as follows:

1 Pre-adaptation: palatable species which for quite other reasons are rare, will in some circumstances gain mimetic forms and develop them into accurate mimics, when mimics can only with difficulty be established in common species; those whose patterns are used as signals in courtship and male–male interaction will tend to develop sex-limited mimicry by a similar 'watershed' effect; those which have appropriately linked clusters of genes controlling their colour patterns will tend to become polymorphic.

2 Group selection: for us to think of polymorphism and sex-limitation as 'adaptations' it is further helpful that rare species be more prone to extinction than common ones, and that a reduction in the rate of predation on the butterflies increases the population density.

3 Adaptation: it will then follow that Batesian mimics, being rare species, will have a higher rate of extinction than the average butterfly, but that those which have or develop further 'dodges' that make the mimics less 'apparent' and hence reduce the rate of predation will lower their rate of extinction again. There seem to be two kinds of mechanism by which apparency can be reduced:

(i) Becoming as similar as possible to the model, both in colour and behaviour, including emergence date. This is favoured by individual selection, which has the further peculiarity that the mimic necessarily evolves faster than the model, so that the resemblance is steadily improved. The adaptationist argument is incorrect in predicting first that the emergence dates will diverge, and second that this divergence would benefit the mimic as well as the model.

(ii) Reducing the actual number of mimics in the population, which because selection is frequency dependent produces a disproportionate gain in protection. Polymorphic mimicry of several models achieves this, and is, again because of frequency dependence, favoured by individual selection. It also requires a fortunate linkage of the appropriate genes if it is to develop to any degree. Suppression of mimicry in the males achieves the same effect, but is clearly not individually favoured by natural selection. It almost certainly is favoured in certain species by individual sexual selection.

It is entirely open to debate whether the rate of extinction of rare butterflies on the one hand and on the other the increase in population density from the lowered apparency of mimics are great enough to make this anything more than an interesting intellectual exercise. A puritanical adherence to individual selective mechanisms, while it is less fun, leaves the mind less cluttered.

WARNING COLOUR AND KIN SELECTION

Co-evolution of model and predator

So far the argument has proceeded in a form which would, I believe, satisfy the most fundamentalist and mechanistic neo-Darwinist. The only process invoked has been individual selection. None of the features of mimicry actually required group selection for its explanation. The origin of aposematism, that is to say unpleasantness accompanied by warning colour, is another matter.

The bright advertisement or road-sign colouring of many poisonous and toxic insects is a biological commonplace. What is not so widely known is that some, and we do not know how many, cryptically coloured insects are distasteful. The buff-tip moth (*Phalera bucephala*) has a warningly coloured, distasteful larva; the adult moth, although itself toxic, is an exquisite cryptic mimic of a silver birch twig (Fisher 1930). There is no necessary connection between unpleasantness and road-sign colouring.

There is no difficulty in believing that distastefulness can be advantageous to the individual: experiments have shown that unpleasant insects can be be released, largely uninjured, by birds (e.g. Wicklund & Järvi 1982). Toxicity and road-sign colouring do present problems. Most aposematic insects derive at least some of their protection from the possession of the toxic secondary defence chemicals of their host plant (or at least from a similar compound), and yet toxicity by itself is of neutral selective value to the individual: as it cannot be delivered up whole like Jonah, the consumed insect gains nothing when the bird vomits. Road-sign colouring is in an even worse case: a distasteful insect may be released unharmed, but there can be only a selective *disadvantage* to the individual in attracting the predator's attention by being conspicuous in the first place, as all attacks will involve some risk of injury. Better to have both defences like the buff-tip moth, to be missed entirely by most predators and rejected, uninjured by a proportion of those which do notice you, then to attract the attention of every passing bird by wearing bright red spots.

Fisher (1930) was led by a consideration of this problem to propose what is now known as the theory of kin selection: '... although with the adult insect the effect of increased distastefulness upon the actions of the predator will be merely to make that individual predator avoid all members of the persecuted species, and so ... to confer no advantage upon its genotype, with gregarious larvae the effect will certainly be to give the increased protection especially to one particular group of larvae, probably brothers and sisters of the individual attacked. The selective potency of the avoidance of brothers will of course be only half as great as if the individual itself were protected; against this is to be set the fact that it applies to the whole of a possibly numerous brood. There is no doubt of the real efficacy of this form of selection ...' Rather oddly, Fisher applied this argument only to the evolution of distastefulness, where he admitted individual selection is also possible, and did not consider bright coloration, where the case for kin selection is much stronger.

Harvey *et al.* (1982) have recently examined this problem in some detail, and have given the necessary conditions for the evolution of bright colour in a distasteful species: the prey must live in family groups, rather widely scattered so that not too many occur within the home range of any one predator; the bright colour must be more readily learned as the mark of a nasty taste than is the original cryptic pattern of the prey, and it must not make the prey excessively conspicuous. These conditions do tend to be met. As Fisher noted, many warningly coloured caterpillars are gregarious, and this even applies to some warningly coloured adults, which have limited home ranges and aggregate in communal roosts at night (Benson 1971; Turner 1975). There is experimental evidence (below) for the more rapid learning of bright colours, and a number of workers have noted that because of optical blending effects the apparently conspicuous warning patterns can become quite well camouflaged when viewed from a distance (Endler 1978, Rothschild 1981). (An eccentric version of this theory states the opposite: that warning patterns and aggregation behaviour are designed to exploit optical blending by making the insect *maximally* conspicuous at a distance [Hinton 1977]. At this point the accusation of just-so story-writing comes uncomfortably near the mark: *both* theories can hardly be right.)

These findings raise as many questions as they answer. The question of the balance between conspicuousness and concealment could do with much more investigation. Papageorgis (1975) suggested quite plausibly that the five different warning patterns in South American rain forest butterflies, which she found to fly at different modal heights above the ground, were partly protected in flight by a flicker effect which made them hard to follow with the eye, and that the different patterns were adapted to the different distributions of light and shade found at different heights in the forest. But if this were the sole satisfactory explanation, as has recently been claimed (Hiam 1982) then there would be a continuous gradation of warning patterns to match the continuously graded dapple pattern. That there are only five patterns among numerous butterfly species must surely be explained in terms of mimicry: by the tendency of warning patterns to converge towards one another if they are sufficiently similar but not to converge if they are already so different that predators never mistake one for the other (Turner 1977, Sheppard *et al.* 1983).

The general evolutionary questions begged are (a) why *should* bright colours be easier to learn? and (b) why do distasteful species become gregarious? I believe that answering these questions is now particularly hard, as we are dealing, as with models and mimics, with a system of co-evolution: predators and insects have been co-evolving since the Carboniferous period, and the invoked changes in the predator may substantially mislead us as to the initial conditions. It is beneficial to predators to find what is poisonous distasteful, and to have an innate tendency to avoid what is likely to be harmful. Once there are substantial numbers of poisonous insects which are brightly coloured, the invoked instinct of the predators to find them distasteful and to avoid bright colours, will create an individual selective advantage for toxicity and for road-sign colouring that did not exist at the beginning. The matter of starting the process off is a somewhat separate one from keeping it going.

Suggestions for why warning colours are easily learned, must include innate avoidance on the part of the predators. Schuler (1982) has shown that young starlings do have an innate avoidance of the black and yellow stripes of wasps. It must now, after so many millions of years of possible co-evolution, be impossible to know whether in addition to a co-evolved avoidance there is some inbuilt quirk in the vertebrate central nervous system which finds the cool colours of sky and grass restful, and warm colours disturbing or stimulating (would you rather look at a diazo lecture slide printed in red or blue?). Certainly there are quirks that are not easily explained: why should goldfish find it easier to learn to 'go' on red and 'stay' on green, than to obey the traffic lights as we understand them? (Bisping *et al.* 1974).

Unless there is, and was always, an innate avoidance of bright colours, then it appears that road-sign colouring must, at least originally have evolved through kin selection, as a result of the colours being more readily learned. The reasons for the faster learning which have been suggested are as follows:

1 Innately faster learning, as a quirk of the vertebrate nervous system, either of bright colours as such, or of colours which contrast with the background which, given the predominantly cool colour of vegetation, would have the same result (Gibson 1980).

2 Avoidance of confusion with predominantly palatable cryptic forms: 'to be recognized as unpalatable is equivalent to avoiding confusion with palatable species' (Fisher 1930).

3 Ease of association with the effects of poisoning (Turner 1983b). Vertebrates can learn avoidance of poisonous food, even when there is an appreciable delay between eating and discomfort (Garcia, McGowan & Green 1972). But if a series of prey items was taken at roughly the same time, it will be the one that stands out in some way from the others that is remembered. If you are sick after dinner, you will blame the exotic fish cocktail, not the familiar roast beef.

4 Encouragement of a faster rate of attack (Gittleman & Harvey 1980). Even with kin selection, this must at the most be a marginal advantage. The slower rate at which distasteful, cryptic prey are learned will be offset by the lower rate at which they are attacked, and little if any advantage will accrue to brightly coloured, distasteful prey.

It is certainly the case that bright colours and/or those which contrast with the background are the more rapidly associated with negative reinforcement, not just with distastefulness but with the rapid escape of the prey (Gittleman & Harvey 1980; Gibson 1980), but the relative importance of brightness, contrast, innate learning differences and external factors is still far from clear.

Gregariousness is often individually advantageous on account of the 'selfish herd' effect (Hamilton 1971); as predators tend to pick off the peripheral members of a group, a solitary individual (which is peripheral when viewed from any direction) is at a decided disadvantage compared with a member of a herd. However this principle ceases to apply when the prey is small compared with the predator, which on discovering one member of the herd may be able to down the whole lot. Small palatable species will usually be dispersed. Once unpalatability has developed, this no longer applies: a large predator picking off one member of the herd will leave the whole herd alone. Aggregation ceases to be a disadvantage with respect to large predators, and may even be an advantage, as it concentrates the prey within the home ranges of a smaller number of individual predators. Selfish herd selection, exercised by small predators, will not be countered by selection for dispersion against large predators, and distasteful prey will be expected to aggregate. Once this stage is reached, other features of the aposematic life style, particularly warning colour, will be free to evolve through kin selection.

I therefore suggest the following scenario for the evolution of warning colour in insects. Toxicity first appears as an incidental result of feeding on toxic foodplants: the mulch of leaves in the gut makes the insect toxic. In a classic experiment, Eltringham (1909, 1910) showed that the otherwise palatable larvae of *Gonodontis bidentata* were unacceptable to lizards after feeding on ivy (the same is reported of humans and ivy-fed snails). This sets the stage for the evolution of distastefulness, both by active evolution on the part of the prey (leading eventually in some to the independent synthesis of the defensive compounds) (e.g. Rothschild, von Euw & Reichstein 1972) and by co-evolution of the predators toward finding what is poisonous also distasteful. Distastefulness leads to aggregation, and this in turn permits kin selection for warning colour. Once this becomes prevalent, co-evolution of the predators toward innate avoidance of these colours accelerates their evolution, across the whole range of prey, both unpalatable and perhaps sometimes palatable, through individual selection. However, neither innate avoidance nor individual selection can be paramount or we would find warningly coloured species which were not Batesian mimics of anything, but which were complete 'frauds', being entirely palatable but protected by the innate avoidance reaction to the bright colour. Predators obviously retain a high degree of flexibility. The only such fraud is 'flash-coloration', which exploits the innate avoidance without giving the predator time to unlearn it in the particular context.

Kin selection and advergence

Unstated assumptions about kin selection lie behind the prediction by Rothschild (1971) and Huheey (1980a) that models will fly before mimics. Both models and mimics will be selected to avoid periods when predators are most actively feeding, whether this be the late afternoon, the early morning when birds are hungry, or the late spring when the world is full of inexperienced fledgelings experimenting with various prey (Jeffords, Waldbauer & Sternburg 1980; Waldbauer & Sheldon 1971). Sometimes the model is unable to avoid flying at this time: the life cycle of bees and wasps will not readily allow them to vanish in late spring and they are present throughout the temperate growing season. On the other hand, any of their mimics which have a more limited flying season may be selected more easily to avoid the peak of predation, as was suggested by Waldbauer & Sheldon (1971), who found that dipteran bee and wasp mimics were indeed scarcer at that time of year. A peak of abundance of the mimics which occurred *before* the fledging season they attributed to the memories of wasps which older birds retained from last season. The effect is shown diagramatically in Fig. 15.11. Clearly this kind of seasonality in mimics could be produced either by pre-adaptation (species emerging outside the fledging season tend to become mimics) or by direct modification of the flying season of the mimics.

What happens if we combine a peak of predator activity with a model which is not perennial, but which itself has a limited flying season? Perhaps the easiest case to think about is the diurnal flight rhythm. Selection on the model will balance the disadvantage of flying during the dawn peak of bird activity against any other advantages of flying early. For the mimic, there might be a simple selective advantage in flying later in the day: although a mimic flying later than the model would be at a disadvantage because more readily discriminated, it might be that even this could be converted to an advantage by a great enough drop in the absolute rate of predation. However, this means that the mimic is merely using another method of predator avoidance; we might expect the mimicry to deteriorate.

Now consider the selective balance on a model and a mimic which are both prevented from moving their flying period forward into an otherwise favourable season (either diurnal or annual) by a peak of predator activity. In the model, kin selection may mitigate the individual selective disadvantage of early emergence; there will be no such mitigation for the mimic. Hence the model will be able, by moving forward in formation, to advance its flying season at a faster rate than the mimic.

It might therefore be that kin selection can slow down the rate of advergence not simply for emergence date, but for other characters as well, by reducing individual selection against deviant models.

EVOLUTION BY JERKS—A JUST-SO STORY?

Once we have uncovered a satisfactory evolutionary pathway to a particular end-point, 'adaptation' becomes a not very interesting semantic problem.

Can something be an adaptation, which improves survival but has not been directly selected? Do populations gain adaptations by accident via individual selection, or do we have to be convinced that their prevalence has been reinforced by group selection? Does it really matter? The substantive question is whether we can discover properly documented sequences through which features of organisms will arise.

At this point the opposition may feel that the range of selective mechanisms available is so wide that any phenomenon could be explained. This criticism is frequently levelled at sociobiology, which can explain behaviours through a hierarchy of individual selection, kin selection and reciprocal altruism. Neo-Darwinism, it is claimed, (Gould 1980) is so flexible it will explain *anything*. The sincerity, if not the force of this criticism is somewhat diminished though, when the same author (Gould & Lewontin 1979; also Gould 1982, p. 54) objects that neo-Darwinists are inflexible and not sufficiently 'pluralist' in their use of available evolutionary theories!

But for at least this one rather complex case we can give a good Darwinist account, backed extensively by experimental evidence on the behaviour of predators and the genetics of the mimics. This assuredly takes mimicry out of the realm of 'just-so stories.' But the alternative macro-mutational theory of evolution is seen, in this case at least, as a bald and unconvincing narrative!

Can the study of mimicry throw any light on the latest version of the macro-mutation theory, which has been set up as an alternative to the selectionist programme as a way of explaining evolutionary diversification? It is important to distinguish between observation and theory in macroevolution. The observation that evolution does not proceed at a uniform speed, sometimes even appearing to alternate long periods of stasis with short periods of rapid change, I shall call *punctuated equilibrium* (Gould & Eldredge 1977). The theory which attempts to put together elements of the theories of Lamarck, Cuvier and Goldschmidt into an alternative to neo-Darwinism (Stanley 1979; Gould 1980; Gould in Goldschmidt 1940), I shall call *evolution by jerks*. The question is, can neo-Darwinism explain punctuated equilibrium? Clearly it can (Charlesworth, Lande & Slatkin 1982). Rapid changes, too fast to be detectable as anything other than instantaneous in the fossil record, can be produced in populations even by quite weak natural selection. To change a distinct phenotype from 10^{-3} of the population to $1 - 10^{-3}$ requires roughly $\sqrt{(10^3)}/s$ generations, largely independently of the genetic architecture of the characteristic under consideration (J.R.G. Turner & M. Mukherjee, unpublished). With one generation a year, a geologically 'instantaneous' change lasting a mere 50×10^3 years could be produced by a selection coefficient (s) of little more than 5×10^{-4}. In fact only with long generation times and very small selective differences could 'gradual' changes be expected in the fossil record. Existing evolutionary theory has no difficulty in explaining punctuational events. But gradual changes would be expected to occur when the population adapted continuously to a slowly changing environment. Hence, at first sight, periods of complete stasis are surprising. But an explanation is forthcoming if, as seems likely, a major part of adaptive evolution is a response not to the physical environment but to other organisms. As ecosystems reach evolutionary equilibrium, there is a strong tendency for the niche space to become fully saturated. Once this state of full 'occupancy' (to resurrect the concept used by Charles Lyell) is reached, only very limited evolution will occur, in response to slow changes in the physical environment and perhaps to unchecked arms races of the gazelle–cheetah type, and of unrestrained male–male competition for dominance. Most characters of most species will be stabilized by interspecific competition and by arms races with stable outcomes like the model–mimic system. Periods of rapid change will tend to occur as a response to sudden changes of the physical environment such as the change in lake level at Lake Turkana (Williamson 1981) but perhaps more often to the emptying of part of the niche space by the extinction of the species that occupies it. A rapid adjustment of the remaining species would then occur, as some of them successfully evolved to exploit the empty niche. As Wright (1949) pointed out, a further but rare cause of rapid change is the opening up of entirely new niches, such as aerobic respiration, the land or the air. Given that only a few aspects of the phenotype are fossilized, many of these periods of change will remain totally undetected in the fossil record. A series of apparently long periods of equilibrium, punctuated only by that small percentage of periods of rapid change which are detectable in fossils, is therefore not unexpected in conventional evolutionary theory.

Mimicry in *Heliconius* butterflies seems to illustrate this kind of punctuated evolution rather well. Certain species show a considerable variation between the colour patterns of their various geographical races, so much so that most were, until very recently, classified into several different allopatric species. We have studied two such variable species, the parallel Muellerian mimics *Heliconius erato* and *H. melpomene* in some detail (Sheppard *et al.* 1983) and, as we saw above, have shown that the highly divergent geographical races differ at a number of major gene loci (Fig. 15.6). Divergence of this kind, in the face of the normalizing selection inflicted on Muellerian mimics, is best explained as the result of the pattern 'switching' to the mimicry of the locally best protected pattern that could be adequately imitated by a single mutation. While such changes of pattern possibly occur all the time in evolution, the correlation of the distribution of the races with centres of evolution deduced for birds, lizards, mammals and plants in South America strongly implies that most of the race formation in the butterflies took place when the neotropical rain forests were fragmented during the cool, dry periods of the Pleistocene (reviews in Prance 1982). During these periods of fragmentation, progressive extinction of elements of the fauna and flora in each of refuges, and of a rather different set of species in each refuge, would produce progressive differentiation of their ecologies and as the remaining species evolved to occupy the vacated niches, trigger evolutionary changes of the type already described. The butterfly patterns have evolved in a kind of photo-negative of this performance, not so as to occupy an empty mimetic niche (i.e. a pattern which no longer exists) but the pattern of whatever species has been caused, by the considerable local re-adjustments of the ecosystem, to become the most abundant.

Because of the comparative lack of dependence of the speed of evolutionary change on the genetic architecture of the trait being selected, the speed with which *Heliconius* responded to this ecological change will not have been atypical of such evolutionary changes, even if the use of major mutations

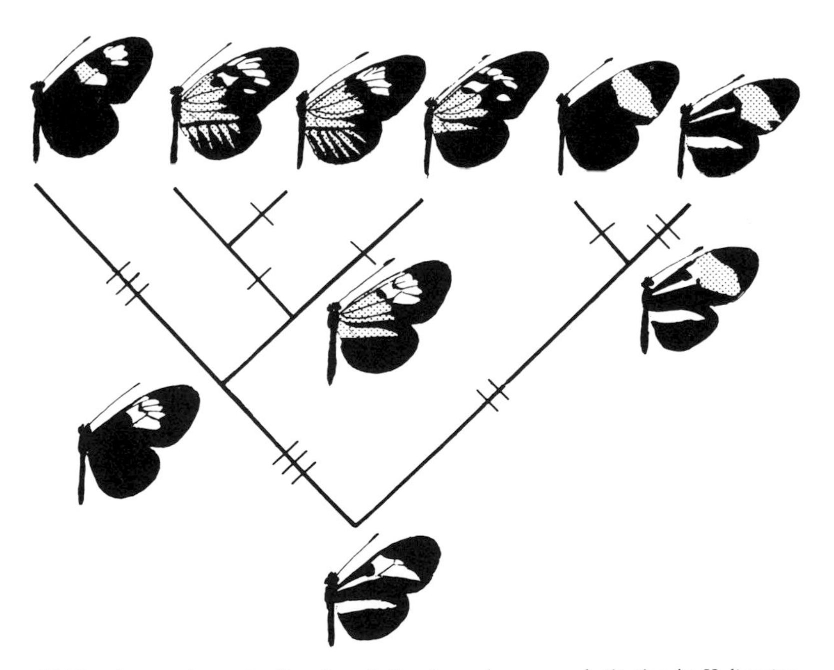

FIG. 15.12. Apparent punctuational evolution by major gene substitution in *Heliconius melpomene*. The cladogram has been constructed by the method of Farris, Kluge & Eckardt (1970); cross bars, placed conventionally at the centres of the branches, show the substitution of major genes. Butterflies at tips of branches are extant races, those at nodes are reconstructed ancestors. For the 'dominance sieve' method of determining the ancestral condition, and the strikingly similar cladogram for *H. erato*, a Muellerian mimic, cf. Turner (1981, 1983a, b).

is a feature of few adaptive changes other than mimicry. How long the changes took we cannot know, but with around ten generations a year, a moderate selection coefficient initially, and the positive frequency dependence inevitable with a Muellerian mimic, most of the major gene substitutions should have altered the population from having the new pattern at one in 1000 to having the old pattern at one in 1000 within a period of 500 years.

Our genetic data (Fig. 15.6) have been used to construct cladograms for both species (one is shown in Fig. 15.12). How near the truth these cladograms are is unimportant for the present argument: in any reasonably parsimonious scheme, the few hundred years of each evolutionary change (a cross bar presents a major gene substitution) must have been interspersed with long periods of evolutionary stasis, as the trees presumably span no less a time than the 30 000 to 50 000 years of the last glacial cycle (Turner 1983a).

In short, if we could see the evolution of these two butterflies in playback, or find a good series of fossil patterns, we would have a typical case of punctuated equilibrium.

It is therefore entirely clear that conventional neo-Darwinism can account for punctuated equilibrium, and what is more account for it, in this case, in terms of adaptive evolution. At this point the proponents of evolution by jerks would I think, raise two objections: (a) as speciation has not occurred, these are not punctuational changes as they understand them, but merely adaptive adjustments *within* a single species; (b) the question of a separate evolutionary process has not been discussed.

To this there are two answers. If the sequence of rapid gene substitutions and long periods of stasis in *Heliconius* were found in fossils, a butterfly palaeontologist would be forced to regard them as evidence of speciation, just as butterfly collectors until recently regarded the highly divergent races as separate species. That is how, willy nilly, species are determined in the fossil record. Second, we have evidence that no further processes are involved in the diversification of *Heliconius* patterns above the species level. The forty-odd species in the group display a great many mimetic patterns (Brown 1981) most of which can be rather easily seen to share homologous elements with other, related species in the group. For instance, *H. ethilla* is closely related to *H. melpomene*, but has a pattern which appears to be very distinct, belonging to a quite different mimicry ring. Yet most of its pattern can be constructed from elements of pattern found in various races of *H. melpomene*, all known to be produced by single major genes (Fig. 15.13); it is a reasonable supposition that were we to synthesize this genotype by the appropriate crosses, a few generations of selection of the modifying genetic background would produce a very passable mimic of *H. ethilla*. Just such an artificial mimic has been manufactured, quite by accident, in one of our crosses in

FIG. 15.13. To show that gene mutation within one species of *Heliconius* can readily produce the pattern of another species. (a) The pattern of *H. ethilla* (left) approximated by conbining genes known in various races of *H. melpomene* (right). (b) The pattern of *H. telesiphe* (left) mimicked by an *H. erato* produced by inter-racial hybridization (right).

H. erato, which produced butterflies with a close resemblance to *H. telesiphe* (Fig. 15.13). What is interesting here is the appearance of the peculiar ragged edge to the yellow bar on the hindwing, which is a species characteristic of *telesiphe*, but which is unknown in naturally occurring phenotypes of *erato*. Thus the potential for producing a species-specific pattern, which could rather easily be attributed to macro-mutations at the species level, is on the contrary contained in the normal genetic variation found *within* a related species.

In sum, for mimicry at least, adaptation and evolutionary diversity have a common cause. To account for mimicry we have a theory of selective path-pathways which is both internally consistent and consistent with a wide range of external evidence. In *Heliconius* the evolution of Muellerian mimicry generates a pattern of change which could reasonably be called punctuated equilibrium, but which is produced by no additional process other than adaptive gene substitution. The spectacular pattern of adaptive radiation and convergence at the species level in these butterflies (cf. Brown 1981) is no different in principle from the same pattern of radiation and divergence which occurs between races within the more polytypic species, and is produced by genes of the same type. No additional process is required to account for diversity at the species level. This is in direct contradiction to the theory of evolution by jerks, which maintains that evolutionary diversity is produced not by selection and adaptation as conventionally understood, but by an additional process occurring at the level of the species.

Fifty *Heliconius* do not admittedly make a summer, but in science we can only argue from known cases, not from hypothetical ones which we might discover if we could. Evolutionary radiation has been accounted for by the processes of individual selection and kin selection (and perhaps, very weakly, by selection at the population level), with the use of real examples. The theory of evolution by jerks, on the other hand, can produce no real cases, nor any consistent account of the alleged additional process. It falls back instead on vague statements about the *milieu intèrieur* and developmental canalization.

The Darwinian synthesis of adaptation and diversity stands, with natural selection as a common cause (although very obviously not the sole cause—Turner 1983d). What is extremely hazardous is to try to convert adaptation from an *outcome* into a *cause*, rather in the manner of an Aristotelian final cause. For want of more detailed modelling much of sociobiology and of evolutionary ecology depends upon this kind of argument. But the trouble with arguing that organisms will evolve what is adaptive for them is that what is adaptive for a cuckoo or a cheetah is manifestly not adaptive for a dunnock or a gazelle: if there are winners there must also be losers. To overcome this difficulty, Dawkins (1982) appeals to another adaptationist principle: the asymmetrical arms race, predicting that the dunnock or the gazelle becomes the loser because selection acts less strongly on it than on the winner. The danger of applying this argument can be seen when Dawkins tries to show why the dunnock and gazelle lose: the explanation is found in two principles—the 'life–dinner principle' (that the prey is more strongly selected because it runs for its life, while the predator runs only for its dinner) and the 'rare predator (or parasite) principle' which states that if, say, cuckoos are less numerous than dunnocks, all cuckoos must encounter dunnocks but not all dunnocks are parasitized by cuckoos: hence selection is stronger on cuckoos. These principles are attractive and deceptive, making us think we have an explanation, when in fact we have explained nothing because we have predicted nothing. Do gazelles 'win' because they run for their lives, or 'lose' because they are commoner than cheetahs?

If we try to apply these two principles to Batesian mimicry, then both turn out not to be reliable. With random dispersal, the rare parasite effect can be shown to operate, in that common mimics can have difficulty in adverging to their models (p. 338), but there is a further asymmetry in the system which ensures that advergence will usually not cease until the mimic is many more times adundant than the model. And it is this asymmetry of the selection patterns on the model and the mimic (Fig. 15.9) that actually causes the advergence: the comparative rarity of the mimic is merely a further permissive factor. A more general principle of asymmetry, rather like the life–dinner principle, suggests that the mimic should 'win' because it gains more from the mimicry than the model losses. Comparing the rates of predation in model and mimic with their respective controls in Fig. 15.2 suggests that this is true, but it is likely that here we will be able to find situations in which a very common mimic gains less than its much-suffering model loses, and yet advergence still occurs. In short, it is not the relative gain and loss to the species, or even to the individual model and mimic, but the *overall* pattern of selection *within* each species which determines the outcome of the arms race. Appealing as may be the idea of adaptive evolution biased by the assymetry of the conflict, this argument is likely to prove difficult to apply reliably and unambiguously when it comes to predicting the outcome of particular cases. Starting off with natural selection is surely more profitable.

But adaptationist arguments can form an invaluable starting point. Rothschild's argument connecting scarcity, sex-limitation and polymorphism in mimics, originally phrased in adaptationist terms, constitutes a profound insight into the evolutionary dynamics of Batesian mimics. As Berry (1982) says, high quality natural history is as indispensible for evolutionary theory now as it was in the days of Charles Darwin.

There are assuredly processes in evolution which occur at a higher level than is accounted for in classical population genetics (we have caught a glimpse of them in the possibility that species with certain types of mimicry are more prone than others to go extinct), but there is no indication from the study of mimicry that these are systemic speciational events of the type postulated by Goldschmidt and in the theory of evolution by jerks. Like the punctuational events postulated by Kipling himself ('Make me different from all other animals by five this afternoon'), these constitute, much more than the concept of adaptation, a *Just-So Story*.

REFERENCES

Barrett J.A. (1976) The maintenance of non-mimetic forms in a dimorphic Batesian mimic species. *Evolution*, **30**, 82–85.

Benson W.W. (1971) Evidence for the evolution of unpalatability through kin selection in the Heliconiinae (Lepidoptera). *American Naturalist*, **105**, 213–226.

Benson W.W. (1972) Natural selection for Müllerian mimicry in *Heliconius erato* in Costa Rica. *Science*, **176**, 936–939.

Benson W.W. (1977) On the supposed spectrum between Batesian and Müllerian mimicry. *Evolution*, **31**, 454–455.

Berry R.J. (1982) *Neo-Darwinism* (*Studies in Biology No. 144*). Edward Arnold, London.

Bertram H. (1966) Genetik des mimetischen Polymorphismus bei Schmetterlingen, *Zeitschrift für zoologische und Evolutions for schung*, **5**, 333–397.

Bisping R., Benz U., Boxer P. & Longo N. (1974) Chemical transfer of learned colour discrimination in goldfish. *Nature*, **249**, 771–773.

Brower J.VZ. (1960) Experimental studies of mimicry. IV. The reactions of starlings to different proportions of models and mimics. *American Naturalist*, **94**, 271–282.

Brower L.P. & Brower J.V.Z. (1972) Parallelism, convergence, divergence, and the new concept of advergence in the evolution of mimicry. In *Growth by Intussusception. Ecological Essays in Honor of G. Evelyn Hutchinson* (Ed. by E.S. Deevey), *Transactions of the Connecticut Academy of Arts and Sciences*, **44**, 57–67.

Brown Jr. K.S. (1981) The biology of *Heliconius* and related genera. *Annual Review of Entomology*, **26**, 427–456.

Bullini L., Sbordoni V. & Ragazzini P. (1969) Mimetismo mulleriano in popolazione italiane di *Zygaena ephialtes* (L.) (Lepidoptera, Zygaenidae), *Archivio Zoologico Italiano*, **44**, 181–214.

Carpenter G.D.H. (1946) Mimetic polymorphism. *Nature*, **158**, 277.

Charlesworth D. & Charlesworth B. (1976) Theoretical genetics of Batesian mimicry. II. Evolution of supergenes. *Journal of theoretical Biology*, **55**, 305–324.

Charlesworth B., Lande R. & Slatkin M. (1982) A neo-Darwinian commentary on macroevolution. *Evolution*, **36**, 474–498.

Clarke C.A. & Sheppard P.M. (1960) The evolution of mimicry in the butterfly *Papilio dardanus*. *Heredity*, **14**, 163–173.

Clarke C.A. & Sheppard P.M. (1962) The genetics of *Papilio dardanus*, Brown. IV. Data on race *ochracea*, race *flavicornis*, and further information on races *polytrophus* and *dardanus*. *Genetics*, **47**, 909–920.

Dawkins R. (1982) *The extended phenotype. The gene as the unit of selection*. Freeman, Oxford.

Dawkins R. & Krebs J.R. (1979) Arms races between and within species. *Proceedings of the Royal Society of London*, **253**, 483–498.

Duncan C.J. & Sheppard P.M. (1965) Sensory discrimination and its role in the evolution of Batesian mimicry. *Behaviour*, **24**, 269–282.

Ebinuma H. & Yoshitake N. (1982) The genetic system controlling recombination in the silkworm. *Genetics*, **99**, 231–245.

Eltringham H. (1909) An account of some experiments on the edibility of certain *Lepidopterous* larvae. *Transactions of the Entomological Society of London 1909*, 471–478.

Eltringham H. (1910) Edibility of lepidopterous larvae. *Proceedings of the Entomological Society of London 1910*, xxxi–xxxii.

Endler J.A. (1978) A predator's view of animal color patterns. In *Evolutionary Biology* (Ed. by M.K. Hecht W.C. Steere & B. Wallace), vol. **11**, pp. 319–364. Plenum Publishing Corp, New York.

Farris J.S., Kluge A.G. & Eckardt M.J. (1970) A numerical approach to phylogenetic systematics. *Systematic Zoology*, **19**, 172–191.

Fisher R.A. (1930) *The Genetical Theory of Natural Selection*. Clarendon Press, Oxford.

Ford E.B. (1971) *Ecological Genetics*, 3rd ed., Chapman and Hall, London.

Garcia J., McGowan B.K. & Green K.F. (1972) Biological constraints on conditioning. In *Classical conditioning. II: Current research and theory* (Ed. by A.H. Black & W.F. Prokasy), Appleton-Century-Crofts, New York.

Gibson D.O. (1980) The role of escape in mimicry and polymorphism: I. The response of captive birds to artificial prey. *Biological Journal of the Linnean Society of London*, **14**, 201–214.

Gilbert L.E. (1983) Coevolution and mimicry. In *Coevolution* (Ed. by D.J. Futuyma, & M. Slatkin). Sinauer, Sunderland, Mass.

Gittleman J.L. & Harvey P.H. (1980) Why are distasteful prey not cryptic? *Nature*, **286**, 149–150.

Goldschmidt R.B. (1940) *The Material Basis of Evolution*. Yale University Press, New Haven. Reprinted (1982) with an introduction by S.J. Gould.

Goldschmidt R.B. (1945) Mimetic polymorphism, a controversial chapter of Darwinism. *Quarterly Review of Biology*, **20**, 147–164, 205–230.

Goodale M.A. & Sneddon I. (1977) The effect of distastefulness of the model on the predation of artificial batesian mimics. *Animal Behaviour*, **25**, 660–665.

Gould S.J. (1980) Is a new and general theory of evolution emerging? *Paleobiology*, **6**, 119–130.

Gould S.J. (1982) *The Panda's Thumb*. Norton Paperback Editions, London.

Gould S.J. & Eldredge N. (1977) Punctuated equilibria: the tempo and mode of evolution reconsidered. *Paleobiology*, **3**, 115–151.

Gould S.J. & Lewontin R.C. (1979) The spandrels of San Marco and the Panglossian paradigm: a critique of the adaptationist programme. *Proceedings of the Royal Society of London B*, **205**, 581–598.

Hamilton W.D. (1971) Geometry for the selfish herd. *Journal of Theoretical Biology*, **31**, 295–311.

Harvey P.H., Bull J.J., Pemberton M. & Paxton R.J. (1982) The evolution of aposematic coloration in distasteful prey: a family model. *American Naturalist*, **119**, 710–719.

Hiam A.W. (1982) Airborne models and flying mimics. *Natural History*, **91(4)**, 42–49.

Hinton H.E. (1977) Subsocial behaviour and biology of some Mexican membracid bugs. *Ecological Entomology*, **2**, 61–79.

Huheey J.E. (1976) Studies in warning coloration and mimicry. VII. Evolutionary consequencies of a Batesian-Müllerian spectrum, a model for Müllerian mimicry. *Evolution*, **30**, 86–93.

Huheey J.E. (1980a) The question of synchrony or 'temporal sympatry' in mimicry. *Evolution*, **34**, 614–616.

Huheey J.E. (1980b) Batesian and Müllerian mimicry: semantic and substantive differences of opinion. *Evolution*, **34**, 1212–1215.

Ikin M. & Turner J.R.G. (1972) Experiments on mimicry: Gestalt perception and the evolution of genetic linkage. *Nature*, **239**, 525–527.

Jeffords M.R., Waldbauer G.P., & Sternburg J.G. (1980) Determination of the time of day at which diurnal moths painted to resemble butterflies are attached by birds. *Evolution*, **34**, 1205–1211.

Johnson M.S. & Turner J.R.G. (1979) Absence of dosage compensation for a sex linked enzyme. *Heredity*, **43**, 71–77.

Kimler W.C. (1983) Mimicry: views of naturalists and ecologists before the Modern Synthesis. In *Dimensions of Darwinism*. (Ed. by M. Grene) Cambridge University Press, New York.

Kimura M. & Ohta T. (1971) *Theoretical Aspects of Population Genetics*. Princeton University Press, Princeton.

Morrell G.M. & Turner J.R.G. (1970) Experiments on mimicry: I. The response of wild birds to artificial prey. *Behaviour*, **36**, 116–130.

Nicholson A.J. (1927) A new theory of mimicry in insects. *Australian Zoologist*, **5**, 10–104.

Nur U. (1970) Evolutionary rates of models and mimics in Batesian mimicry. *American Naturalist* **104**, 477–486.

Owen D.F. & Chanter D.O. (1972) Polymorphic mimicry in a population of the African butterfly, *Pseudacraea eurytus* (L.). (Lep. Nymphalidae). *Entomologica scandinavica*, **3**, 258–266.

Papageorgis C. (1975) Mimicry in neotropical butterflies. *American Scientist*, **63**, 522–532.

Pasteur G. (1982) A classificatory review of mimicry systems. *Annual Review of Ecology and Systematics*, **13**, 169–199.

Platt A. P. & Brower L.P. (1968) Mimetic verus disruptive coloration in intergrading populations of *Limenitis arthemis* and *astyanax* butterflies. *Evolution*, **22**, 699–718.

Poulton E.B. (1912) Darwin and Bergson on the interpretation of evolution. *Bedrock*, **1(1)**, 48–65.

Poulton E.B. (1931) Two specially significant examples of insect mimicry, *Transactions of the Entomological Society of London*, **79**, 395–398.

Prance G.T. (1982) *Biological Diversification in the Tropics*. Columbia University Press, New York.

Punnett R.C. (1915) *Mimicry in Butterflies*. Cambridge University Press, Cambridge.

Rettenmeyer C.W. (1970) Insect mimicry. *Annual Review of Entymology*, **15**, 43–74.

Rothschild M. (1971) Speculations about mimicry with Henry Ford. In *Ecological Genetics and Evolution* (Ed. by E.R. Creed), pp. 202–223. Blackwell Scientific Publications, Oxford.

Rothschild M. (1980) Mimicry, butterflies and plants. *Symbolae Botanicae Upsaliensis*, **22(4)**, 82–99.

Rothschild M. (1981) The mimicrats must move with the times. *Biological Journal of theLinnean Society of London*, **16**, 21–23.

Rothschild M., von Euw J. & Reichstein T. (1972) Some problems connected with warningly coloured insects and toxic defense mechanisms. In *Impulse eines Landes extremer Bedingungen für die Wissenschaft*, pp. 135–158. Mitteilungen der Basler Afrike Bibliographien.

Sargent R.C. & Turner, J.R.G. (1981) Gestalt perception by birds as individuals and in flocks. *American Naturalist*, **117**, 99–103.

Sbordoni V., Bullini L., Scarpelli G., Forestiero S. & Rampini M. (1979) Mimicry in the burnet both *Zygaena ephialtes*: population studies and evidence of a Batesian–Müllerian situation. *Ecological Entomology*, **4**, 83–93.

Schuler W. (1974) Die Schutzwirkung künstlicher Batesscher Mimikry abhängig von Modellähnlichkeit und Beuteangebot. *Zeitschrift für Tierpsychologie*, **36**, 71–127.

Schuler W. (1982) Zur Funktion von Warnfarber: die Reation junger Stare auf wespenänlich schwarz-gelbe Attrapen. *Zeitschrift für Tierpsychologie*, **58**, 66–78.

Sheppard P.M. (1961) Recent genetical work on polymorphic mimetic Papilios. In *Insect Polymorphism*. (Ed. by J.S. Kennedy) pp. 20–29. Royal Entomological Society of London.

Sheppard P.M. (1975) *Natural Selection and Heredity*. 4th ed. Hutchinson, London.

Sheppard P.M. & Turner J.R.G. (1977) The existence of muellerian mimicry. *Evolution*, **31**, 452–453.

Sheppard P.M., Turner J.R.G. Brown jr. K.S., Benson W.W. & Singer M.C. (1983) Genetics and the evolution of Muellerian mimicry in *Heliconius* butterflies. *Philosophical Transactions of Royal Society of London B* in press.

Silberglied R. (1983) Visual communication and sexual selection among butterflies. In *Butterfly Biology*. (Ed. by R.I. Vane-Wright & P.R. Ackery) Academic Press. London.

Smiley J.T. (1978) The host plant ecology of *Heliconius* butterflies in northeastern Costa Rica. PhD. thesis, University of Texas at Austin, Texas.

Stanley S.M. (1979) *Macroevolution. Pattern and process*. Freeman, San Francisco.

Turner J.R.G. (1975) Communal roosting in relation to warning colour in two heliconiine butterflies (Nymphalidae). *Journal of the Lepidopterists Society*, **29**, 221–226.

Turner J.R.G. (1977) Butterfly mimicry: the genetical evolution of an adaptation. In *Evolutionary Biology* (Ed. by M.K. Hecht, W.C. Steere & B. Wallace) vol. 10, pp. 163–206. Plenum Publishing Corp., New York.

Turner J.R.G. (1978) Why male butterflies are non-mimetic: natural selection, sexual selection, group selection, modification and sieving. *Biological Journal of the Linnean Society*, **10**, 385–432.

Turner J.R.G. (1979) Genetic control of recombination in the silkworm. I. Multigenic control of chromosome. 2. *Heredity*, **43**, 273–293.

Turner J.R.G. (1980) Oscillations of frequency in batesian mimics, hawks and doves, and other simple frequency dependent polymorphisms. *Heredity*, **45**, 113–126.

Turner J.R.G. (1981) Adaptation and evolution in *Heliconius*: a defense of neoDarwinism. *Annual Review of Ecology and Systematics*, **12**, 99–121.

Turner J.R.G. (1983a) Mimetic butterflies and punctuated equilibria: Some old light on a new paradigm. *Biological Journal of the Linnean Society*, **20**.

Turner J.R.G. (1983b) Mimicry: the palatability spectrum and its consequences. In *The Biology of Butterflies* (Ed. by R.I. Vane-Wright & P.R. Ackery). Academic Press, London.

Turner J.R.G. (1983c) The evolutionary dynamics of batesian and mullerian mimicry: similarities and differences. *Colloques Internationaux du Centre National de la Recherche Scientifique* (in press).

Turner J.R.G. (1983d) 'The hypothesis that explains mimetic resemblance explains evolution': the Gradualism Saltationism schism. In *Dimensions in Darwinism* (Ed. by M. Grene) Cambridge University Press, New York.

Vane-Wright R.I. (1980) Mimicry and its unknown ecological consequences. In *The Evolving Biosphere* (Ed. by P.L. Forey). pp. 157–168. British Museum and Cambridge University Press, London.

Vane-Wright R.I., Ackery P.R., & Smiles R.L. (1977) The polymorphism, mimicry, and host plant relationships of *Hypolimnas* butterflies. *Biological Journal of the Linnean Society* **9**, 285–297.

Waldbauer G.P. & Sheldon J.K. (1971) Phenological relationships of some aculeate hymenoptera, their dipteran mimics, and insectivorous birds. *Evolution*, **25**, 371–382.

Wickler W. (1968) *Mimicry in Plants and Animals*. Wiedenfield and Nicolson, London.

Wicklund C. & Järvi T. (1982) Survival of distasteful insects after being attacked by naive birds: a reappraisal of the theory of aposematic coloration evolving through individual selection. *Evolution*, **36**, 998–1002.

Wiens D. (1978) Mimicry in plants. In *Evolutionary Biology* (Ed. by M.K. Hecht, W.C. Steere & B. Wallace), vol. **11**, pp. 365–403. Plenum Press, New York.

Williamson P.G. (1981) Palaeontological documentation of speciation in Cenozoic molluscs from Turkana Basin. *Nature*, **293**, 437–443.

Wright S. (1949) Adaptation and selection. In *Genetics, Paleontology and Evolution*. (Ed. by G.L. Jepson, E. Mayr & G.G. Simpson). Princeton University Press, Princeton.

16. GENETIC DIVERSITY AND ECOLOGICAL STABILITY

G.S. MANI
Schuster Laboratory, Department of Physics, University of Manchester, Manchester M13 9PL

INTRODUCTION

The overwhelming diversity of life, at the population, as well as at the species level, provide ample evidence for the existence of a large amount of individual variability. Biological evolution is the result of forces acting upon the existing variability. The nature of the variability at any point in time depends on the earlier history and on the type of environmental interaction present at that time.

Biological diversity is exhibited both at the intrapopulation (individual) level as well as at the interpopulation (geographic) level. The latter case refers to differences among individuals representing separate populations and a measure of the variability is obtained by comparing the averages of the variation taken over single populations. The intrapopulation variation of a character is usually either quantitative or polymorphic. The pattern is referred to as quantitative when there exists a continuous or graded sequence of morphological types, such as size or weight variation. Polymorphic variation is said to occur when a few distinct classes of easily identifiable morphs exist within the population, such as distinct blood groups. Often the distinction between these two types are blurred and there can exist large regions of overlap. Traditionally, experimental biologists have studied the variation in the population at the morphological level in terms of the broad classifications mentioned above. The observed variation is at the phenotypic level, which encompases the morphological, physiological, ecological and behavioural attributes of the individual at any or at all stages of its life cycle. Although the existence of a large amount of variation within the population had been recognized, the explanation for the maintenance of this variation interms of the underlying genetics of the system has often proved difficult. Part of this difficulty stems from the fact that the observed morphological variation is often polygenic in character.

The difficulties associated with polymorphism observed at the morphological (macroscopic) level, namely, its relationship to the underlying genetic structure is to some extent circumvented by studying the polymorphism at the enzyme level. In the past two decades this has been made possible through advances in electrophoretic techniques and this has led to an information explosion for enzyme polymorphism. The data spans over a large variety of species, from *Drosophilla* to man. The study of polymorphism at the enzyme level has been very popular, primarily because the method yields better, unambiguous quantitative data compared with the information obtainable at the macroscopic level. Further, since a particular enzyme is associated with a single gene at a single locus, the relation between the enzyme and the DNA substructure is much simpler than the equivalent relation between morphological polymorphism and molecular genetics.

The vast amount of data on enzyme polymorphism indicate the following general features (cf. Nevo 1978; Nevo *et al.* 1983 for a review of the data):

1 All species studied exhibit considerable genetic diversity with polymorphism averaging around 25% for any population.

2 The mean heterozygosity per locus varies between 4 and 25%. For large population sizes, the heterozygosity seems to attain the limiting value of around 25%.

3 Between 20 and 50% of all loci studied in any population are heterozygous.

4 The distribution of allelic frequencies at any locus is often U-shaped, with relatively high frequencies occuring in the neighbourhood of zero and unity.

5 For a given population, the average number of alleles at any locus varies from one to four. In comparing geographically separated populations, on a global scale, anywhere between ten and thirty alleles can be recognized at every polymorphic loci. Most geographically distinct population have often one common allele occuring with relatively high frequency while the set of alleles with smaller frequencies show very little correlation among the different populations. Further, the values of genetic similarity extracted from the data indicate quasi randomness in relation to spatial distribution.

6 Often, small populations, such as those on small islands, or laboratory populations, show as much variation as large continental populations.

7 The data shows some degree of correlation with ecological parameters, such as habitat, environmental conditions, resources etc.

From what has been said till now, it is evident that considerable diversity is observed both at the macroscopic and at the microscopic level. In fact, recent studies of the DNA substructure indicates that the variability can be continued down to the DNA level. Two fundamental questions can be asked.

(i) What mechanism generates diversity?

(ii) Through what mechanism is the diversity maintained?

There is general consensus to the answer to the first question. There exists only one mode for generating diversity, namely, mutation. No doubt mutation is not a simple or single mechanism; it can be a point mutation when a single base pair in the DNA is changed, or it can affect the structure of the chromosome through the processes of inversions, translocations, duplications and deletions. In the case of enzyme polymorphism, point mutation is the main process for generating diversity. There is much less agreement among biologists concerning the second question. The opinion regarding the maintenance of diversity is divided into two main camps, the selectionists and the neutralists. There exists no uniform consensus even within either of these two groups. I shall now attempt to briefly summarize the basic contentions of both the selectionists and the neutralists and some of the difficulties encountered in both the models. The rest of the chapter is then devoted to the development of a model based on ecological genetics, which attempts in some way to bridge the gap between the extreme selectionists and the neutralists. The model could also provide some insight into the problems of ecological stability.

The starting point for the selectionists is the Darwinian argument that evolution proceeds as a result of natural selection based on a struggle for existence and on the survival of the fittest. In the Darwinian context, the evolutionary force is generated both through the phenotypic variation in the population and through the interaction with other components in the ecosystem. On the other hand, the selectionists' models, imbedded in population genetics, have often neglected extremely important ecological factors. One of the main modes in the selectionists' models for maintaining polymorphism is through balancing selection. The formal aspect of balancing selection can be written as the following equation.

$$\Delta q = \frac{q(1-q)}{2\bar{W}}\frac{d\bar{W}}{dq}. \tag{1}$$

Here q is the frequency of the allele under consideration, Δq the change in the frequency per generation and $\bar{W}$ is the mean fitness for the whole population. Though eqn (1), as written, refers to the case of two alleles at a single locus, the extension to n-alleles at a single locus is straight forward and yield equations with very similar structure. The mean fitness $\bar{W}$ is obtained by ascribing relative fitness values W_{ij} to the genotypes A_iA_j. The fitness values W_{ij} for the genotypes are taken in some way to describe the relative number of viable offsprings produced per generation by the parental genotypes. It bears little or no relation to the basic Darwinian idea of natural selection, since it by-passes the fact that the struggle for existence implies a struggle against the external ecological world. Thus to ascribe a constant fitness parameter to each genotype yields a poor representation of the real system. Also in eqn (1), one ignores the fact that populations are finite in size and that they are constantly under mutation pressure. As mentioned earlier, it is through mutation that diversity is generated; on the other hand, random drift generated through sampling errors in finite populations tend to reduce the variability. Inspite of all these shortcomings, eqn (1) with various cosmetic changes has been used to generate a wide and rich variety of scenarios. The various limitations to the model have been recognized and discussed over the past few decades, but there has been very little attempt at a detailed treatment of the subject. This has, in part, led to the claim that the selectionists' models are not quantitative and have no predictive value (Kimura 1979).

Referring to eqn (1), one can easily demonstrate that the condition for maintaining polymorphism through balancing selection, with constant fitnesses, is the selective advantage of the heterozygotes over the homozygotes. How does the heterozygote advantage arise from so many allele combinations, including multiple alleles? A possible answer (Sheppard 1975) is that the heterozygote advantage arises through the modification of the phenotypic expression as a result of selection of modifiers. This argument is an extension of Fisher's Theory of the Evolution of Dominance. Experimentally, Ford (1971) has demonstrated that such modification of expression can occur within the span of a few generations. It can be demonstrated that changes in the frequency of alleles with modifying effect, situated at unlinked loci, producing the large amount of polymorphism observed, can only occur if the genome contains a ready supply of selectively neutral genic variability, (Ewens 1979). If, on the other hand, the modifier is closely linked to the locus it modifies (O'Donald & Barrett 1973; Charlesworth & Charlesworth 1976), then the question arises of how common are such close linkages to be expected.

The difficulty of the evolution of heterozygote advantage may rest in the fact that the model, as represented by eqn (1), has not taken mutation into account. In a series of detailed calculations which include both the effects of mutation and the effects of finite population, I have been able to show that it is possible to maintain a large amount of polymorphism. In this case, when there is no constraint on the fitnesses of the genotypes, the polymorphic alleles exhibit heterozygote advantage. The problem is thus no longer to produce plausible mechanism for the evolution of heterozygote advantage; it is more concerned with the fact that most of the polymorphism observed in nature rarely exhibit heterozygote advantage.

There are two mechanisms suggested for the maintenance of polymorphism without the need for heterozygote advantage. It is now well established that in the presence of migration, polymorphism can occur without the need for heterozygote advantage. In a recent series of papers (Cook & Mani 1980; Mani 1980, 1982), it has been demostrated that the observed distribution of polymorphism of the melanic morphs of the moth species *Biston betularia* over England and Wales can be adequately explained on the basis of balance between migration and selection, without the need for invoking heterosis. It is equally clear that polymorphism can be maintained through frequency- and density-dependent selection, a view which has been forcibly put forth in a recent paper by Clarke (1979). Clarke advocates that frequency-dependent selection may be the cause of the maintenance of a large amount of genetic diversity.

Recent detailed calculation of various types of selection models (G.S. Mani, unpublished), which take into account the effects of finite population and mutation, indicate that such models can produce large amounts of polymorphism. On the other hand, they fail to reproduce in detail many features of the polymorphism observed at the enzyme level, especially the shape of the allelic frequency distribution and the distribution of heterozygosity. Also they predict heterozygote advantage, even in the presence of frequency-dependent selection, a fact that is not observed at the enzyme level. Further, the parameters in the model are ad hoc and bear little relation to the Darwinian concept of fitness and struggle for existence. A further difficulty with such population genetics models of selection is the so-called 'load problem'. If selection is to act at every locus, then the system with thousands of loci would require an almost impossible reproductive effort, provided the total selective component for the whole system is obtained through a multiplicative model (Lewontin 1974). Various models, such as the truncation models (Wills 1981) have been proposed to avoid the load problem, but all these models have a certain built in artificiality. It is also not very clear how good these models are in describing the features of enzyme polymorphism. The problem of 'load' is discussed further below.

The observation of such extensive amounts of polymorphism at the enzyme level motivated Kimura (1968a, b) to suggest the possibility of a very large class of effectively neutral alleles. The birth of the neutralist controversy started in the following year with the publication by King & Jukes (1969) of their paper with the inflamatory title 'Non-Darwinian evolution'. The extreme neutralist's position is that a major part of the observed polymorphism arises from alleles that are selectively neutral, all the deleterious mutations having already been removed through purifying selection. The polymorphism is maintained through a balance between mutation and drift. This extreme neutralist model is very appealing since it yields a large number of predictions based on only two parameters, namely the effective population size N_e and the mutation rate v. With these two parameters alone, the model predicts the following:

1 The effective number of alleles at a single locus is given by

$$n_e = M + 1, \quad M = 4N_e v \tag{2}$$

2 The expected heterozygosity at a single locus in an equilibrium population is given by:

$$H = M/(M + 1) \tag{3}$$

with the variance

$$V_H = 2M/\{(M + 1)^2(M + 2)(M + 3)\} \tag{4}$$

3 The expected number of alleles in the frequency interval x and $x + dx$ (the allele frequency distribution) is given by:

$$\Psi(x)dx = M(1 - x)^{M-1}x^{-1}dx \tag{5}$$

All the above relations are for the case when an infinitely large number of alleles are possible at any locus (the infinite allele model). The expressions have to be suitably modified for a finite number of possible alleles. Since the two parameters, N_e and v, always occur in the combination $4N_e v$, one has effectively a one-parameter model.

It is not my intention, due to lack of space, to give a detailed appraisal of the neutralist's position. Interested readers could consult many of the review articles written on the subject, especially the monograph by Kimura & Ohta (1971), the book devoted to molecular genetics by Nei (1975) and the article defending the neutralist position by Nei at the Vito Volterra Symposium, 1980. In a series of papers Nei and his collaborators (Nei, Fuerst & Chakraborty 1976, 1978; Fuerst, Chakraborty & Nei 1977; Chakraborty, Feurst & Nei 1978, 1980) have shown that a large part of the data on enzyme polymorphism can be explained in terms of the neutralist model, provided one makes the following additional assumptions, some of which having some experimental justification:

1 The mutation rate varies among the loci. This is needed to explain the observed correlation between the mean heterozygosity and its variance as well as to fit the observed allele frequency distribution better.

2 A majority of polymorphic alleles are slightly deleterious compared with a typical allele and in large populations a mutation-selection balance is established. This assumption was due to Ohta (1973) and was introduced to explain that the observed heterozygosity is always less than 25% or so, even for very large populations. Referring to eqn (3) it is easily seen than in the extreme neutralist model, when $M = 4N_e v$ is very large, the heterozygosity is almost close to unity. Nei (1980) has objected to this suggestion on the ground that though it could reproduce the data concerning heterozygosity,

it does not yield constant rate of gene substitution for a wide range of organisms. Also the allele frequency distribution will not be U-shaped for large populations.

3 The bottle-neck hypothesis. To explain the small value of heterozygosity in large populations, it is assumed that the populations have gone through a bottle neck in some recent past. The effect of this would be to drastically reduce the heterozygosity at the time of the bottle neck and what is being observed today is the extremely slow recovery, yielding the low values for heterozygosity. If a sequence of past bottle-necks are responsible for the low values of heterozygosity exhibited by large populations, then why do cosmopolitan deep-sea invertibrates, for example, whose population sizes need not be very small, show heterozygosity of the order of 20%? In this case one expects no large-scale geological upheavals to create bottle-necks. Secondly, for organisms differing widely in their population sizes and in their generation time, how is it that a single bottle-neck, or multiply recurring bottle-necks are so finely adjusted as to yield a reasonably smooth variation of heterozygosity with population size, reaching the limiting value of around 25%?

Thus, though the neutral model started as a one parameter model, many additional, and in some cases ad hoc, assumptions had to be introduced to explain the observed data at the enzyme level. What the neutral model has demonstrated is that, in general, the genes in populations exhibit very weak or no selection and that the strong selection situation is not common. The main objection by selectionists to the neutral model stems from the fact that the latter is seen to be non-Darwinian in spirit. Though the selectionists give lip service to the Darwinian principle of natural selection and the struggle for existence, in reality the neo-Darwinian models rarely take into account the principles originally stated by Darwin. I suspect that a large part of the 'neutral-selection' controversy arises through a misunderstanding of what selection or neutrality implies. It is not an inherent property of the gene to be either neutral or selective. The gene exhibits either of these properties depending on the environment into which it is placed. What is neutral or near neutral in one environment can become extremely selective with respect to other genes in another environment. Also, neutral theory does not claim the genes to be functionless. In a particular environment that is reasonably stable (here I define the environment to encompass the interactions from the rest of the ecosystem on the individual or population), it is quite conceivable that at any locus a large number of genes each producing slightly different enzymes, could be supported and that these genes are so selected through the constraints of the whole system that the inherent differences do not alter materially the morphological adaptation. When a sudden change takes place, where the time constant of the change is much smaller than the intrinsic time constant of the system, selection would operate and the system, if left alone, would attain a new state of near neutrality. Thus the central question to my mind is whether the constraints of the rest of the ecosystem could produce, from the source of mutations containing widely differing selective values, the set of nearly neutral or weakly selective alleles? Would such a model reproduce the data obtained in the study of enzyme polymorphism? Would such a model be capable of including the effects of strong selection under certain circumstances? In the next two sections I describe such a model. The ecosystem being an exceptionally complex system, any model that one constructs is necessarily a feeble representation of the real world. What will be attempted is to show that the influence of the external system can produce the weak selection limit that is commonly observed at the enzyme level. The model is robust, and small changes in the assumptions do not radically alter the conclusions. The main aim of this work is to emphasize that the mere observation of a large number of apparently near neutral alleles maintained in a population need not indicate a non-Darwinian mechanism.

THE ECOLOGICAL GENETIC MODEL

No population exists in isolation, with unlimited resources and unchanging environment. To include in any theory the total effects of the rest of the environment on the population is near impossible. I shall assume for this simple model that the effects of the rest of the ecosystem can be approximated into a single effect of population regulation. This is a very drastic approximation and some ideas for building a more realistic model will be discussed later.

Consider a diploid, sexually mating population with n alleles $A_1 \ldots A_n$ at a single locus. There would be $n(n+1)/2$ distinct genotypes represented by A_iA_j. Let N_{ij} be the population size for the genotype A_iA_j and let N_T be the total population. Let R_{ij} be the intrinsic growth rate and K_{ij} the carrying capacity for the genotype A_iA_j. Here the carrying capacity K_{ij} is taken to represent the effects of the rest of the environment. Then we can define the absolute selective value W_{ij} in terms of the individual's contribution to the population growth as follows (Roughgarden 1979):

$$W_{ij} = F(R_{ij}, K_{ij}, N_T) \tag{6}$$

where $F(.)$ is a density-dependent population regulating function. Many forms for the function $F(.)$ have been suggested in the ecological literature (cf. for example, May 1976 and the references given there). I have chosen two most commonly quoted forms for $F(.)$, namely

$$F(R_{ij}, K_{ij}, N_T) = 1 + R_{ij}\left\{1 - \frac{N_T}{K_{ij}}\right\} \tag{7}$$

$$= \text{Exp}\left\{R_{ij}\left(1 - \frac{N_T}{K_{ij}}\right)\right\} \tag{8}$$

inorder to see if the model is very sensitive to the form of the function used.

In terms of the genotypic selective value W_{ij}, we can write the following equations for the evolution of the gene frequency and of the total population:

$$p_i(t) - p_i(t-1)W_i\cdot(t-1)/\bar{W}(t-1) \tag{9}$$

$$N_T(t) = \bar{W}(t-1)N_T(t-1) \tag{10}$$

$$W_i\cdot(t) = \sum_j p_i(t)W_{ij}(t) \tag{11}$$

$$\bar{W}(t) = \sum_i \sum_j p_i(t)p_j(t)W_{ij}(t) \tag{12}$$

where $p_i(t)$ is the frequency of the allele A_i in generation t and $\bar{W}$ is fitness averaged over the whole population.

Roughgarden (1979) has argued that R, the intrinsic growth rate and K, the carrying capacity have to be negatively correlated. His argument can be stated as follows. A large value of R implies that the organism rapidly allocates the energy it has aquired to the production of offspring. On the other hand, the individual with high K would defer allocation of energy for reproduction till later, and thus would utilize the energy in surviving under crowded conditions. Since energy and time in the growing season is limited, the organism cannot simultaneously have high R and K. It has either high R (and low K) or high K (and hence low R), but not both. To take this into account, we assume the following simple linear relationship between R and K. It must be stressed that this linear form is chosen for convenience and may not represent the situation in the field. As will be seen later, the conclusions of the model are insensitive to the exact form of the correlation between R and K; also I feel that more sophisticated forms are just a luxury for the simple model discussed here.

$$K_{ij} = \{\beta R_{ij} + 1.0\}K_0 \tag{13}$$

where for negative values of β, R and K are negatively correlated in accordance with Roughgarden's arguments given above. For $\beta = 0$, from eqns (6)–(8) we see that one obtains the nearly neutral limit. In eqn (13), K_0 is a constant corresponding to the carrying capacity when $R_{ij} = 0$, that is, when the intrinsic growth rate is at the replacement level. For reasons that will become clearer later, we define K_0 to be the population size.

As mentioned earlier, for finite populations one has to include the effect of genetic drift arising from sampling errors. I shall not give here the mathematical details of how this can be done, except to point out that the effects of mutation and drift can be calculated using the diffusion approximation; the equations that one uses are the so-called Kolmogorov Backward Equation. Details of the diffusion approximation are well doccumented in Crow & Kimura (1970). The equations are solved numerically on a computer using a method due to Itoh (1979).

The concept of the population size is not very transparent. The model, both through the eqn (9) and through the method of handling the question of genetic drift, assumes a perfect random mating situation. Thus the population size that enters into the model cannot be the total population of the species distributed over a very wide geographical range. Also it is not very evident what size the population has had during its evolutionary history. One avoids these embarassing questions in theoretical models by considering an 'effective population size' which is defined to be an equivalent population with perfect random mating that would yield the same results as real populations. Defined this way, from spatial consideration alone, one can say that the effective population size must be related to the populations that are found within a geographical area that corresponds in some way to the migration aspects of the individuals in the population. Though the dispersal distance indicates some measure of the population size to be used in the model, it still does not yield an unambiguous value for two reasons. First, because there can be non-random modes of mating within the population, the effective size has to be smaller by some factor compared with the actual size. Second, the present size of the population within a dispersal distance may have no bearing on the population size in evolutionary time scale. This fundamental question concerning population size is avoided in the model by calculating the effects of the model for various effective population sizes. For convenience, we define K_0 of eqn (13) to be our population size though the total population N_T might differ from this value. In most cases N_T is not very different from K_0.

The next problem is the choice of the mutation rate. For point mutations this is estimated to be roughly of the order of 10^{-7} per locus per year. Taking into account all types of mutations, this figure could easily be one to two orders of magnitude higher. In the present case I have chosen as a compromise the value of 5×10^{-6} per locus per generation. Since the generation times differ by one to two orders of magnitude, a few calculations were done with lower values of mutation rate. These are discussed later.

The computer simulation proceeds as follows. The initial conditions at

$t = 0$ is determined by chosing two alleles at a single locus in a diploid population, with one of the alleles being at mutant frequency. The initial total population is arbitrarily chosen to be $N_T = 0.1K_0$, for a given choice of population size K_0; it was found that the long-term evolutionary characteristics are not influenced by the choice of N_T. The intrinsic growth rates. R_{ij} for the three initial genotypes are chosen from a random distribution. The random distributions used are discussed later. The value of the correlation parameter β, (eqn 13), is an input quantity and thus K_{ij}'s are determined from the R_{ij}'s. For each generation the sequence of events is given by selection–mutation–segregation. For a population size N_T, a single mutation has a frequency of $1/2N_T$. Hence the occurrence of mutations in any generation, for a mutation rate of v, is treated stochastically. As explained earlier, the effective population sizes are small compared to the total global population. To obtain mean values of observables and variances for the population, an ensemble of 100 populations, each with different starting conditions and each with different stochastic pathways, is considered. Each population is allowed to evolve to 100 000 generations and the observables for each population are calculated as the mean over the last 5000 generations. The mean and the variances of the observables are then evaluated from the ensemble of 100 populations.

It is well known that the linear density-dependent regulation that is imbedded in eqn (10) can produce in the total population dynamics exhibiting cycles and chaos, depending on the value of the parameter R. Also, for $F(.)$ given by eqn (7), R_{ij} must be within the interval $(-1, 3)$ to ensure that the population size N_T is positive at all times. The type of dynamics obtained for various values of R is summarized in Table 16.1 (May 1976). As explained earlier, the values of R_{ij} for the genotypes of each mutant is obtained from a random distribution. Two forms of random distribution were used, the uniform distribution and the distribution with correlated heterozygotes.

(i) *Uniform distribution.* In this case the values of R_{ij}, both for the

TABLE 16.1. Stability conditions for density-dependent regulation given by eqns (7) and (8)

Equation (7)	Equation (8)	Stability conditions
$R < 0$	$R < 0$	$N = O$ stable node
$0 < R < 1$	$0 < R < 1$	$N = K$ stable node
$1 < R < 2$	$1 < R < 2$	$N = K$ stable focus
$2 < R < 2.449$	$2 < R < 2.526$	2-point cycle
$2.449 < R < 2.570$	$2.526 < R < 2.692$	Higher 2^k bifurcations
$2.570 < R < 3.000$	$2.692 < R$	Chaotic behaviour

homozygotes and the heterozygotes were chosen from a uniform distribution in the interval $(-1, 3)$ for eqn (7) and $(-1, 8)$ for eqn (8).

(ii) *The correlated heterozygote distribution.* In this case the values R_{ii} for the homozygotes are chosen from a random uniform distribution as above. The R_{ij} values for the heterozygotes were obtained from the following equation:

$$R_{ij}(i \neq j) = \tfrac{1}{2}\{R_{ii} + R_{jj}\}\{1 + \lambda(0, \sigma^2)\} \quad (14)$$

where $\lambda(0, \sigma^2)$ is a normally distributed random variable with mean zero and variance σ^2.

The population sizes K_0 used were 100, 500, 5000, 50 000 and 500 000. Two types of calculations were done. In the constant environmental model, the values of R_{ij} and K_{ij}, once chosen, were kept constant throughout the rest of the calculations. In the variable environmental model, the effects of the fluctuations in the environment are supposed to be reflected in the values of K_{ij} through the following equation:

$$K_{ij}(t) = K_{ij}(0)\{1 + \eta_{ij}(0, \sigma_e^2)\} \quad (15)$$

where $\eta_{ij}(0, \sigma_e^2)$ is a normally distributed random variable with zero mean and variance σ_e^2 and $K_{ij}(0)$ is the initial choice of K_{ij} when the mutant first appeared. The subscripts (ij) are used for η to emphasize that the model assumes that the genotypic fluctuations are uncorrelated, but all have the same variance σ_e^2, representing the fluctuations in the environment.

I shall now discuss some of the results obtained from this model and its relevance to ecology and genetics. Only a small fraction of the results obtained will be discussed. A detailed discussion of all the results will be published elsewhere.

WHY IS CHAOTIC MOTION RARE IN NATURAL POPULATIONS?

From the available litterature, Hassel *et al.* (1976), have analysed the data for twenty-four field populations and four laboratory populations of seasonally breeding anthropods with discrete, non-overlapping generations. The population dynamics was analysed using the two parameter equation:

$$N(t+1) = \lambda N(t)(1 + aN(t))^{-\beta} \quad (16)$$

This is defined to be a two parameter equation, since the stability conditions depend only on the two parameters λ and β. Discussion of the stability criteria can be found in May (1976). For the field populations, they find that around 88% of the populations lie in the region of monotonic damping, around 8% in the region of damped oscillations and only one of the twenty-four populations yield values of λ and β, indicating a two-point stable limit cycle dynamics. There was no field population that showed a dynamics of oscillations with higher than two-point cycle or chaos. On the other hand, there was one population of the four laboratory populations studied that indicated chaotic dynamics. Stubbs (1977), using a different form for the density regulating function, arrived at the same type of conclusion from a similar compilation of populations. More recently, Bellows (1981) had subsumed from the data used by Hassel *et al.* (1976) and by Stubbs (1977), fourteen best documented populations and analysed them with density-regulating function slightly different from that used by the previous two workers and arrived at very similar conclusions. Thomas, Pomerantz & Gilpin (1980) analysed the populations of twenty-seven species of *Drosophila* at two different temperatures using the following two parameter model:

$$N(t+1) = N(t)\,\mathrm{Exp}\{R(1 - \{N(t)/K\}^{\theta}\} \tag{17}$$

They observed that *all* the populations appear to have a stable fixed point. They argue that this tendency towards stability is a direct result of natural selection.

From the experimental data presented above, it may be tempting to make the generalization that all natural populations exhibit stable behaviour and that cyclic or chaotic behaviour is rare in wild populations. May (1976) gives two reasons why caution should be exercised in making such generalizations. He points out that there may exist inherent bias in the data. Chaotic populations fluctuate rapidly and may thus be very rare in most years. These then stand a higher chance of being excluded in the population studies than the populations that are reasonably stable from year to year. Also, no population is isolated and the interactions with other populations would not be included in the simple models used in the analysis of the data. Inspite of these reservations one could say that a large sample of natural populations do exhibit some degree of stability and enquire whether the ecological genetic model presented above predict similar features. This will not provide an absolute proof of stability for natural populations but would at least indicate the effects of constraints postulated, on the population dynamics.

Referring to eqn (10), the population dynamics is described by

$$N(t+1) = N(t)\{1 + \langle R\rangle - \langle R/K\rangle N(t)\} \tag{18}$$

where the average of any variable x_{ij} corresponding to the genotype A_iA_j is given by

$$\langle x\rangle = \sum_{ij} x_{ij} p_i p_j \tag{19}$$

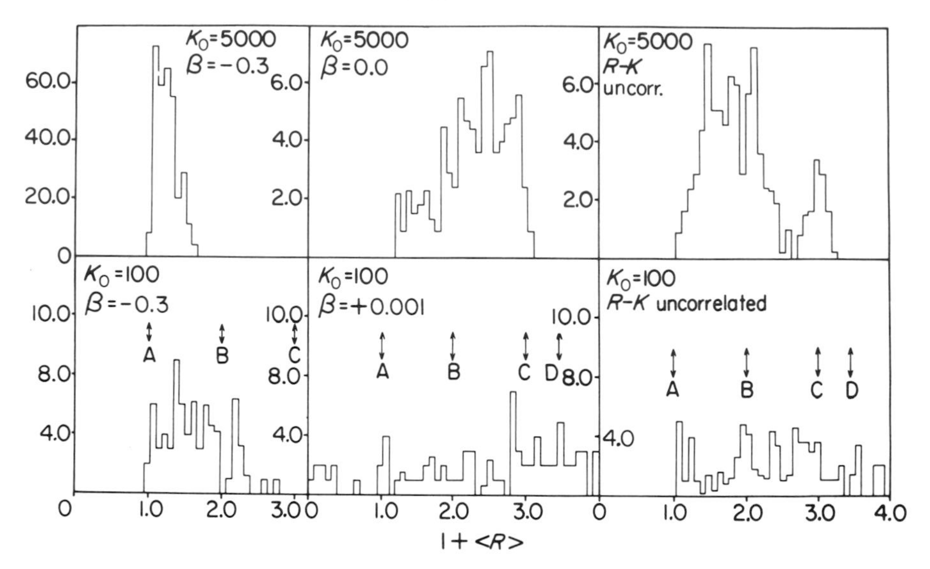

FIG. 16.1. The distribution of $1 + \langle R \rangle$ for population sizes $K_0 = 100$ and 5000 using the density-regulating function given by eqn (7). Three cases are shown for each population: two cases are when R and K are correlated with $\beta = -0.3$ and 0.0 and one case with R and K uncorrelated. The arrows marked A, B, C and D determine the stability criteria listed here; 0.0–A, population goes to extinction; A–B, stable node; B–C, stable focus; C–D, stable two-point cycle; above D, higher order cycles and chaotic dynamics.

The form of eqn (7) is used in eqn (18). A similar equation would result if one used the form of eqn (8). The stability criteria for eqn (18) are determined by the value of $\langle R \rangle$ alone, as shown in Table 16.1.

The distribution of $\langle R \rangle$ for the 100 populations studied, for population sizes of 100 and 5000, with no environmental fluctuation, is shown in Fig. 16.1. In this figure, the form of eqn (7) was used. In Fig. 16.2, the results for the exponential form given by eqn (8) are shown. In comparing these two figures one sees:

1 The exact form for the density-dependent regulation is not important, both types yielding very similar results.

2 For small populations, there is a larger probability for the population to enter into the chaotic regime.

Since the two forms of eqns (7) and (8) yield very similar results, in what follows we shall only discuss the results obtained using eqn (7). The variation of $\langle R \rangle$ with β is shown in Fig. 16.3. It is seen from this figure that the values of $\langle R \rangle$ are very insensitive to the value of β chosen except in the neighbourhood of $\beta = 0$. The distribution of $\langle R \rangle$ with 10% environmental fluctuation per generation, $(\sqrt{\sigma_e^2} = 0.1)$, for population sizes of 100 and 5000 are shown

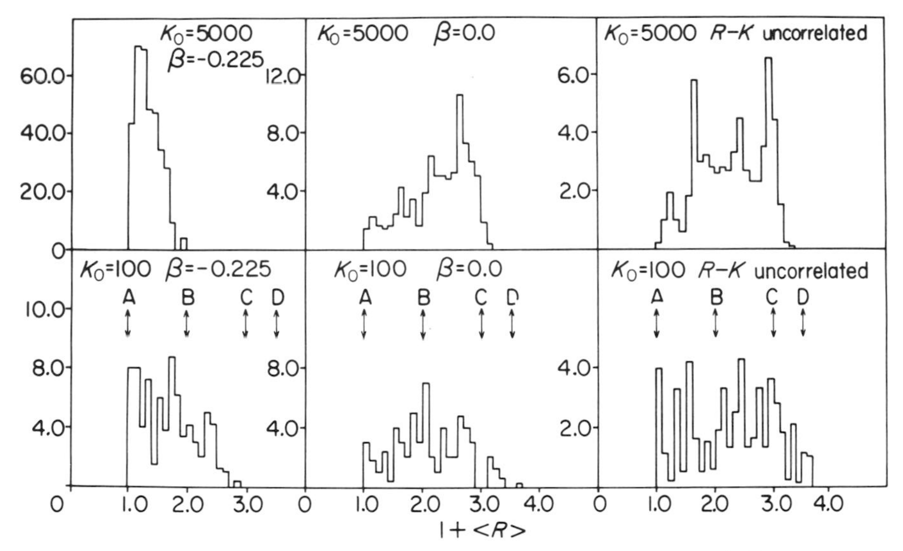

FIG. 16.2. The distribution of $1 + \langle R \rangle$ for population sizes $K_0 = 100$ and 5000 using the form factor (8). Two cases with R and K correlated and with $\beta = -0.225$ and 0.0 and one case for R and K correlated are shown. The significance of the arrows marked A, B, C and D are same as in Fig. 16.1.

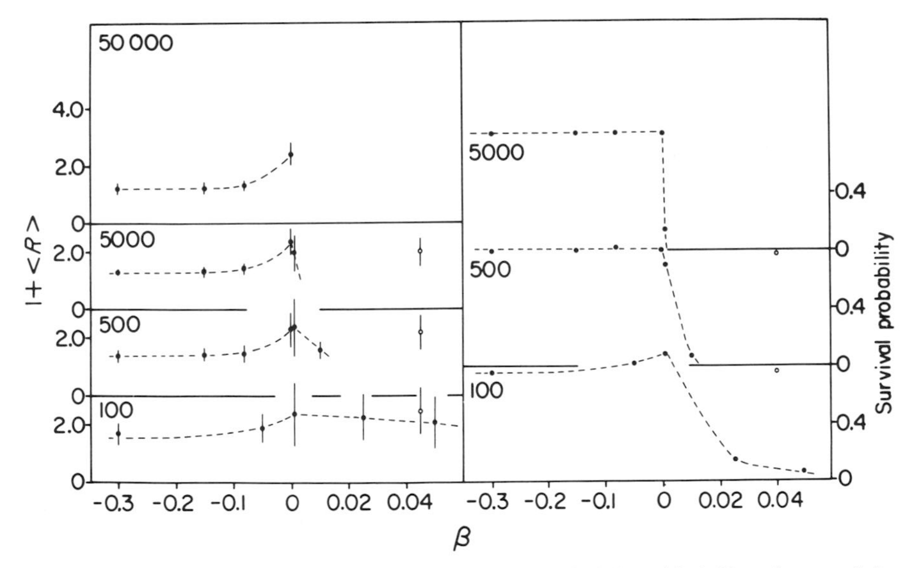

FIG. 16.3. The variation of $1 + \langle R \rangle$ and the survival probability with β. Note the expanded scale for $\beta > 0$. The R–K uncorrelated case is indicated by points marked ○ or ϕ. The survival probability for $K_0 = 50\,000$ goes to zero extremely rapidly above $\beta = 0$ and hence not shown. The broken lines are arbitrarily drawn to give to give some visual impression of the variation with β.

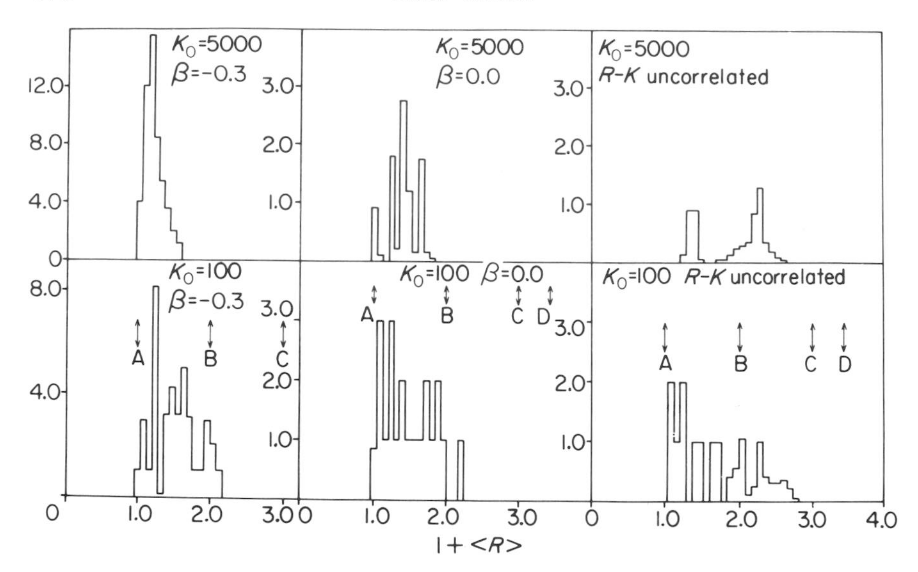

FIG. 16.4. The distribution of $1 + \langle R \rangle$ for form factor (7) and in the presence of fluctuating environment with $\Sigma_e = 0.1$. The values of K_0 and β and the significance of the arrows marked A, B, C and D are same as in Fig. 16.1.

in Fig. 16.4. This figure demonstrates that the mean value of $\langle R \rangle$ decreases with increasing environmental disturbance and thereby reducing substantially the probability for the population to become chaotic. Thus chaotic dynamics is not favoured by natural selection for large populations and for populations in fluctuating environment. Since laboratory populations are very small and are usually maintained in controlled conditions, one would expect such populations to exhibit chaotic motion with much higher probability than natural populations which are usually large in size and which experience more or less rapid changes in the environment. What is demonstrated here is that in the presence of density-regulating mechanisms and environmental fluctuations, natural selection forces a high degree of stability on the populations. This then could provide a partial explanation for the data presented at the begining of this section.

The probability for the population to go into the unstable phase increases as β increases from a negative value to zero; increasing β into the positive regime reduces again the probability for instability. If K_{ij} and R_{ij} are chosen such that they are uncorrelated, then one obtain the maximum probability for the population to experience chaos. In all these situations, increasing the population size or the amount of environmental fluctuation tend to dampen instability.

A few examples of the types of cyclic or chaotic motions that could be obtained under conditions described above are shown in Fig. 16.5.

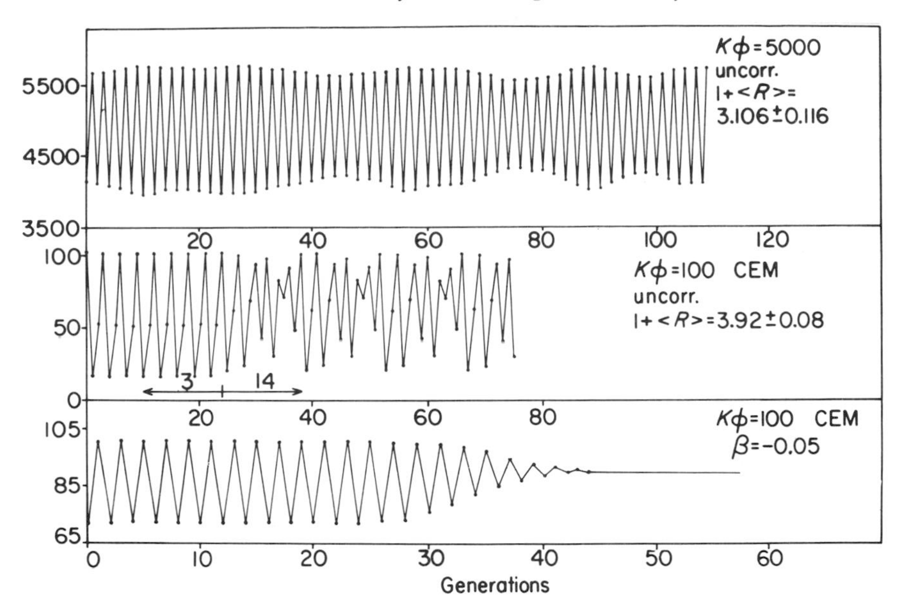

FIG. 16.5. Three examples of populations exhibiting cyclic or chaotic behaviour. The top curve shows a two-point cycle for $K_0 = 5000$ and R–K uncorrelated. The modulation of the curve is due to the continual input of mutants which slightly alter the value of the total population. The period of the modulation is around twenty generations which is the average number of generations for the introduction of a new mutant. Note that the zero in the figure corresponds to 99 500 generations. The middle curve is for $K_0 = 100$ and R–K uncorrelated. There is an initial three–point cycle which goes over to a fourteen-point cycle with the introduction of a new mutant. The three-point cycle stretches backwards to a few 10 000 generations. This particular population went extinct after 450 000 generations. Over half the populations that exhibited such chaotic behaviour, on the other hand, persisted for 1 000 000 generations. The bottom curve shown an example of a two-point cycle eventually going over to a stable point behaviour with the introduction of a new mutant.

POPULATION EXTINCTION PROBABILITY

Another aspect of stability of populations is their persistence to survival through the evoltionary period. Here one can ask two questions:

1 What is the probability for the population to go to extinction in a constant environment as the value of β is changed?

2 With what probability does the population go to extinction in the presence of environmental fluctuations, for a given value of β?

There are two reasons why a population could go to extinction. First, the stochastic process could have driven the population into the domain where all R_{ij}'s are less than zero. Second, the fluctuations, choatic or environmental or both, could accidently drive the population to extinction.

The survival probability is defined to be the fraction of the 100 populations studied that survive for 100 000 generations. The survival probability as a function of β is shown in Fig. 16.3. These results were obtained using the form factor of eqn (7). For values of $\beta < 0$, the survival probability is independent of the population size, being around 80%. Since the values of R_{ij} are chosen within the interval $(-1, 3)$, there is 25% probability that the values of R_{ij} chosen lie below zero and this would explain the observed survival probability. The populations that do go to extinction, do so within 10–500 generations. Also, for the exponential form factor of eqn (8), when the interval of R_{ij} values is set much larger, the survival probability for $\beta < 0$ correspondingly increases. For positive values of β, the survival probability rapidly decreases with population size. For positive β, from eqn (13) one sees that K_{ij} is an increasing function with R_{ij}. Consequently, K_{ij} becomes very small compared to N_T when R_{ij} goes below zero. Since the population regulation forces the value of N_T to be close to K_0, for negative R_{ij}'s the selective values become large as seen from eqn (7). This forces the population to move towards the domain of negative R_{ij}'s and consequently the population goes to extinction. The number of generations required for the population to go to extinction is much larger than when $\beta < 0$, by one to two orders of magnitude. The extinction probability can be reduced by allowing β to be negative in the region $R_{ij} < 0$. Such procedure is arbitrary and has little justification. Further, such populations become very unstable in the presence of environmental fluctuations, since they have a high probability of being in the chaotic region. Also, when β is chosen to be negative for $R_{ij} < 0$ and positive for $R_{ij} > 0$, the genetic properties of the system do not agree with observations.

The linear relation of eqn (13) may be an over simplification to what exists in the field but I think it is a reasonable approximation. If this be true, then the results on survival probability indicates that Roughgarden's (1979) prescription of negative correlation between R and K based on biological reasoning is also borne out by theory.

I now turn to the effects of environmental changes on the survival probability of populations. The fluctuations in the environment are introduced through eqn (15) and the measure of fluctuations will be referred to in what follows in terms of $\Sigma_e = \{\sigma_e^2\}^{\frac{1}{2}}$. The survival probability as a function of Σ_e for various population sizes and for various values of β are shown in Fig. 16.6. The survival probability is seen to be fairly insensitive to the values of β less than zero. On the other hand, $\beta = 0$ exhibits a very different behaviour. For $\beta < 0$, larger the population the better they withstand severe environmental disturbances, in conformity with the very widely held view.

When $\beta = 0$, the situation is the reverse. For small populations, the survival probability is not very different from the case when $\beta < 0$. On the other hand, large populations go extinct with greater probability when

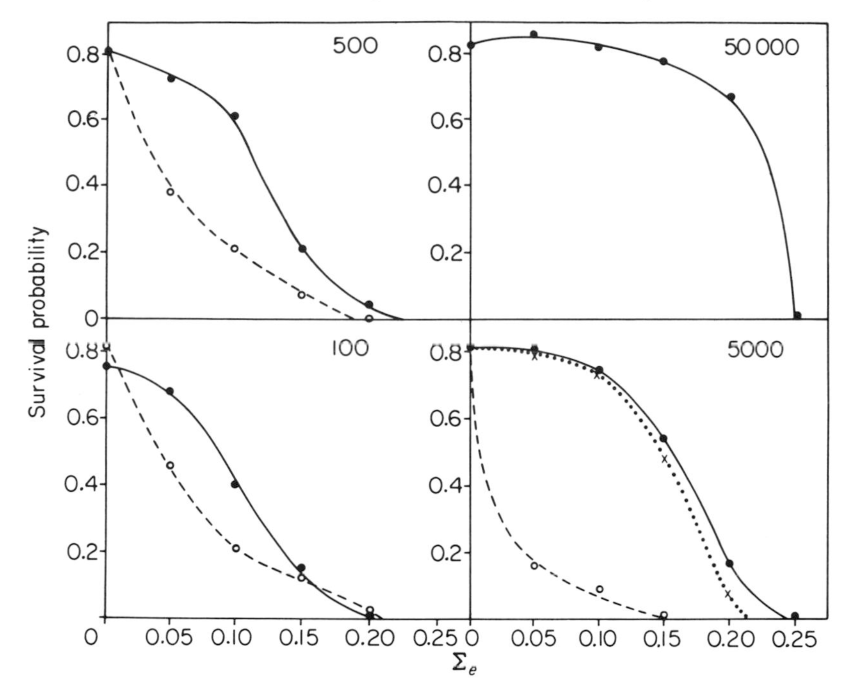

FIG. 16.6. The variation of survival probability with the magnitude Σ_e of environmental fluctuation for different population sizes and for different values of β. Since the curves $\beta = -0.08$ are very close to $\beta = -0.30$ curves, only one case is shown. ●—●, $\beta = -0.30$; ×---×, $\beta = -0.08$; 0---0, $\beta = 0.0$.

subjected to even small amount of environmental changes. Referring to Fig. 16.3, it is seen that the mean value of $\langle R \rangle$ rapidly increases in the neighbourhood of $\beta = 0$. In the presence of environmental fluctuations, the mean value of $\langle R \rangle$ is pushed towards the chaotic regime, thereby introducing instabilities. As discussed in the earlier section, evolutionary stability through natural selection requires that the mean value of $\langle R \rangle$ be nearer to unity.

It was remarked above (cf. p. 371) that $\beta = 0$ corresponds to the extreme neutralist's limit, especially for large populations. Thus the above result indicates that the extreme neutral model is structurally unstable. A similar conclusion was arrived at by G.S. Mani (unpublished) from a different view point. I must hasten to add that the present advocates of the neutral theory do not subscribe to the extreme or strictly neutral view point.

When $\beta > 0$ or when R_{ij} and K_{ij} are uncorrelated, the extinction probability increases by a large amount in the presence of even small environmental changes. For $\beta > 0$, the instabilities appear even when β is arbitrarily set to be < 0 when R_{ij} goes negative. These results reinforce the assumption of negative correlation between R_{ij} and K_{ij}.

PREDICTIONS OF THE MODEL AT THE GENETIC LEVEL

In the above two sections the ecological consequences of the model were discussed. In this section I present some of the genetic consequences of the model.

Probability density for the allele frequency distribution

The allele frequency distribution, as given by eqn (5) for the neutral model, is shown in Fig. 16.7. In this figure the mutation rate was taken to be 5×10^{-6}.

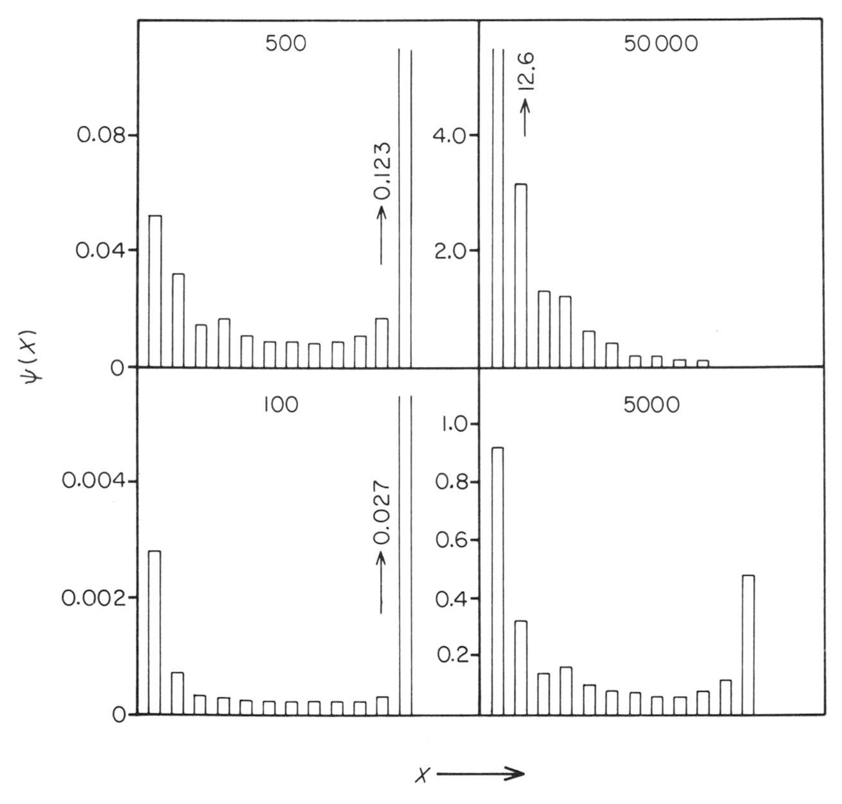

FIG. 16.7. The allele frequency distribution predicted by the infinite allele neutral model is shown. The curves were obtained by integrating the expression for the gene frequency distribution given by eqn (5). The various values of population size used are shown in the figure. The first three boxes in each figure correspond to the frequency intervals < 0.01, 0.01–0.05 and 0.05–0.10 respectively. The rest of the boxes are for successive intervals of 0.1, from 0.1 to 1.0. These intervals were chosen to enable comparison with the results of Fuerst, Chakraborty & Nei (1977).

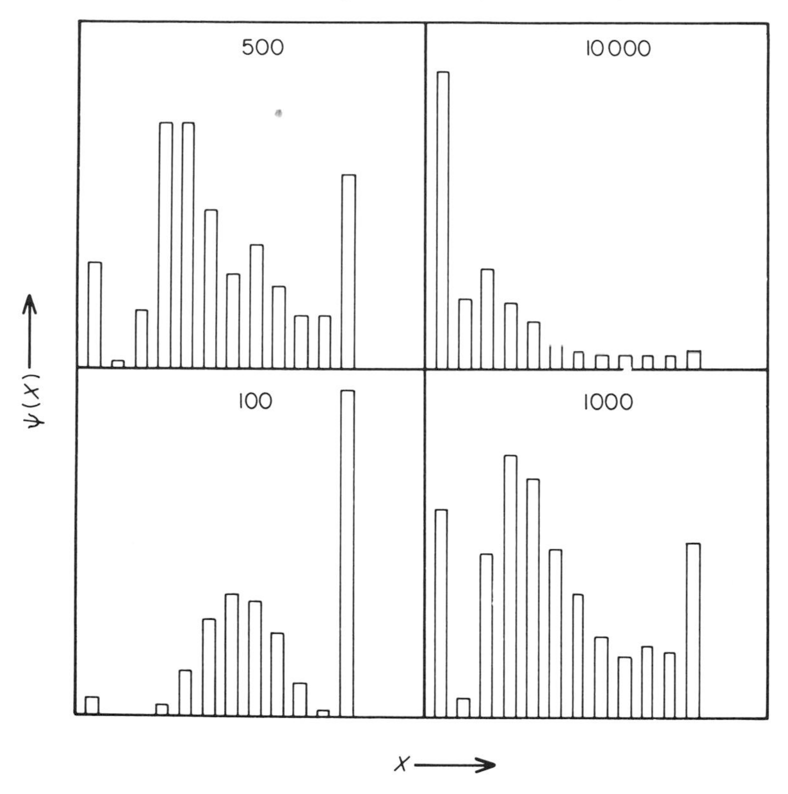

FIG. 16.8. The allele frequency distribution predicted by the strong selection model. The frequency ranges used are the same as in Fig. 16.7. These results are from calculations by Mani (unpublished).

As seen from the figure, for population sizes between 100 and 5000, one has a more or less U-shaped distribution. The arm of the U near zero or unity is larger depending on whether the population size is nearer 5000 or 100 respectively. For very large populations, greater than 50 000, one observes an L-shaped distribution with the probability density concentrated in the neighbourhood of zero.

In the case of the strong selection models, a large fraction of the probability density is peaked around the frequency of 0.5 apart from the increased probability near zero or unity. Some typical examples are shown in Fig. 16.8.

The predictions of the ecological model for $\beta = -0.3$ and 0.0, and for the case when R_{ij} and K_{ij} are uncorrelated are shown in Fig. 16.9. This figure is for the case of constant environment.

In general, as discussed earlier, enzyme polymorphism yield U-shaped distributions while in the very few examples of strong selections observed

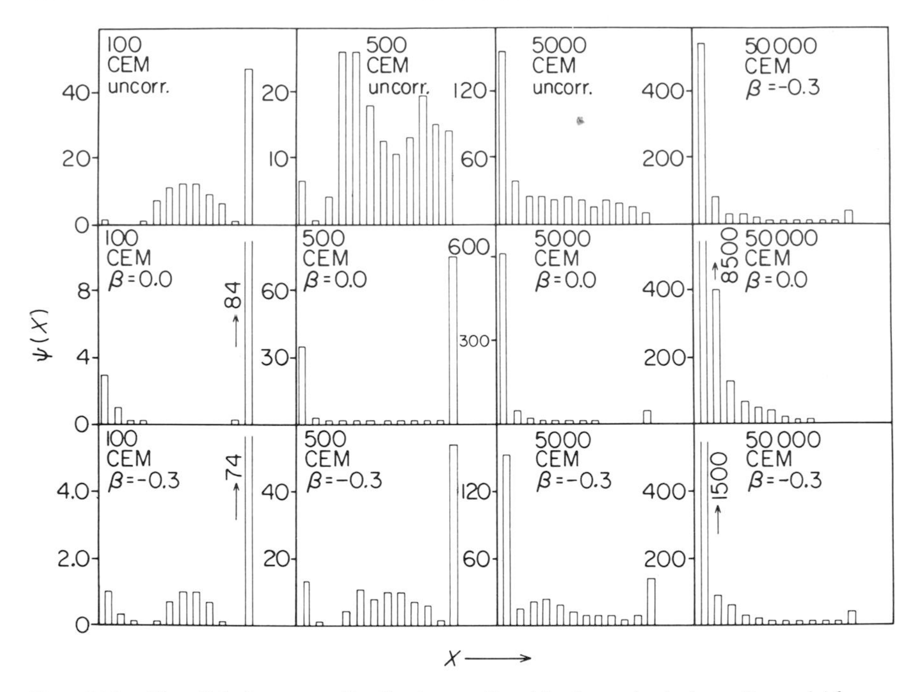

FIG. 16.9. The allele frequency distribution predicted by the ecological genetic model for various values of K_0 and β are shown. The cases with R–K uncorrelated are also included. Note that the distributions due to uncorrelated R and K closely follow the strong selection limit while the R–K correlated cases approach the neutral limit. These results correspond to the case when there is no environmental fluctuation. The frequency ranges used are the same as in Fig. 16.7.

in the field, the distributions indicate a peak between zero and unity. The model shows, (Fig. 16.9), that in general when R and K are correlated, one has the weak selection limit and the results agree both with the neutral model and with the observations at the enzyme level. In the presence of very strong selection, the selection would dominate and the population regulation would tend to be weakened. Thus one would have a situation not dissimilar from the uncorrelated case. As seen from Fig. 16.9, the uncorrelated case do approach the strong selection limit.

When environmental fluctuation is superimposed, selection operates with increased strength, especially for small population sizes and for large negative β's. This is understandable since for those populations that survive, the genetic composition is such as to withstand external disturbances. As seen earlier, small populations are less stable in the presence of environmental fluctuations than large populations and hence the selective effects are enhanced for small populations. The effects of environmental disturbance

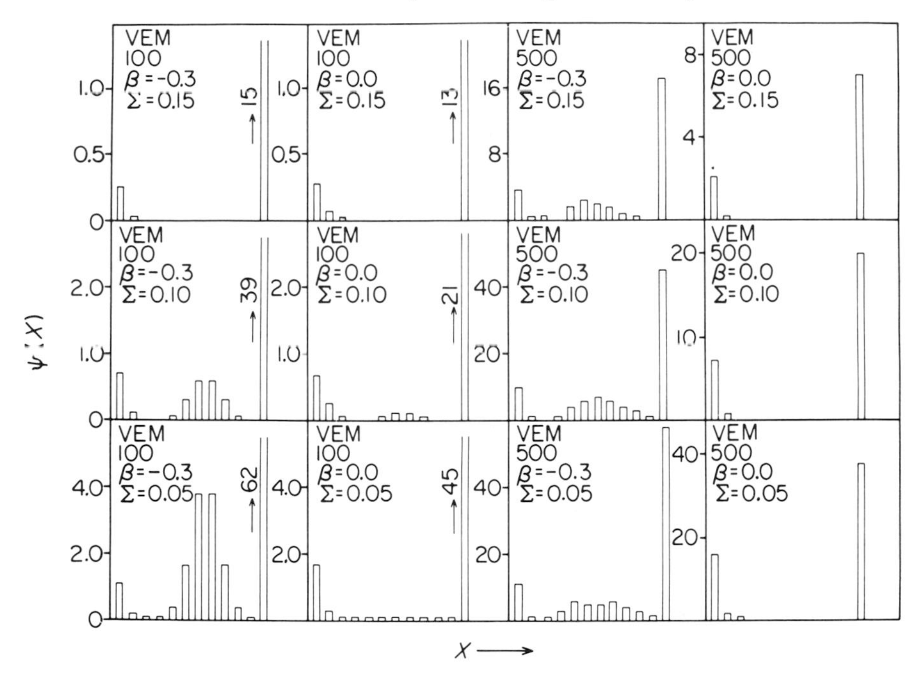

FIG. 16.10. The allele frequency distribution predicted by the ecological genetic model in the presence of varying magnitude of environmental fluctuations. The frequency ranges used are same as in Fig. 16.7.

on population size and on the value of β, is shown in Figs 16.10 and 16.11. The interesting aspect as shown by these figures is that when the fluctuations are large or when $\beta = 0$, the distributions tend to indicate a weak selection. What is happening in these cases is that the selective differences are not sufficient to adjust to the strong environmental disturbances; most of the populations go extinct and the few that survive are not very stable. In fact if one continued the evolutionary path way to a million generations, most of these populations would also go extinct.

Single locus heterozygosity

We have already seen that the enzyme data yield heterozygosity of a few percent to a maximum of 25%. The distribution of the heterozygosity shows a very large peak around zero with must smaller peaks between zero and unity, especially around 0.5. The neutral model fails to predict the fact that the heterozygosity is limited to the value of 25% for large populations. Further assumptions of slightly deleterious mutants and bottle-necks of population sizes have to be invoked to reduce the predicted values of the

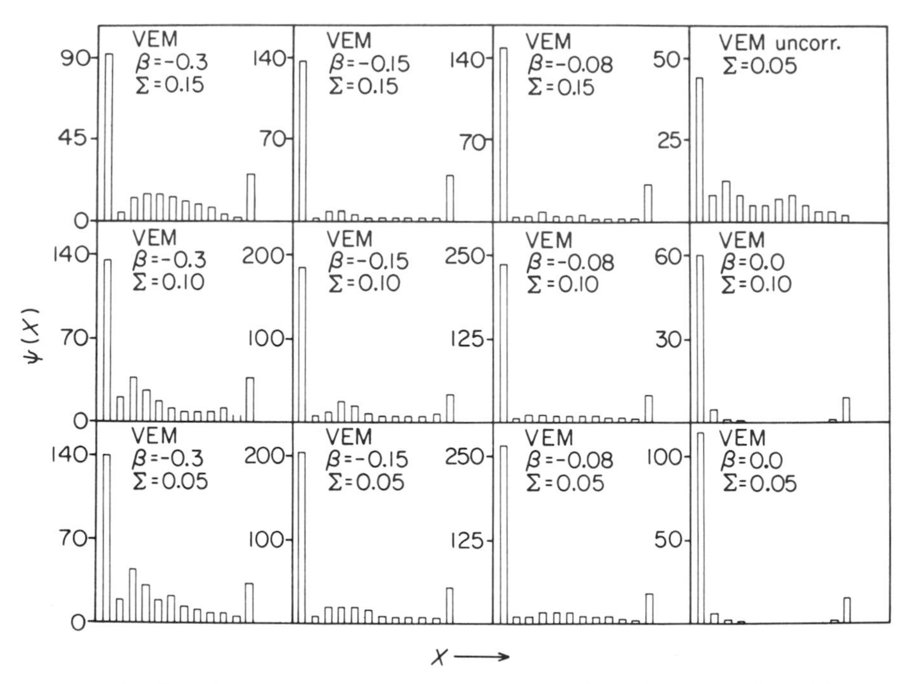

FIG. 16.11. The allele frequency distribution predicted by the ecological genetic model in the presence of varying magnitude of environmental fluctuations for $K_0 = 5000$. The frequency ranges used are the same as in Fig. 16.7.

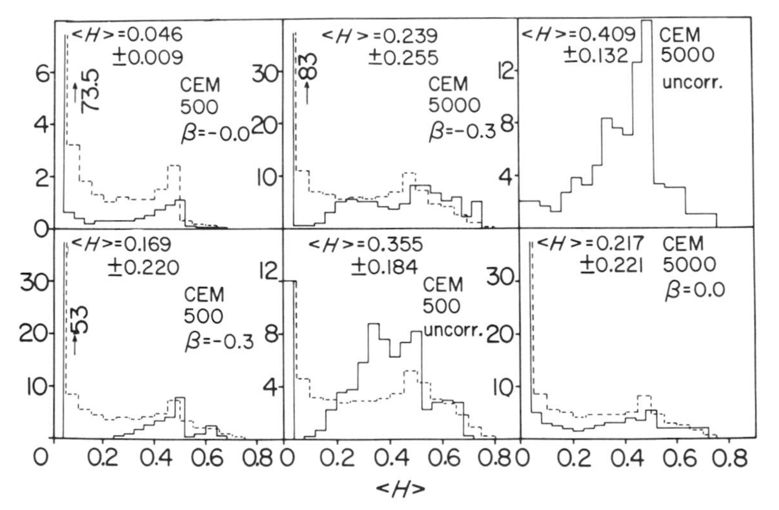

FIG. 16.12. A few examples of the heterozygosity distribution predicted by the ecological genetic model. The broken-line curve corresponds to the predictions of the neutral model. Note the occurrence of the large peak around 0.5 for the case when R and K are uncorrelated, reminiscent of the strong selection model. —, S; ----, N.

neutral theory to the observed heterozygosity of 25% for large populations. In the strong selection models, the heterozygosity reaches a limiting value of 50–60% for very large populations. Also the heterozygosity distribution is peaked around 0.5 rather than at the origin.

Some typical examples of the distribution of heterozygosity calculated on the ecological genetic model are shown in Fig. 16.12. These are compared to the neutral model, in these figures. The heterozygosity distribution for the neutral model is obtained using the method of Stewart quoted by Feurst, *et al.* (1977). In the same paper they show that the theoretical distribution resembles with the observed distribution, especially the appearance of the dominant peak near zero and small peaks between zero and unity. On the other hand, the details of the distribution between near zero and unity show some differences. From Fig. 16.12, one sees that the predictions of the ecological model for $\beta < 0$, are in reasonable agreement with the neutral model and hence agree with experiments to the same extent. In the case of uncorrelated R and K, the heterozygosity distribution approaches the strong selection limit as expected. The distributions do not materially alter in the presence of environmental fluctuations.

The next question is how sensitive the mean value of heterozygosity $\langle H \rangle$ is to changes in the value of β. This is shown in Fig. 16.13. We again arrive at the conclusion that for $\beta < 0$, the predictions of the model are not very sensitive to the value of β.

One of the important comparisons between experiment and theory used

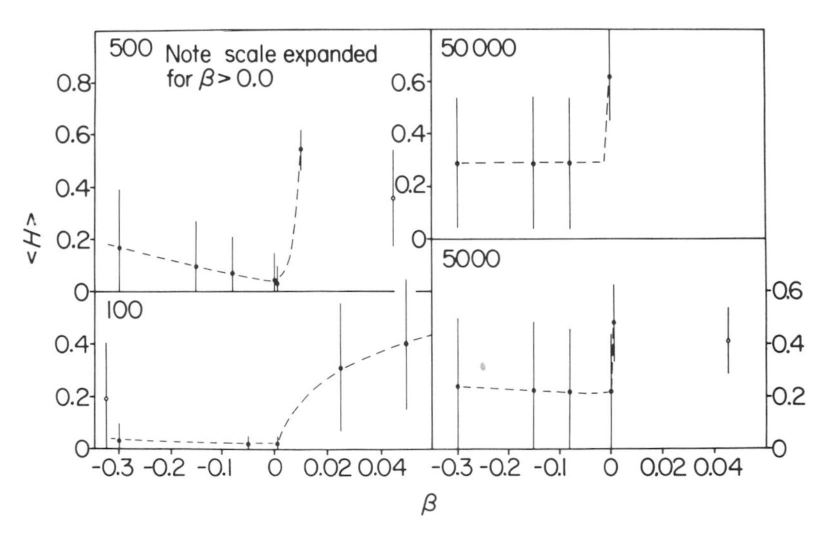

FIG. 16.13. The variation of mean heterozygosity with β. Note the expanded scale for $\beta > 0$. The figure demonstrates the insensitivity of β for $\beta < 0$.

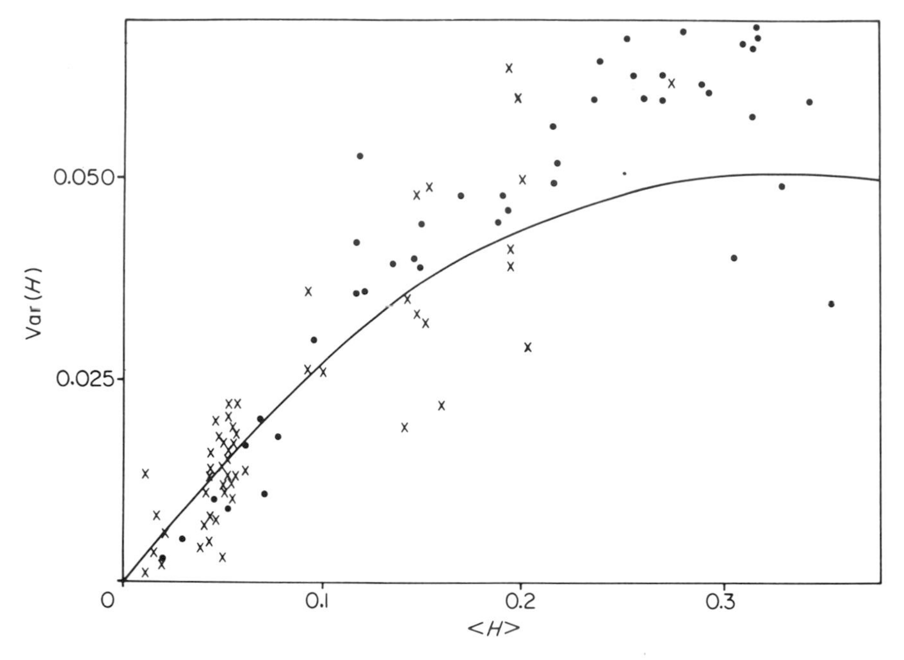

FIG. 16.14. The observed and predicted relation between the variance and the mean of heterozygosity. ——, The strictly neutral infinite allele model; X, experimental data taken from the compilation by Fuerst *et al.* (1977); ●, the prediction from the ecological genetic model. See text for detailes.

by Feurst *et al.* (1977) is to compare the relationship between the variance of H, (var (H)), and the mean value of heterozygosity, $\langle H \rangle$. In their paper, they show that the extreme neutral model agrees, but not very well, with the data between var(H) and $\langle H \rangle$. Introducing the assumption of varying mutation rate with loci, they obtain a better agreement, considering the wide fluctuations in the data. In Fig. 16.14, the predictions of the present model for the relation between var(H) and $\langle H \rangle$, together with the observed data and the extreme neutral predictions are shown. The predictions of the ecological genetic model was obtained by pooling all cases studied without imposing any bias. The R–K uncorrelated cases are not included in the figure for obvious reasons. Comparing the results of the present model with those of the modified neutral model of Feurst *et al.* (1977, Figs 1 and 2), one sees that the present model agrees with the data to the same degree as the modified neutral model.

The variation of heterozygosity with population size as predicted by the present model is shown in Fig. 16.15. In this figure the predictions of the neutral model and the upper bound of the experimental data estimated by

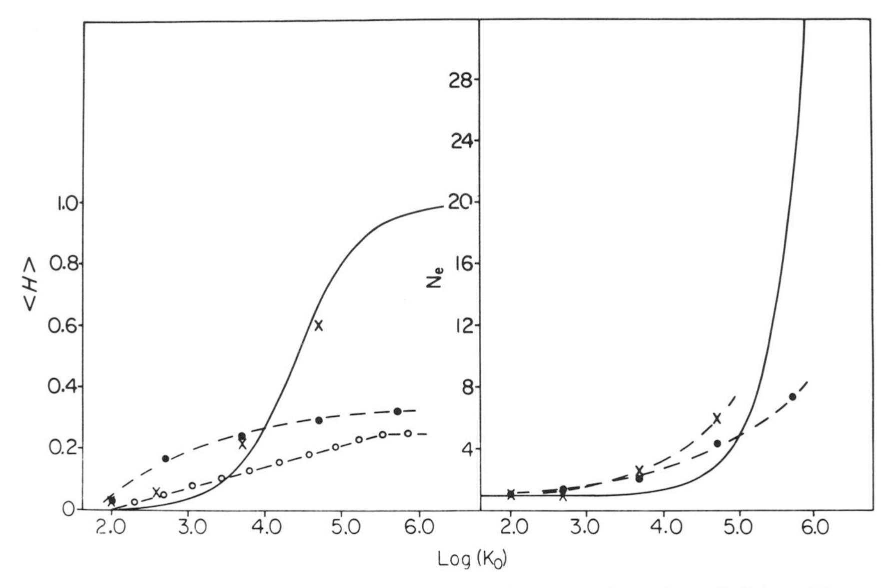

FIG. 16.15. The variation of heterozygosity and the expected number of alleles with population size.—, The infinite allele neutral model; ● the ecological genetic model with = − 0.30. Values of $\beta < 0$, are all very close to these points; X, the ecological genetic model predictions with $\beta = 0.0$. o–o, the upper bound of the experimental data extracted by Soulé (1976).

Soulé (1976) are also shown. Nei (1980) has argued that Soulé has probably underestimated the population size. If this be true, then the agreement with the present model would become even better. The reason for this agreement in the present model will become clear in the following subsection. Since for $\beta < 0$, the results are insensitive, the departure from experiment only occurs in the neighbourhood of $\beta = 0$. This is also shown in Fig. 16.15.

The fitness distribution for homozygotes and heterozygotes

Some typical examples of the fitness distribution for homozygotes and heterozygotes are shown in Fig. 16.16. The model yields the following general features:

1 There is a slight heterozygote advantage of the order of a few per cent when $\beta = -0.3$. The heterozygote advantage decreases with increasing population size, varying from about 12% for $K_0 = 100$ to 2% for $K_0 = 50\,000$. The reason for the peaking of the heterozygosity distribution in the presence of environmental fluctions for small populations discussed in the previous subsection arises from this difference between the fitnesses of homozygotes

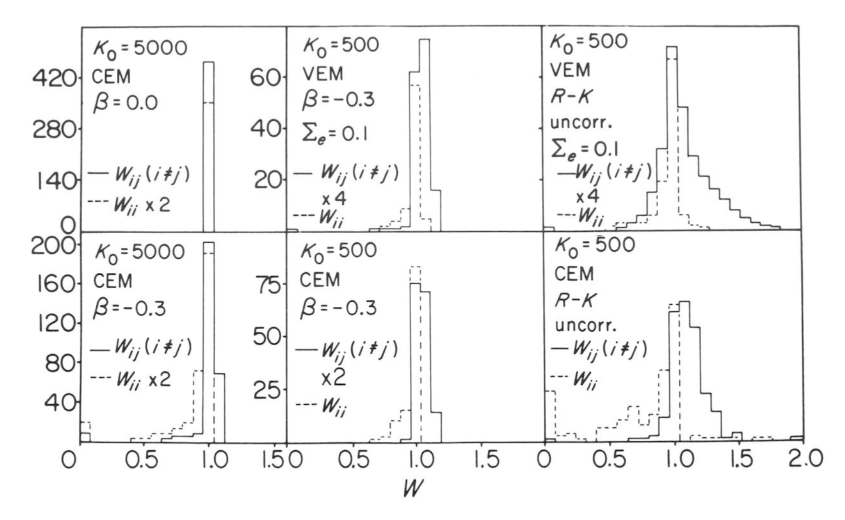

FIG. 16.16. The distribution of the fitnesses of the heterozygotes and the homozygotes predicted by the ecological genetic model. The full line histograms are for the heterozygotes and the broken lines for the homoygotes. Note the very large heterozygote advantage for the uncorrelated cases.

and the heterozygotes in the case of small population size. As the value of β moves towards zero, the heterozygote advantage decreases, in some cases even becoming negative. When $\beta = 0$, the mean fitness of the heterozygotes equals that of the homozygotes.

2 The populations contain slightly deleterious alleles. The frequency of the deleterious alleles decrease rapidly with fitness difference with the typical alleles. Thus one obtains the Ohta model in a natural way. It is due to these deleterious alleles that the heterozygosity is damped for large population sizes.

3 The mean fitness $\bar{W}$ over the whole population is very close to unity differing by about $10^{-4} - 10^{-5}$. This indicates that there is no 'load problem'.

The variation of the heterozygote advantage with population size and with β is shown in Fig. 16.17. I have used the following equation as a measure of heterozygote advantage:

$$\Delta W = \frac{\langle W_{ij} \rangle - \langle W_{ii} \rangle}{\langle W_{ii} \rangle} \tag{20}$$

where $\langle W_{ij} \rangle$ ($i \neq j$) and $\langle W_{ii} \rangle$ are the mean fitness values averaged over all the homozygotes and the heterozygotes respectively. Note that the heterozygote advantage, ΔW, increases rapidly for $\beta > 0$. In the same figure is also shown the mean values of the intrinsic growth rates $\langle R_{ij} \rangle$ ($i \neq j$) and $\langle R_{ii} \rangle$ for the heterozygotes and the homozygotes, respectively. It is seen that for $\beta \leqslant 0$, $\langle R_{ij} \rangle$ and $\langle R_{ii} \rangle$ are almost equal while for $\beta > 0$, they rapidly diverge.

Number of polymorphic alleles at a single locus

The variation of the number of alleles at a single locus with population size is shown in Fig. 16.15. The predictions of the extreme neutral model is also shown for comparison. The experimental data is much less certain for this case compared to heterozygosity. In general, the data indicates that for any single population, there exists between one and five alleles at any locus. This is true for populations with very widely differing sizes. Thus the present model is in closer accord with data than the strict neutral model.

Rate of gene substitution

The neutral model predicts that the rate of gene substitution should be constant with respect to the generation time, being equal to the mutation rate per generation and should be independent of population size. This

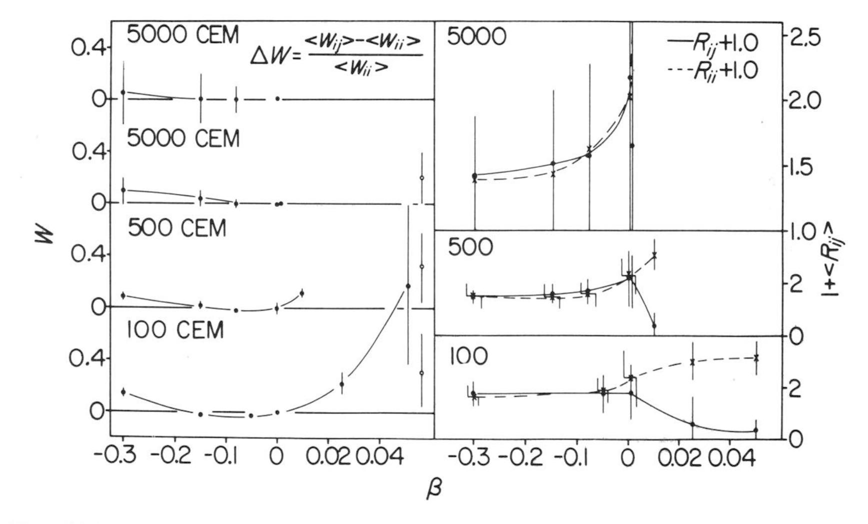

FIG. 16.17. The left-hand curve shows the variation of the magnitude of the heterozygote advantage with β and population size. The heterozygote advantage is measured through ΔW defined by eqn (20). Note that ΔW becomes very slightly negative under some situations, indicating heterozygote disadvantage in these cases. The right-hand side curve compares the mean value of R_{ij} for the heterozygotes and homozygotes respectively. The full line corresponds to heterozygotes and the broken to homozygotes. Note that the scale for $\beta > 0$ is expanded. The mean values of R_{ij} for the homozygotes and the heterozygotes are very close to each other for $\beta < 0$ but diverge for $\beta > 0$.

follows from the fact that the probability for fixation of a mutant gene is $1/2N$ and the number of mutations per generation is $2Nu$, where u is the mutation rate per generation. Thus the substitution rate per generation equals u. In practice, the neutralists would argue that the substitution rate need not be a constant with respect to the generation time due to the occurence of bottle-necks. After a bottle-neck it would take tens of thousands of generations before neutral mutations accumulate sufficiently for the rate of fixation to equal the mutation rate. Thus the constancy of gene substitution is unlikely to be observed in periods shorter than the geological epoch. Given these uncertainties, the agreement claimed between the neutral model and the molecular-evolution data is not very convincing. It must also be remarked that the present state of the data at the molecular level is not very clear cut to make detailed comparison with theory. In their review paper on biochemical evolution, Wilson, Carlson & White (1977) have shown that (a) the *average* substitution rate is probably constant over the evolutionary period; (b) at least for the data on the mammals ranging from small rodents with many generations per year to large mammals, such as elephants, with many years per generation, the data is more consistent with constancy over absolute time rather than over generation time. It becomes difficult to understand how these populations, so diverse both in population size and in generation time, could have mutation rates and bottle-necks so adjusted as to yield a constant rate of substitution of neutral genes over absolute time.

Nei (1980) using some specific models of selection, has shown that such selective models yield too small values for heterozygosity and percentage of polymorphism for a given rate of substitution when compared with the neutral model and enzyme data. He thus concludes that selective models cannot explain the data at the molecular level consistently.

I shall now discuss some preliminary calculations for gene substitution that I have carried out using the ecological genetic model, More detailed results will be published elsewhere. The rate of gene substitution for a mutation rate of 5×10^{-6} and for a population size of 500 is found to be about 10^{-5} per generation for 100 000 generations and $2–3 \times 10^{-5}$ per generation for 300 000 generations. When the population size is varied by two orders of magnitude, from 500 to 50 000, these values alter by less than a factor of two. When the mutation rate is increased by a factor of 10, the substitution rate increases by more than a factor of five. From the very brief discussion given above, it is evident that these results are as consistent with the data as the neutral model. In comparing the neutral model and the present results, it is seen that the present model indicates that there is less constancy with generation time and that there exists a very slight dependance on population size. The status of the experimental data at the present time is not sufficiently good or consistent to chose between the two models. The predictions of thc present model are not inconsistent with the neutral model modified to include slightly deleterious mutants as suggested by Ohta (1973).

The correlated heterozygote model

All the results till now were based on the uniform distribution assumption. I shall very briefly discuss the results when the correlated heterozygote approximation, (see eqn [14], is used. The calculations were done with $\sigma^2 = 0.01$ and 0.0001. The results using this approximation are very similar in nature to the uniform distribution case. As the value of σ^2 is decreased, the mean heterozygosity and the mean number of alleles are slightly reduced, and one moves more towards the extreme neutral model.

DISCUSSION

My aim in writing this chapter was not to present yet another discussion on the rather futile argument between neutralism and selectionism. It is now an undisputed fact that most of the data on enzyme polymorphism indicates that the selection at this level is weak. No selectionist can ignore this fact. On the other hand, the enzyme data is incompatible with the strict neutral approach and most neutralists today concede that some selective components have to be included in their model. Also, the advocates of the neutral hypothesis cannot ignore the fact that there are many instances when selection is dominant and that the very motive force in evolution is natural selection and the 'struggle for existence'. The advocates of either theories argue concerning the fraction of the alleles or loci that exhibit more or less strong selection, as if selection is a specific property residing within the genes. The selectionists claim this fraction to be large while the neutralists maintain that it is small. The real question, on the other hand, is what aspect of the environmental interaction and perhaps of the genome structure that yields such wide spread polymorphism with apparently weak selection? What changes in the external conditions induce observable selective differences at the genetic level?

In this chapter, I have demonstrated that weak selection is a natural consequence of resource allocation and of competition and hence, in some sense, a 'struggle for existence'. In the presence of strong selection, such as industrial pollution or insecticide applications, such constraints are modified. When the selective force or impulse is removed, the population would slowly revert to the neutral norm (though it may occupy a different neutral state) under the force of external constraints. Nature is very complex and the interactions involved are very varied. Thus the above is a very simplistic view of the real world. What such a simple model demonstrates is that one cannot ignore the ecological aspects in describing the populations at the genetic level.

I have shown in this chapter that the stability of single species system depends on natural selection and the resultant genetic diversity. It is true that the complex interactions that are involved between any species and its external ecological environment cannot be described by the simple equations used in the present model. But I believe that the plasticity produced by genetic diversity as a result of ecological interactions is an important factor in maintenance of persistence in ecosystems. The problem of complexity versus stability can only be understood and resolved by including genetic aspects into the ecological theories. Such studies are contemplated for the future.

One of the criticism that has been levelled against the model is the multiplicity of parameters that enter into it. The parameter set used in the model are $\{K_0, v, R_{ij}, \beta\}$ which, apart from β are the same set used in any selection model. We have already seen that the model is insensitive to the exact choice of β as long as $\beta < 0$. A negative value of β has biological justification. Further we have seen that the simple one parameter neutral model, for all its attractiveness, does not fit the data unless other assumptions that destroy its basic simplicity, are invoked. I maintain that the present model, with less or equal number of parameters, is able to provide results both at the ecological and at the genetic level. The model is a predictive model and it further shows how the strong selection limit can be included. Thus it bridges the gap between selectionists and neutralists.

There is still much work to be done in the understanding of the relation between the ecological or macroscopic world and the genetic or the microscopic world. All the results quoted here are for a single locus system. Extension to many loci is being currently undertaken. The model can be used in the study of the nature and evolution of niches in competitive systems, parasite–host interactions for the maintenance of genetic diversity (Clarke 1979) and for the evolution of sex (Hamilton 1980).

ACKNOWLEDGEMENTS

A nuclear physicist entering into the totally alien environment of biology is confronted by what appears to him to be a complex and chaotic situation. My deep felt thanks to L.M. Cook, B.C. Clarke, R.J. Wood and the late

J. Bishop for showing me the order and beauty that exist in biological systems.

REFERENCES

Bellows T.S. Jr (1981) The descriptive properties of some models for density dependence. *Journal of Animal Ecology* **50**, 139–156.

Crow J.F. & Kimura M. (1970) *An Introduction to Population Genetics Theory*. Harper & Row, New York.

Chakraborty R., Fuerst P.A. & Nei M. (1978) Statistical studies on protein polymorphism in natural populations. II. Gene differentiation between populations. *Genetics*, **88**, 367–390.

Chakraborty R., Fuerst P.A. & Nei M. (1980) Statistical studies on protein polymorphism in natural populations. III. Distribution of allele frequencies and number of alleles per locus. *Genetics*, **94**, 1039–1063.

Cook L.M. & Mani G.S. (1980) A migration-selection model for the morph frequency variation in the peppered moth *Biston betularia* over England and Wales. *Biological Journal of the Linean Society*, **13**, 251–262.

Clarke B.C. (1979) The evolution of genetic diversity. *Proceedings of Royal Society (London) Series B* **205**, 453–474.

Charlesworth D. & Charlesworth B. (1975) Theoretical genetics of Batesian mimicry. II. Evolution of supergenes. *Journal of Theoretical Biology*, **55**, 305–324.

Ewens W.J. (1979) *Mathematical Population Genetics*. Springer, Berlin.

Fuerst P.A., Chakraborty R. & Nei M. (1977) Statistical studies on protein polymorphism in natural populations. I. Distribution of single-locus heterozygosity *Genetics*, **86**, 455–483.

Ford E.B. (1971) *Ecological Genetics*. Chapman and Hall, London.

Hamilton W.D. (1980) Sex versus non-sex versus parasite. *Oikos*, **35**, 282–290.

Hassel M.P., Lawton J.H. & May R.M. (1976) Patterns of dynamical behaviour of single-species populations. *Journal of Animal Ecology*, **45**, 471–486.

Itoh Y. (1979) *Random collision Process of Oriented Graph*. Institute of Statistical Mathematics (Japan) Research memorandum no. 154, 1–20.

Kimura M. (1968a) Evolutionary rate at the molecular level. *Nature*, **217**, 624–626.

Kimura M. (1968b) Genetic variability maintained in a finite population due to mutational production of neutral and nearly neutral isoalleles. *Genetical Research*, **11**, 247–269.

Kimura M. & Ohta T. (1971) *Theoretical Aspects of Population Genetics*. Monographs in Population Biology, vol. 4 Princeton University Press, Princeton.

Kimura M. (1979) The neutral theory of molecular evolution. *Scientific American*, **241**, 94–104.

King J.L. & Jukes T.H. (1969) Non-Darwinian evolution. *Science*, **164**, 788–798.

Lewontin R.C. (1974) *The Genetic Basis of Evolutionary Change*. Columbia University Press, New York.

Mani G.S. (1980) A theoretical study of morph ratio clines with special reference to melanism in moths. *Proceedings of the Royal Society (London) Series B*, **210**, 299–316.

Mani G.S. (1982) A theoretical analysis of the morph frequency variation in the peppered moth over England and Wales. *Biological Journal of the Linean Society*, **17**, 259–267.

May R.M. (1976) Models for single populations. In *Theoretical Ecology: Principles and applications* (Ed. by R.M. May) Blackwell Scientific Publications, Oxford.

Nei M. (1975) *Molecular Population Genetics and Evolution*. North-Holland, Amsterdam.

Nei M., Fuerst P.A. & Chakraborty R. (1976) Testing the neutral mutation hypothesis by distribution of single-locus heterozygosity. *Nature*, **262**, 491–493.

Nei M., Fuerst P.A. & Chakraborty R. (1978) Subunit molecular weight and genetic variability of proteins in natural populations. *Proceedings of the National Academy of Science (U.S.A)*, **75**, 3359–3362.

Nei M. (1980) Stochastic theory of population genetics and evolution. In *Proceedings of Vito Volterra Symposium on Mathematical Models in Biology*, Lecture Notes in Biomathematics, Vol. 39. Springer, Berlin.

Nevo E. (1978) Genetic variation in natural populations: Patterns and theory. *Theoretical Population Biology*, **13**, 121–177.

Nevo E., Beiles & Ben-shlomo R. (1983) The evolutionary significance of genetic diversity: Ecological, demographic and life-history correlates. Paper presented at the symposium on '*The Basis of Genetic Diversity*' March, 1983, Manchester, U.K. (to be published).

O'Donald P. & Barrett J.A. (1973) Evolution of dominance in polymorphic Batesian mimicry. *Theoretical Population Biology*, **4**, 173–192.

Ohta T. (1973) Slightly deleterious mutant substitutions in evolution. *Nature*, **246**, 96–98.

Roughgarden J. (1979) *Theory of Population Genetics and Evolutionary Ecology: an Introduction*. Macmillan, New York.

Sheppard P.M. (1975) *Natural Selection and Heridity*. Hutchinson, London.

Solué M. (1976) Allozyme variation: Its determinants in space and time. In *Molecular Evolution* (Ed. by J. Ayala), Sinauer Associates Inc., Sunderland, Mass.

Stubbs M. (1977) Density dependence in the life cycle of animals and its importance in *k*- and *r*-strategies. *Journal of Animal Ecology*, **46**, 677–688.

Thomas W.R., Pomerantz M.J. & Gilpin M.E. (1980) Chaos, asymmetric growth and group selection for dynamical stability. *Ecology*, **61**, 1312–1320.

Wills C. (1981) *Genetic Variability*. Clarendon Press, Oxford.

Wilson A.C., Carlson S.S. & White T.J. (1977) Biochemical evolution. *Annual Review of Biochemistry*, **46**, 573–639.